T0324106

Current Trends in Estuarine and Coastal Dynamics

Estuarine and Coastal Sciences Series

Current Trends in Estuarine and Coastal Dynamics

Observations and Modelling

Volume 4

Edited by

Xiao Hua Wang

The Sino-Australian Research Consortium for Coastal Management, School of Science, University of New South Wales, Canberra, ACT, Australia

Lulu Qiao

College of Marine Geosciences, Ocean University of China, Key Lab of Submarine Geosciences and Prospecting Techniques, Ministry of Education, Qingdao, China

Series Editors

Steve Mitchell

Michael Elliott

ELSEVIER

Elsevier
Radarweg 29, PO Box 211, 1000 AE Amsterdam, Netherlands
125 London Wall, London EC2Y 5AS, United Kingdom
50 Hampshire Street, 5th Floor, Cambridge, MA 02139, United States

Notices
Knowledge and best practice in this field are constantly changing. As new research and experience broaden our understanding, changes in research methods, professional practices, or medical treatment may become necessary.

Practitioners and researchers must always rely on their own experience and knowledge in evaluating and using any information, methods, compounds, or experiments described herein. In using such information or methods they should be mindful of their own safety and the safety of others, including parties for whom they have a professional responsibility.

To the fullest extent of the law, neither the Publisher nor the authors, contributors, or editors, assume any liability for any injury and/or damage to persons or property as a matter of products liability, negligence or otherwise, or from any use or operation of any methods, products, instructions, or ideas contained in the material herein.

ISBN: 978-0-443-21728-9
ISSN: 2589-6970

For information on all Elsevier publications visit our website at
https://www.elsevier.com/books-and-journals

Publisher: Candice Janco
Acquisitions Editor: Maria Elekidou
Editorial Project Manager: Rupinder Heron
Production Project Manager: Kumar Anbazhagan
Cover Designer: Christian Bilbow

Typeset by TNQ Technologies

Working together to grow libraries in developing countries
www.elsevier.com • www.bookaid.org

Contents

List of contributors

Gavin Birch School of Geosciences, University of Sydney, Sydney, NSW, Australia

Zhixin Cheng College of Environmental Science and Engineering, Dalian Maritime University, Dalian, China; The Sino-Australian Research Consortium for Coastal Management, School of Science, University of New South Wales, Canberra, ACT, Australia

Pingxing Ding State Key Laboratory of Estuarine and Coastal Research, East China Normal University, Shanghai, China

Zhaopeng Du College of Environmental Science and Engineering, Dalian Maritime University, Dalian, China

Jun Du First Institute of Oceanography, Ministry of Natural Resources of China, Coastal Zone Science and Ocean Development Strategy Research Center, Qingdao, China; First Institute of Oceanography, Ministry of Natural Resources, Qingdao, China

Chenhui Fan Ocean College, Zhejiang University, Zhoushan, China

Guandong Gao Key Laboratory of Ocean Observation and Forecasting, Key Laboratory of Ocean Circulation and Waves, Institute of Oceanology, Chinese Academy of Sciences, Qingdao, China; CAS Key Laboratory of Ocean Circulation and Waves, Institute of Oceanology Chinese Academy of Sciences, Qingdao, China; Laoshan Laboratory, Qingdao, China

Kai Gao Ocean College, Zhejiang University, Zhoushan, China

Jianzhong Ge State Key Laboratory of Estuarine and Coastal Research, East China Normal University, Shanghai, China

Wenyun Guo College of Ocean Science and Engineering, Shanghai Maritime University, Shanghai, China

Xia Ju First Institute of Oceanography, Ministry of Natural Resources of China, Coastal Zone Science and Ocean Development Strategy Research Center, Qingdao, China; First Institute of Oceanography, Ministry of Natural Resources, Qingdao, China

Li Li Ocean College, Zhejiang University, Zhoushan, China

Zhibing Li University of New South Wales, School of Science, Canberra, ACT, Australia

Yunhuan Li   Dalian University of Technology, Panjin, China

Fanglou Liao   Ocean Institute, Northwestern Polytechnical University, Taicang, China

Yue Ma   Wuhan University, School of Electronic Information, Wuhan, China

Wei Mao   College of Ocean Science and Engineering, Shanghai Maritime University, Shanghai, China

Lulu Qiao   College of Marine Geosciences, Ocean University of China, Key Lab of Submarine Geosciences and Prospecting Techniques, Ministry of Education, Qingdao, China

Moninya Roughan   School of Biological Earth and Environmental Science, University of New South Wales, Sydney, NSW, Australia

Youn-Jong Sun   The Sino-Australian Research Consortium for Coastal Management, School of Science, University of New South Wales, Canberra, ACT, Australia

Yi Wan   Ocean College, Zhejiang University, Zhoushan, China

Liu Wan   Ocean University of China, College of Oceanic and Atmospheric Sciences, Qingdao, China

Xiao Hua Wang   The Sino-Australian Research Consortium for Coastal Management, School of Science, University of New South Wales, Canberra, ACT, Australia

Yongzhi Wang   First Institute of Oceanography, Ministry of Natural Resources of China, Coastal Zone Science and Ocean Development Strategy Research Center, Qingdao, China; First Institute of Oceanography, Ministry of Natural Resources, Qingdao, China

Wen Wu   Ocean University of China, College of Oceanic and Atmospheric Sciences, Qingdao, China

Ziyu Xiao   CSIRO Environment, Brisbane, QLD, Australia

Rushui Xiao   National Marine Environment Monitoring Center, Dalian, China

Nan Xu   Hohai University, College of Geography and Remote Sensing, Nanjing, China

Gang Yang   Nanjing University of Information Science and Technology School of Marine Sciences, Nanjing, China

Yueying Zha   Ocean College, Zhejiang University, Zhoushan, China

Haifeng Zhang   Bureau of Meteorology, Docklands, VIC, Australia

Changle Zhang   First Institute of Oceanography, Ministry of Natural Resources of China, Coastal Zone Science and Ocean Development Strategy Research Center, Qingdao, China; First Institute of Oceanography, Ministry of Natural Resources, Qingdao, China

Introduction

1

Xiao Hua Wang[1] *and Lulu Qiao*[2]
[1]The Sino-Australian Research Consortium for Coastal Management, School of Science, University of New South Wales, Canberra, ACT, Australia; [2]College of Marine Geosciences, Ocean University of China, Key Lab of Submarine Geosciences and Prospecting Techniques, Ministry of Education, Qingdao, China

1. Background

The ever-evolving dynamics of estuarine and coastal environments are critical components of the Earth's intricate systems, influencing both natural processes and anthropogenic activities. In the pursuit of understanding and managing these dynamic coastal systems, our monograph, titled "Recent Trends in Estuarine and Coastal Dynamics: Observations and Modelling," serves as a comprehensive exploration of key aspects in the realm of Physical Oceanography, Remote Sensing, and Ocean Modeling.

Coastal environments are key locations for transport, commercial, residential and defense infrastructure and provide conditions suitable for economic growth. They also fulfill important cultural, recreational, and aesthetic needs. The coasts cover a diverse range of ecosystems within marine, estuarine, and freshwater environments and have intrinsic ecosystem service values. They are some of the most heavily populated and visited areas and are also some of the most threatened natural habitats. Therefore, coastal zones are critically important not only to the people who live there but for the health of the planet.

Estuaries are semi-enclosed bodies of water where freshwater rivers or streams meet the ocean. They give rise to unique ecosystems distinguished by their ever-changing salinity patterns. Located on coastal plains estuaries, they display a wide array of geomorphological features and unique salinity distributions. The interaction of factors like river discharge, tidal forces, current velocities, basin shapes, sediment composition, and wave energy contributes to their intricate nature and complex dynamics. This elaborate interplay between river flow, tidal currents, and basin geometry shapes the circulation patterns and distinct salinity variations found in each estuary, ultimately influencing the plant and animal species that thrive within them.

Estuaries and coasts hold immense importance in coastal management and oceanographic studies. Communities grew alongside estuaries and coasts because they provided drinking and irrigation water, sewage disposal, and transport routes. Industrial revolution increased urban and industrial pressures. Cities grew, urbanization mushroomed, clean water, and clean beaches provided for recreation water, waste and heat removal. Nowadays compromise is no longer possible for improving coastal environment health versus human development.

Current Trends in Estuarine and Coastal Dynamics. https://doi.org/10.1016/B978-0-443-21728-9.00001-6

Therefore, a good understanding of the current state of these marine environments and lessons learned from human influences would be extremely valuable to restore and protect these habitats and ecosystems from further environmental degradation and even catastrophe. We need to understand flushing time and safe limits of waste and heat disposal; coastal and estuarine ecosystem dynamics, that is, circulation, mixing characteristics and biogeochemical processes; and interaction with offshore environment. In addition, large-scale oceanographic processes are often mirrored in small-scale estuarine and coastal environments. In this book, we will document recent trend and development in observing and modeling of estuaries and coastal oceans with a focus on sediment transport dynamics and hydrodynamics of these environments. As such, the monography will identify some key coastal management issues and offer strategies to address these environmental problems facing coastal oceans worldwide.

2. Scopes

The chapters in this monograph are meticulously curated to provide a holistic view of the multifaceted interactions within estuarine and coastal regions. The compilation seamlessly integrates insights from in situ observations, satellite remote sensing, and advanced modeling techniques, offering a balanced and thorough examination of the complexities inherent in these environments.

The chapters are organized to cover diverse topics, ranging from in-situ observations of sea surface temperature (Chapter 2) to the intricacies of coastal bathymetry inversion using satellite remote sensing (Chapter 3). We delve into the significance of coastal eddies as transient yet impactful features (Chapter 4) and explore the effects of stratification asymmetries on residual flows in the Sydney Harbor Estuary, Australia (Chapter 5). Furthermore, the monograph delves into contemporary methods for estuarine sediment provenance (Chapter 6) and hydrologic and sediment investigations in coastal zones (Chapter 7).

Our scope extends to historical changes in hydro and sediment dynamics due to coastline alterations in Hangzhou Bay, China (Chapter 8) and modeling studies on hydrodynamics and sediment dynamics in estuarine environments (Chapter 9). The book also scrutinizes the response of tidal dynamics to the construction of Yangshan Harbor in Shanghai, China (Chapter 10) and recent progress in the studies of boundary upwelling (Chapter 11). Additionally, we explore the influences of extreme rainfall events on nutrient and chlorophyll dynamics in coastal regions (Chapter 12) and review theories and evaluation methods for coastal ecological and environmental management under multiple anthropogenic pressures (Chapter 13).

With its comprehensive coverage, this monograph is poised to serve not only as a valuable reference for researchers and practitioners in the field but also as a textbook for advanced undergraduate courses in oceanography. It aspires to bridge the gap between theoretical knowledge and practical applications, fostering a deeper understanding of the intricate dynamics governing estuarine and coastal systems.

skin SST being usually slightly cooler, typically by a few tenths of a degree, than the water temperature immediately below, the sub-skin SST, due to direct heat loss within the skin layer from the water to the atmosphere (Fairall et al., 1996; Minnett et al., 2011; Saunders, 1967; Zhang et al., 2020). The sub-skin temperature can be approximated to the measurement made by a spaceborne or airborne passive microwave radiometer operating in the 6−11 GHz frequency range at depth of ∼1 mm.

- Depth SST (SST_{depth}): all SST measurements beneath the sub-skin SST, and this is where SST DV may exist. During the daytime, SST from below the surface to a few meters depth could be warmer than the foundation SST (see definition below) by up to several degrees due to strong solar heating or low wind-induced mixing or a combination of both. This is referred to as the SST DV. This signal often largely vanishes over the nighttime (Fairall et al., 1996; Zhang et al., 2016).

- Foundation SST (SST_{fnd}): the temperature free of DV, that is, the temperature at the first time of the day when the heat gain from the solar radiation absorption exceeds the heat loss at the sea surface (approximately 10 m water depth under calm conditions).

In situ SST measurements have been a crucial component of the Global Ocean Observing System (GOOS, https://www.goosocean.org/accessed on September 4, 2023 (Moltmann et al., 2019; Tanhua et al., 2019)), and they are of particular importance around Australia's coastline due to their historical scarcity. Compared to satellite SSTs, which are strictly confined to the sea surface (measuring only SST_{skin} or $SST_{sub-skin}$) and become less optimal close to the coast, in situ observations can be obtained at both the surface and at depth. Additionally, the calibration and validation (cal/val) of satellite retrievals are highly dependent on good quality in situ measurements. Over the past few decades, the accuracy of in situ SST sensors has significantly improved with emerging new technologies. The spatial and vertical coverage of in situ SSTs have also greatly expanded with increasing numbers of internationally coordinated programs and projects. Historically, in Australian and adjacent waters, high-quality in situ SSTs were relatively low in number. For example, the Indonesian seas, Gulf of Carpentaria, Bass Strait, the Great Australian Bight, and the Southern Ocean have been known to lack drifting buoy data (Beggs et al., 2012). However, this situation has been substantially enhanced over the past 15 years or so since the launch of the Integrated Marine Observing System (IMOS, https://imos.org.au/accessed on September 4, 2023). The IMOS project is a large-scale collaborative observing system that is carried out by national facilities and operated by partner institutions around Australia. Each facility manages a particular type of observing platform. By bringing together 10 different organizations, including universities and marine agencies from all over the country, IMOS aims to facilitate marine and climate research with an open-data approach. It has now become the cornerstone of Australia's contribution to GOOS and plays a leading role in the development of observing systems in the Southern Hemisphere (Hill et al., 2010; Lara-Lopez et al., 2016; Lynch et al., 2014; Proctor et al., 2010). More information regarding SST-related facilities and sub-facilities will be provided in the following sections.

For the remainder of this chapter, we will discuss in situ SST measurements obtained from different types of platforms with more attention focused on those available around Australian adjacent waters. The following sections will be divided on a platform basis with detailed introductions on ships and coastal moorings and largely neglecting those platforms primarily designed for the open ocean, such as Argo floats, regardless of their critical role in GOOS. For more information on open ocean platforms, readers are referred to several excellent previous review studies (Centurioni et al., 2019; Kent et al., 2010; Kent & Kennedy, 2021; Legler et al., 2015; O'Carroll et al., 2019; Send, 2006; Ravichandran 2011). Development history and features for each observation platform will be highlighted in the following sections. Projects that deliver in situ SST measurements around Australia will be presented. Data transmission, management, visualization, and acquisition are described toward the end.

2. SST from ships

Sea surface temperatures observed from ships have the longest history extending back more than a century. The earliest reliable time series of SST measurements began when the Maritime Conference was held in Brussels in 1853 (Rayner et al., 2006). Over the years, a range of different types of ships have been employed to provide SST measurements, including cargo vessels, passenger vessels, and research vessels (RVs) through programs such as the international Voluntary Observing Ships (VOS) and the Ship of Opportunity Program (SOOP). The observing methods, or sensors, for shipborne SSTs have been growing more accurate over time, leading to increasingly better-quality data. Shipborne SSTs are an important data source in coastal waters as they can get closer to the coastline than many other platforms. In this section, programs such as the VOS and SOOP will be first introduced, followed by instrumentation, and then ship SST projects in Australia are separately discussed.

2.1 VOS and SOOP

The VOS program is one critical component of the World Meteorological Organization (WMO, https://public.wmo.int/en, accessed on September 7, 2023)—Intergovernmental Oceanographic Commission (IOC) Ship Observations Team (SOT) (http://sot.jcommops.org/vos/vos.html, accessed on August 23, 2023; Smith et al., 2019). The VOS program became more established at the start of the twentieth century when wireless telegraphy was introduced, and Port Meteorological Officers (PMOs) were appointed to service the ships (Legler et al., 2015). In 1974 at the International Maritime Organization (IMO) Safety of Life at Sea (SOLAS) convention, governments were encouraged to participate in the VOS scheme and provide high-accuracy meteorological and oceanographic equipment to selected ships that are recruited into national fleets. Consequently, following this convention, the number of VOS ships peaked in the 1980s. In March 1984, there were more than 7800 ships reporting SST measurements to Global Telecommunication System (GTS). From then on, the size of the fleet steadily declined until around 2015 when it started to stabilize at around 2000 ships (Fig. 2.2 panel a). However, the amount of SST observations reported to GTS has been relatively stable throughout the 1980s and 1990s and has

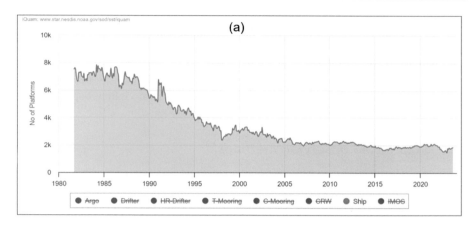

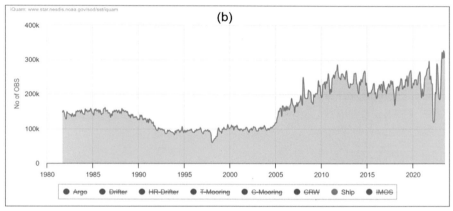

Figure 2.2 Number of ships and ship SSTs. (a) Number of ships reporting SST from September 1981 to September 2023; and (b) number of ship SST observations.
From NOAA In Situ SST Quality Monitor - iQuam; https://www.star.nesdis.noaa.gov/socd/sst/iquam/?tab=2&dayofmon=monthly&qcrefsst=_qcrey&qcrefsst=_qccmc&qmrefsst=reyn#tims, accessed on September 11, 2023; Xu, F., & Ignatov, A. (2014). In situ SST quality monitor (iQuam). Journal of Atmospheric and Oceanic Technology, 31(1), 164−180. https://doi.org/10.1175/JTECH-D-13-00121.1.

even showed a climbing trend in the past 2 decades with a peak at > 300,000 per month in August 2023 (Fig. 2.2 panel b). This is largely because of the frequent employment of the Shipborne Automatic Weather Stations (S-AWSs), which provide hourly measurements instead of the traditional four observations per day at the synoptic reporting hours (0000, 0600, 1200, and 1800 UTC).

Another critical component of the WMO SOT is the SOOP (https://www.ocean-ops.org/sot/soop/, accessed on August 23, 2023). Currently, there are at least 20 agencies from around the world involved in the SOOP. The primary objective of SOOP is to fulfill the Expendable BathyThermograph (XBT) upper ocean data

requirements. However, after the introduction of Argo floats in 1999, XBT temperature profiles are not as widely used in the SST community, although they have been
employed in the studies of near-surface heat content and its variability (Domingues
et al., 2008). In addition to the deployment of XBTs, many SOOP ships also contribute
to other programs such as the Argo float project and the Global Drifter Program. Some
SOOP vessels measure NRT SST and salinity using a ThermoSalinoGraph (TSG), in
addition to many other air—sea interfacial variables (Goni et al., 2010).

2.2   Ship-based observing methods

Thus far, there are five major methods from which an SST measurement can be taken
from a ship, namely from thermometers placed in buckets, or in engine room intake
(ERI) waterpipes, from hull-contact sensors and through-hull water temperature sensors (combined in Subsection 2.2.3), and from shipborne IR radiometers. Note that
several studies have previously reviewed one or more of these methods and/or associated data uncertainties (Folland & Parker, 1995; James & Fox, 1972, pp. 0—336; Kennedy et al., 2011; Kennedy, 2014; Kent et al., 2017; Kent & Kennedy, 2021; Kent &
Taylor, 2006; Matthews, 2013). This chapter will briefly introduce each method with
foci on the sensor evolvement and its general data quality. Also note that there are
other methods to measure SST from a ship that are not included in this study due to
scope limitation, such as the floating thermistors attached to the end of an umbilical
towed along the surface (a.k.a., sea snakes) or CTD (Conductivity, Temperature,
Depth) casts.

2.2.1   Buckets

From the mid-1850s until the 1940s, it is believed that almost all SST data were
derived from bucket samples, with wooden buckets gradually being phased out and
replaced by "uninsulated" canvas buckets (Kent & Taylor, 2006). The first issue of
rubber buckets was in 1957, and these mostly replaced the canvas buckets by the
1970s. From the 1970s, nearly all the bucket SST measurements were obtained
from insulated rubber or plastic buckets (Kent et al., 2017). Overall, the portion of
bucket SSTs in ship SST measurements has been decreasing since the 1970s, which
is mostly due to the increasing number of ERI measurements. Zhang and Ignatov
(2021) found that selecting 2 years from 2016—2017 as an example, bucket SSTs
only account for ~ 2.8% of all ship measurements collected within the International
Comprehensive Ocean—Atmosphere Data Set (ICOADS) NRT release R3.0.1, and
the United States Global Ocean Data Assimilation Experiment (GODAE (Bell et al.,
2009) Fleet Numerical Meteorology and Oceanography Center (FNMOC, https://
www.metoc.navy.mil/fnmoc/fnmoc.html, accessed on September 26, 2023) data set
(Fig. 2.3). Spatially, bucket SSTs are most common in the Atlantic due to the preference of some European countries for this type of sampling method.

Historically, the quality of bucket SSTs have been less than optimal. A few sources
of error are expected. For example, the water might be cooled after being pulled up due
to direct heat loss to the often colder air, or due to precipitation, or evaporation from

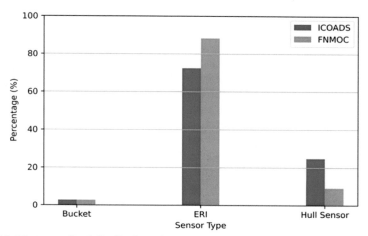

Figure 2.3 The normalized distribution of the shipborne SST measurements obtained from different sensor types, that is, bucket, engine room intake (ERI), and hull-contact sensors, for ICOADS (*blue*) and FNMOC (*orange*) data from 2016−17.
From Zhang, H., & Ignatov, A. (2021). A completeness and complementarity analysis of the data sources in the noaa in situ sea surface temperature quality monitor (Iquam) system. Remote Sensing, 13(18), 3741. https://doi.org/10.3390/rs13183741.

the surface or from the walls of the bucket under windy conditions, even for insulated instruments (James & Fox, 1972, pp. 0−336). During the earlier days, these samples could also be severely affected by diurnal warming signals during calm and sunny days, yet this was not well quantified (Kennedy, 2014). Moreover, when there is manual intervention in making a report, there is a tendency to round the report, perhaps to a whole number or to half a degree (Kent & Taylor, 2006). Quantitatively, (Kent & Challenor, 2006) found that the random error for bucket SSTs is 1.1 ± 0.3 K using data from 1970 to 1998 on a nearly global scale (45°S-75°N). The number is similar in (Cummings, 2006), which is 1.2 K. Many efforts have also been devoted to the better understanding and correction of uncertainties of ship SSTs (Kennedy, 2014; Kent et al., 1993; Kent & Kaplan, 2006; Matthews, 2013). Note the differences between random and systematic errors here. Random errors typically include misreading of the thermometer, rounding errors, incorrectly recorded values, etc. They tend to cancel out when large number of measurements are averaged. However, systematic errors occur when a particular thermometer is mis-calibrated or poorly sited, which become more pronounced as greater numbers of observations are aggregated (Kennedy, 2014).

2.2.2 Engine room intake

In the 1920s and 1930s, engine-inlet temperatures started to be recorded on many ships. The ERI SST is the temperature of the pumped seawater used to cool the engines. It is not a dedicated method of measurement. Nonetheless, these measurements were considered satisfactory under most circumstances. Early ERI temperatures may have been taken with a mercury thermometer in a well in the intake pipe or from a dial with temperature intervals of several degrees. They are now more likely to be taken with an electronic thermometer. Since the 1970s, the proportion of ERI records increased steadily, and became more numerous than buckets by the early 1980s (Kent & Taylor, 2006; Zhang & Ignatov, 2021). Based on the measurement method indicator flag information contained in the ICOADS data set, the number of ERI measurements only became significant from 1982 following a code change to allow the reporting of the method with the GTS report. In 1995−97, nearly 50% of ship SSTs were recorded as engine-intake data, with the SST method information being included in the National Centers for Environmental Prediction (NCEP, https://www.weather.gov/ncep/, accessed on September 18, 2023) metadata. In a more recent case study, for 2 years from 2016−17, ERI SSTs make up 72.4% of ship measurements based on the ICOADS R3.0.1 data set (Zhang & Ignatov, 2021). Spatially, ERI data are widely spread globally.

Like bucket SSTs, the quality of ERI SSTs can also be more problematic compared to other in situ SST platforms such as drifting buoys or moorings. Error sources for ERI SSTs can be incrustation and fouling, poor exposure of the thermometer well to the flow in the pipe, air pockets inside the thermometer well, and heat conduction along metal supports to the thermometer bulb. Kent and Challenor (2006) found that the random error for ERI SSTs is 1.4 ± 0.3 K using data from 1970 to 1998 on a nearly global scale (45°S-75°N). ERI SSTs are known to usually have a warm bias before the 1990s. Kent and Challenor (2006). identified a negative difference between bucket and ERI SST measurements that is present up to around 1986 and then declines by around 1990. This could be due to that the heating problems with ERI measurements may have been solved by the early 1990s. Note that given the difference in measurement depths (surface for the bucket compared with an average of 10 m for the engine intake), the ERI SST should, if anything, be colder than the bucket SST.

2.2.3   Hull-contact and through-hull sensors

There are two main types of hull sensors, the hull-contact, or hull-mounted, and the through-hull sensors. The hull-contact sensors started to be applied on ships to measure SST from the late 1970s. It is a dedicated SST sensor that measures the water temperature through the hull. The hull-contact sensors were usually installed in the engine room because it is often the only place where the hull can be easily reached (Emery et al., 1997). When installing the instrument, to make sure the sensor is always under water, the minimum and maximum water lines must be located before the installation of the hull sensors can proceed. Another consideration before installing the hull-contact sensors is the location of release valves and outflow pipes for engine room cooling water discharge. Nowadays, hull-contact sensors are still relatively uncommon, but their usage is steadily increasing (Kent & Taylor, 2006). Hull-contact sensor SSTs are usually considered of reasonably good quality. These sensors were found to be relatively unbiased and showed no systematic change of bias with depth (Kent et al., 1993). A total uncertainty of ∼0.6 K for hull-contact SSTs was found in (Cummings,

2006). More recently, Beggs et al. (2012) demonstrated that the hull-contact temperature sensors produce SST data with comparable uncertainties to those available from drifting buoys in the same region.

The through-hull SST sensor is a thermistor that is placed in water intakes close to the outer skin of the vessel and directly measures the water temperature. This SST instrument is typically only installed on RVs. The water intake is for scientific purposes, hence clean and largely free of contamination from sources such as engine cooling water discharge. Through-hull sensors are rare on VOS or other non-RVs because the installation costs can be prohibitive unless the sensor and cabling are fitted when the ship is built (Kent & Taylor, 2006). The SSTs measured by these sensors can have very high quality and be used for DV and cool skin effect studies (Jessup & Brance, 2008; Zhang et al., 2020).

2.2.4 Shipborne IR radiometer

A shipborne IR sensor measures the skin SST at the same depth as a spaceborne IR radiometer (10–20 μm). Due to the often well-known observing conditions (ocean surface mixing, view angle, etc.) and lack of atmospheric interference, shipborne IR sensors can measure skin SST to extremely high accuracy (Donlon et al., 2014). Therefore, they are the optimal option for the cal/val of spaceborne IR SSTs. The first shipborne IR radiometer dates back to 1970 when the Barnes model PRT-5 radiometer was developed (Colacino et al., 1970; Schluessel et al., 1987). Over the last 50+ years, a steady development of shipborne radiometer design has taken place, leading to the present generation of instruments being capable of taking measurements with an accuracy of better than 0.1 K (Donlon et al., 2014). The European Space Agency (ESA) established a project named as the International SST Fiducial Reference Measurements (FRM) Radiometer Network, also known as "Ships4SST" (https://ships4sst.org/, accessed August 25, 2023). The major objective of this project is to collect, process, archive, and quality control (QC) shipborne IR SST data to facilitate the validation of satellite retrievals from new sensors such as sea and land surface temperature radiometer (SLSTR) onboard Sentinel-3 satellite (Coppo et al., 2013; Wimmer et al., 2020). Note it is often recommended that shipborne IR SSTs are not included in the SST retrieval algorithm training data set and should be reserved as an independent source for validation purposes only (Peter Minnett, *personal communication*). Several popular shipborne IR radiometer models are briefly introduced here.

2.2.4.1 SISTeR

The scanning IR sea surface temperature radiometer (SISTeR) is a compact and robust chopped self-calibrating filter radiometer that was designed in 1992 (Barton et al., 2004). All SISTeR measurement sequences contain repeat measurements of its two internal blackbodies. To calculate skin SST, the SISTeR is programmed to make measurements of upwelling radiances from the sea surface and complementary downwelling sky radiances. The SISTeR instrument can operate unattended for extended periods even through bad weather as it is equipped with a rain detector and an electronically signaled weather door, and its fore-optics and electronics enclosures are waterproof. These SISTeR instruments have been deployed since 1996 on a range of RVs and passenger vessels. In the SISTeR longwave channels, the measured noise temperature for a 1s sample at typical SSTs is < 30 mK (Nightingale, 2006).

2.2.4.2 M-AERI

The marine-atmospheric emitted radiance interferometer (M-AERI) is a seagoing Fourier-transform IR spectroradiometer that has now been used on many research cruises ranging from Arctic to equatorial conditions (Luo & Minnett, 2020; Minnett et al., 2001, 2011). It has proven to be sufficiently robust to function at sea and take measurements in conditions where the temperature has ranged from -20–$35°C$ and in wind speeds of up to ~ 17 m/second. The skin SSTs derived from the M-AERI spectra have an uncertainty of ~ 0.04 K (Luo & Minnett, 2020).

2.2.4.3 ISAR

The infrared SST autonomous radiometer (ISAR) is a self-calibrating instrument capable of measuring skin SST to an accuracy of 0.1 K (Donlon et al., 2008). ISAR is a fully autonomous radiometer system that has been developed for satellite SST validation and other scientific programs. Like M-AERI, the ISAR system uses an optical rain detector and storm shutter arrangement that completely seals the instrument from the environment when the atmosphere contains dust or water droplets (from precipitation or ocean spray). Up to 2021, at least 16 ISAR instruments have been built and are in sustained use by institutions around the world, including University of Southampton, University of Miami, Ocean University of China, Royal Navy, Japan Aerospace Exploration Agency, Danish Meteorological Institute, Woods Hole Oceanographic Institute (WHOI), and the Australian Commonwealth Scientific and Industrial Research Organization (CSIRO).

Other shipborne IR radiometers are left out of this study, such as the Calibrated Infrared In situ Measurement System (CIRIMS; Jessup & Brance, 2008), the Heitronics radiometers (https://www.heitronics.com/en/, accessed on October 5, 2023), or the OUCFIRST (FIRST IR radiometer made by the Ocean University of China; Zhang et al., 2018). Interested readers are pointed to the corresponding references above for more information.

2.3 IMOS ship SST

Historically, ship SST measurements around the Australian coasts have been scarce and of low quality. Prior to 2008, SSTs obtained from ERI sensors onboard VOS ships around Australia were substantially noisier than drifting buoy measurements, hence not suitable for the NRT validation of satellite retrievals (Beggs et al., 2012). This situation has been largely improved since 2008 when the Australian IMOS was initiated. Specifically, two IMOS sub-facilities operated by the Australian Bureau of Meteorology (the Bureau), namely the Shipborne SST Sensors for Australian Vessels (http://imos.org.au/sstsensors.html, accessed on August 29, 2023) and the RV Real-time Air-sea Fluxes (https://www.imos.org.au/facilities/shipsofopportunity/airseaflux, accessed on August 29, 2023), have been producing high-accuracy, quality controlled (QC'd) SST measurements (Beggs et al., 2012, 2017). These SSTs are measured at a few meters below the surface. Up to August 2023, there have been 25 ships involved in this project, including multiple VOS, SOOP ships and RVs, with seven ships currently in operation (as of September 2023). Twelve of those vessels are or were equipped with Seabird (https://www.seabird.com/, accessed September 18, 2023) SBE3 or SBE38 digital thermometers located in the water intake, 14 with SBE48 sensors positioned against the inside of the ship's hull, and one vessel (MV, Merchant Vessel, Harbour Master, Rottnest Island ferry in Western Australia) with both SBE38 and SBE48 sensors. RV Investigator has also been equipped with an ISAR and can measure skin SSTs (Zhang et al., 2020, 2023). Tracks of all ships from the beginning of the IMOS sub-facilities in 2007–2023 that have reported SSTs are shown below (Fig. 2.4).

These NRT SST measurements are most valuable after being carefully quality controlled. At the Bureau, the QC process is automatically performed in NRT in a dedicated data processing system. The QC procedure employs a method based on the Shipboard Automated Meteorological and Oceanographic System (SAMOS, http://samos.coaps.fsu.edu/html/, accessed on August 29, 2023). Specific tests consist of checks on physical bounds, time, position, platform speed, exhaust contamination, statistical, and climatological bounds. Please refer to (Schulz et al., 2021/2021) for more details. Following QC these SST measurements are transmitted to the GTS in NRT. These data are also uploaded daily to the Australian Ocean Data Network, AODN (https://portal.aodn.org.au/, accessed on August 29, 2023), as Network Common Data Form (NetCDF) files. Refer to Section 5 for more information on data transmission.

Figure 2.4 All tracks of IMOS ships, including VOS, SOOP ships and RVs, that have provided SSTs from 2008 to 2023.
From Australian Ocean Data Network, AODN https://portal.aodn.org.au/, on August 29, 2023

These ship SST measurements are utilized in many downstream applications. For example, the NRT data are ingested into several analyses at the Bureau, including the Regional Australian Multi-Sensor SST Analysis (RAMSSA; Beggs et al., 2011) and Global Australian Multi-Sensor SST Analysis (GAMSSA (Zhong & Beggs, 2008), which feed into boundary conditions of Numerical Weather Prediction (NWP) and climate models. From the GTS, they are also included in many international SST analyses at various agencies (UK Met Office, NOAA, Canadian Meteorology Centre, the Japan Meteorological Agency, etc.), and global in situ ocean databases. The ISAR skin SSTs obtained from RV Investigator are also periodically reprocessed at CSIRO to ensure their high quality and are then re-published by the Bureau to AODN (Beggs et al., 2017; Zhang et al., 2020). The delayed mode ISAR data serve as excellent ground truth data for validation of satellite SST (Yang et al., 2020; Zhang et al., 2023).

3. SST from coastal moorings

Moorings, such as anchored buoys or an anchored configuration of instruments suspended in the water column, are another key platform type to take high quality in situ SST measurements. The sustained long (multi-year or often multidecadal) time series of SST collected from a mooring at a fixed location allow for the tracking of change in oceanographic variables and are of great value to, for example, climatological studies or validation of model outputs or satellite products. Generally, moorings are divided into two groups depending on their locations: coastal moorings (CMs) and open ocean moorings such as the Global Tropical Moored Buoy Array (GTMBA, https://www.pmel.noaa.gov/gtmba/, accessed on September 18, 2023). In this chapter, emphasis will be put on CMs, especially those with a surface float.

3.1 Mooring history and surface float designs

In the 1940s, before the deployment of moorings, Ocean Weather Stations (OWSs) were used by oceanographers to monitor the ocean. The OWSs were examples of specialized ships introduced during the time of World War II and provided high-quality data at fixed locations in support of NWP operations. These OWSs were gradually phased out after the 1970s and can be considered as the predecessor of moorings (Woodruff et al., 2008). To obtain the best design of a mooring's surface float, the Office of Naval Research (https://www.nre.navy.mil/, accessed on September 18, 2023) conducted a number of studies in the 1960s. As a result, the 12 m discus buoy emerged as the most reliable and survivable (Soreide et al., 2001). While better communications and sensor reliability were achieved and could be advantageous, the escalating costs of the 12 m hull became a major problem. A 10 m discus buoy was then developed as a compromise for survivability and continued reliable performance of the sensors. The 10 m buoy was the first to be used operationally in 1974 to obtain marine weather data by the National Data Buoy Service, later to become the National Data Buoy Centre

(NDBC, https://www.ndbc.noaa.gov/, accessed on September 18, 2023). Later the WHOI developed a 3 m discus design, which was then employed operationally by NDBC (Soreide et al., 2001). More recently, the 6 m NOMAD (Navy Oceanographic Meteorological Automated Device) buoy was designed. It is an aluminum-hulled, boat-shaped buoy, which provides relatively high cost-effectiveness and excellent long-term survivability in severe seas. Like the 3 m discus, they are less likely to corrode and the magnetic effects on the compass are slight. The Data Buoy Cooperation Panel (DBCP, https://www.ocean-ops.org/dbcp/, accessed on September 18, 2023), which is an official joint body of the WMO and the IOC, coordinates over 400 CMs, and ensures the meteorological and oceanographic data are available in NRT to support global forecasts of weather and ocean conditions. Currently, many countries operate CMs and contribute to, or are coordinated by, the DBCP, including the United States of America (USA), Canada, Australia, UK, India, Ireland, Spain, and the Republic of Korea. Note that the DBCP inventory does not represent a complete set of global CMs as it relies on observations sent to the GTS, hence missing those CMs which fail to do so (such as CMs in Australia) (Bailey et al., 2019).

3.2 CM SSTs in satellite cal/val

The SST sensor onboard a CM surface float is typically at a depth of ~ 1 m. The *i*Quam (in situ SST Quality Monitor; https://www.star.nesdis.noaa.gov/socd/sst/iquam/, accessed on 8/9(/2023; Xu & Ignatov, 2014) data show that when compared against the level 4 (L4) SST product from the Canadian Meteorological Center (CMC, https://weather.gc.ca/, accessed on September 21, 2023), CM SSTs constantly show a standard deviation of 0.5−0.7 K, which is better than most ship bucket or ERI SSTs, yet less optimal than other platform types such as drifting buoys or tropical moorings. Here an L4 SST product is a gridded, gap-free data set that represents the foundation SST free of DV signals. Relatively large uncertainties of CMs have also been found when compared to remote sensing retrievals in other studies (Castro et al., 2012; Kennedy et al., 2011; Merchant et al., 2012; Xu & Ignatov, 2016). This is, to some extent, because in coastal waters, there can be large local variations in temperature, which satellites or L4 products cannot resolve. Some moorings along coastlines are located in estuaries and river mouths and are therefore less likely to be representative of open ocean areas (Kennedy, 2014).

3.3 Australian National Mooring Network

As one of the largest facilities in IMOS, the Australian National Mooring Network (ANMN, https://imos.org.au/facilities/nationalmooringnetwork, accessed on September 18, 2023) maintains a series of National Reference Stations (NRSs) and associated regional moorings that provide oceanographic observations along Australia coasts and continental shelves (Lynch et al., 2014). The history of Australian coastal observations dates to 1942 when the first coastal station was established in Port Hacking, New South Wales (NSW). Over the decades, dozens of coastal stations were put into operation but most of them were sustained for only short periods of time. Only a few are still operational at the following locations: Maria Island on Tasmania's east coast, Port Hacking in NSW, and Rottnest Island in Western Australia (Roughan et al., 2022). Records of a suite of physical, chemical, and later biological (temperature, salinity, current, nutrients, dissolved oxygen, etc.) oceanographic parameters generated from these stations form some of the longest sub-surface time series in the world (Roughan et al., 2022).

3.3.1 National reference stations

Designed to build upon the existing time-series stations to ensure their maintenance, the expanded NRS scheme was established as part of the IMOS project in 2008, aiming to also cover other marine provinces around Australia's coastal seas. Critical goals for this NRS network consist of the establishment of a coastal information infrastructure through the development of national data standards and to facilitate the validation of modeling activities and coastal remote sensing products. As of today, there are seven NRS sites scattered within the five IMOS regional nodes of Queensland, NSW, Western Australia, South Australia, and Tasmania (Fig. 2.5). These long-term water temperature time series data have enabled investigation into decadal trends of sea temperatures (Hemming et al., 2023; Ridgway & Ling, 2023).

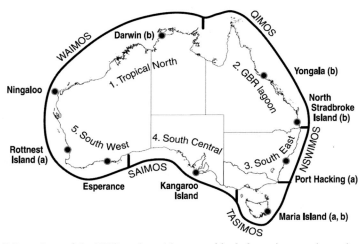

Figure 2.5 Locations of the NRS stations (shown as *black dots*; nine are shown here but the Ningaloo and Esperance ones were decommissioned in 2014), the five regional IMOS nodes, i.e., QIMOS (Queensland node), NSWIMOS (NSW node), TASIMOS (Tasmania node), SAIMOS (Southern Australia node) and WAIMOS (western Australia node); *a* in bracket indicates a long-term site, and *b*, site with telemetry.
From Lynch, T.P., Morello, E.B., Evans, K., Richardson, A.J., Rochester, W., Steinberg, C.R., Roughan, M., Thompson, P., Middleton, J.F., Feng, M., Sherrington, R., Brando, V., Tilbrook, B., Ridgway, K., Allen, S., Doherty, P., Hill, K., Moltmann, T.C.. 2014, 12, 17. IMOS national reference stations: A continental-wide physical, chemical and biological coastal observing system. PLoS One. 9(12), 19326203. https://doi.org/10.1371/journal.pone.0113652.

The NRS sampling program consists of several components. The continuous and robust data collected from moorings at each site is one critical part. In terms of water temperatures, some of the NRS moorings provide only sub-surface observations with the top layer being at a depth of ~ 20 m or deeper, as they lack a surface float. Prior to 2017, out of the seven sites, there were four moorings (Maria Island, Darwin, Yongala, and North Stradbroke Island; Fig. 2.5) equipped with a surface float sensor package and telemetry capability, including a temperature sensor (SBE39) measuring SST at depths in the range of 0.6—1 m. The electronic packages available on these four moorings allowed for the NRT transmission of SST and sub-surface data on an hourly basis (Lynch et al., 2014). In 2017, the NRT data systems at Maria and North Stradbroke Islands were removed due to reductions in operating budgets (David Hughes, *personal communication*). For those sub-surface moorings, data will be collected when the mooring is serviced, which occurs every few months (Bainbridge et al., 2008).

In addition to the continuous moored sensors, the NRS sampling program also incorporates vessel-based sampling and laboratory analyses. Historical bottle samples were usually collected at nominal depths of 0, 10, 20, 30, 40, and 50 m, of which the top two layers within 10 m of the surface can be considered as SST. Since 2008, a series of Seabird CTD profilers have been employed at the NRS sites with the primary purpose to facilitate the biogeochemical sampling of the water column. Frequency of the vessel-based CTD sampling ranges from monthly to quarterly. The temperature sensor has very high resolution (0.0001 K) and accuracy (0.005 K). However, the precision of the readings will be limited and likely significantly less (~ 0.05 K) due to reasons such as the need for transportation (Davies et al., 2020).

3.3.2 Regional array moorings

In addition to the NRS sites, each of the five geographic IMOS nodes is also responsible for the local array moorings. Common physical and biogeochemical parameters are measured at most locations. Depending on their unique oceanographic features, these local moorings may serve a particular purpose. For example, the Great Barrier Reef (GBR) moorings (https://data.aims.gov.au/moorings/index.html, accessed on September 5, 2023) are largely dedicated to monitoring the impact of climate change and other environmental factors on coral reef ecosystems (Bainbridge et al., 2008). Few of these local moorings have surface floats, hence incapable of measuring and transmitting NRT SST observations. In the GBR region, except for one mooring at NRS Yongala (Fig. 2.5), all the others are sub-surface moorings with temperature sensors at typically $\sim$ 15 m depth and below (Simon Spagnol, *personal communication*).

4. SST from other platforms

Other than the moored buoys and ships, coastal in situ SST observations may be obtained from other platforms as well. A selected few are briefly described here as examples.

4.1 Drifting buoys

In the early 1980s, the World Climate Research Program (WCRP) and oceanographic community recognized that a global array of drifters would be invaluable for oceanographic and climate research. The development of a standardized, low-cost, lightweight, easily deployed drifter with a semi-rigid drogue became necessary. By 1993, a design for the Surface Velocity Program (SVP; Lumpkin & Pazos, 2007) had emerged, which combined the holey-sock drogue of the Atlantic Oceanographic and Meteorological Laboratory (AOML, https://www.aoml.noaa.gov/accessed on October 6, 2023) drifters with reinforced tether ends and surface float designs from Scripps Institution of Oceanography (SIO, https://scripps.ucsd.edu/, accessed on October 6, 2023). These satellite-tracked drifting buoys have drogues centered at a depth of 15 m to measure mixed layer currents and to minimize direct wind forcing and Stokes drift. This design became the foundation for future SVP drifter development. Today, the array of SVP drifters is known collectively as the Global Drifter Program (GDP; https://www.aoml.noaa.gov/phod/gdp/, accessed on October 6, 2023), a critical component of GOOS and a scientific project of the DBCP (Fig. 2.6; Centurioni et al., 2019; Lumpkin et al., 2013). More recently, an improved type of SVP drifter was designed and built by the Toward fiducial Reference measUrements of Sea-Surface Temperature by European Drifters (TRUSTED) project (Le Menn et al., 2019; Poli et al., 2019). In addition to the regular SST sensor, a second high-resolution SST (HRSST) sensor is installed on these TRUSTED drifters, which have an SST uncertainty smaller than 0.01 K, making it suitable for the cal/val of high accuracy satellite SSTs from the Copernicus Sentinel-3 SLSTR (Poli et al., 2019). Note that in addition to the SVP family of drifters, there are other models that are widely applied as well. An example is the Coastal Ocean Dynamics Experiment (CODE) drifter, developed by

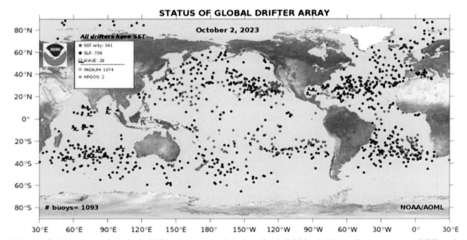

Figure 2.6 Status of global drifter array on October 2nd, 2023. *Red circles* represent SST only drifters. *Blue circles* indicate those which also measure sea level pressure. Wave drifters are indicated by the wavy line below the symbol. Note all buoys measure SST.
From NOAA AOML GDP webpage https://www.aoml.noaa.gov/phod/gdp/index.php, accessed on October 6, 2023.

(Davis, 1985) to measure the currents in the top 1 m of the water column in coastal areas.

In the open oceans, drifting buoys provide the most abundant, high quality in situ SSTs at a depth of ~ 20 cm, due to their good temporal (hourly) and spatial (global) coverages. According to iQuam, in September 2023, the monthly total number of SST observations from drifters ($\sim 770,000$) was dramatically larger than those from ships ($\sim 287,000$), Argo floats ($\sim 12,000$), or tropical moorings ($\sim 18,000$). In terms of data quality, many studies have indicated that drifters have an overall standard deviation of $\sim 0.1-0.2$ K (Gentemann, 2014; Kennedy, 2014; O'Carroll et al., 2008; Xu & Ignatov, 2016). Although principally designed for open waters (Fig. 2.6), some buoys occasionally do drift close to the coastline and then provide valuable coastal SSTs.

4.2 XBT

EXpendable BathyThermographs (XBTs) are instruments that provide a quick and inexpensive means to record temperature profiles to a certain depth, typically 800 to near a 1000 meters. The history of XBTs dates to the 1960s when the first instrument was deployed in the North Atlantic Ocean (Goni et al., 2019). During the 1970s−1990s, XBTs provided the largest source of subsurface temperature data, until the introduction of Argo profiling floats. As part of the SOOP, XBTs are deployed along fixed transects globally and repeated at varying intervals through the year. Many of these transects have been occupied for decades, providing the longest records of upper ocean temperature profiles.

In Australia, the IMOS SOOP XBT sub-facility operated by CSIRO is responsible for the collection of high density data for three XBT transects: IX28 across the Southern Ocean from Hobart to Antarctica, PX30 across the Eastern Australian Current (EAC) from Brisbane to Fiji, and PX34 from Sydney to Wellington, New Zealand. The Bureau also contributes to this sub-facility, managing three frequently repeated XBT transects: IX01 (Fremantle to Singapore), PX02 (between the Flores and Arafura Seas), and PX11-IX22 (from Dampier, WA to 20°N, near Japan). Many studies have taken advantage of these decades of XBT measurements, combined with other observations, to investigate the EAC and the east Auckland Current (Hill et al., 2011; Ridgway & Dunn, 2003; Ridgway et al., 2008). Although XBTs are primarily released in open, deeper oceans rather than coastal regions, they still provide valuable coastal SSTs whenever available. All NRT and delayed mode of XBT data can be accessed from the AODN.

4.3 Gliders

Gliders are AUVs that are capable of measuring temperature profiles in the upper ocean (Rudnick et al., 2004; Rudnick, 2016; Stommel, 1989; Testor et al., 2010). They can move vertically by changing their buoyance and horizontally using their wings. The technology of underwater gliders is essentially a successor to the Argo profiling floats but with the fundamental advantage of sampling horizontally. Currently, there are three widely used operational models of gliders: Slocum Battery manufactured by Webb Research Corp with a maximum depth of 200 m, Seaglider built at the University of Washington with a maximum depth of 1000 m (Eriksen et al., 2001), and Spray built at Scripps Institution of Oceanography with a maximum depth of 1500 m (Sherman et al., 2001).

Funded by IMOS, the Australian National Facility for Ocean Gliders (ANFOG; https://imos.org.au/facilities/oceangliders, accessed on September 13, 2023) operates a fleet of eight gliders, three Slocum and five Seaglider models. The purpose of this facility is to study the major boundary currents that run along the Australian coasts including the EAC and Leeuwin Current. The temperature profiles obtained from these gliders can be easily found from the AODN portal in NRT, mostly for the Slocum models in the shallower coastal regions, or in delayed mode for the Seagliders. They are also available from the IMOS OceanCurrent portal at https://oceancurrent. aodn.org.au/gliders/index.php (accessed on October 2, 2023). NRT glider data are also provided to the Bureau in glider BUFR (Binary Universal Form for Representation of Meteorological Data; see Section 5) format and then uploaded to the GTS.

4.4   Wave buoys

Wave buoys are primarily designed to measure the state of the sea surface, such as wave height and direction. They play a critical role in the development and verification of national- and regional-scale wave models, as well as the cal/val of satellite sensors (Durrant et al., 2014; Greenslade et al., 2018, 2020). Along Australian coasts, wave buoys such as the Waverider models (https://datawell.nl/, accessed on September 5, 2023) are often equipped with a temperature sensor as well. When available, SST measurements are reported along with wave variables to AODN in NRT or delayed mode (Fig. 2.7). Note that SST is not included in the wave observation messages delivered to the GTS via the Bureau (Joel Cabrie, *personal communication*).

Figure 2.7 Locations of wave buoys along the Australian coastline. Different colors indicate different management agencies.
From AODN https://portal.aodn.org.au/ accessed on October 2, 2023

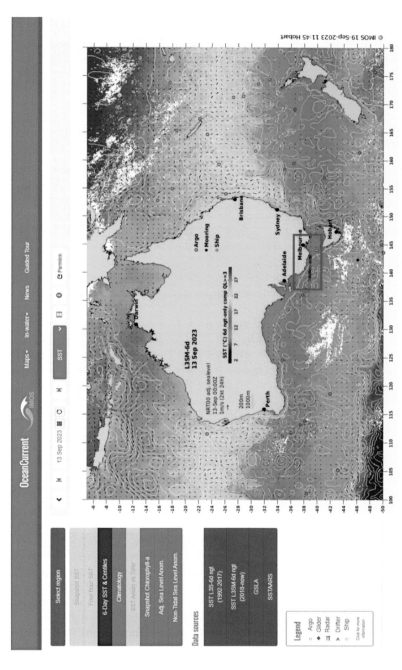

Figure 2.10 IMOS OceanCurrent webpage. A screenshot of the IMOS OceanCurrent webpage showing the 6-day SST centered around September 13, 2023. The 6-day window is used to maximize coverage and night-only satellite SST is used to minimize bias due to diurnal heating. The 6d multi-sensor L3S (level 3 super-located) composites are produced by the Bureau for IMOS using the GHRSST protocol on a 0.02° × 0.02 degrees grid (Govekar et al., 2022). The value of ship SSTs are indicated by the colors inside the black circles in the *red box*. From IMOS OceanCurrent webpage (https://oceancurrent.aodn.org.au/index.php, accessed on September 20, 2023).

been successfully used as a mechanism to improve the data uptake within both scientific and user communities. An example of mapping the 6-day composite average SST (centered on September 13, 2023) is shown in Fig. 2.10. In regional maps of 6-day SST and SST anomaly composites, if any IMOS ship SST observations are collocated in space and time, then these are also displayed (Fig. 2.10). This provides the user with an easily accessible check of the agreement between the satellite SST composite product and in situ SST observations.

6. Discussion and conclusions

Coastal waters are of vital importance both environmentally and socioeconomically, and this is particularly true in Australia given its 36,000 km long coastline, which increases to 60,000 km when all the islands are also considered (Lynch et al., 2014). The boundary currents and interbasin flows along the Australian coastline, including the EAC and Leeuwin Current, exert considerable influences on the nation's weather and climate, fishing activities, tourism (especially at the GBR area), and citizen's recreational activities at the beach. To gain a better understanding of all these impacts, the measurement of coastal SST is essential, which is why there are as many as 18 different methods to measure water temperature across all IMOS facilities (Lara-Lopez et al., 2016).

SSTs obtained from ships and coastal moorings in Australian adjacent waters provide invaluable ground truthing data for the cal/val of modeled outputs or satellite retrievals. One thing worth noting is that nearly all ship SST sensors are placed at a depth of $7-10$ m, and those on mooring float buoys are at ~ 1 m depth. When these data are used, especially during the daytime under warm and calm conditions, extra attention should be paid to the effects of SST DV signals (Zhang et al., 2016). Where available, skin SSTs measured from a shipborne IR radiometer (in the Australian case, the ISAR onboard RV Investigator) are the optimal source for the cal/val of satellite IR skin SST retrievals. Additionally, it should be taken into account that the in situ SST measurements are mostly point observations, such as buoys, or averaged over very small regions (a few meters for some ships), while satellite data or model outputs are usually averages over much larger areas.

Near real-time coastal in situ SSTs are also essential for NWP or climate models as they are ingested into the boundary conditions. Despite their overall small data amount when compared to other platforms such as aircraft or radiosondes, the ships and buoys have a much larger average impact per observation on the forecast skill of the NWP model over Australia (Soldatenko et al., 2018). For the longer term, these ship and CM SST measurements have enabled studies in SST trends and marine climatology where long-term time series are available. Another advantage of these platforms is that a more complete set of essential climate variables (air temperature, wind, insolation, waves, etc.) can be concurrently observed to allow for a comprehensive analysis of the physical or geochemical processes in the upper ocean layers or at the air—sea interface.

The collection of coastal in situ SSTs has not been carried out without issues. Taking the Australian VOS vessels as an example, a ship may change its usual route or occasionally be sold. This poses great challenges on the maintenance of the SST instruments onboard, if they are not entirely lost. During the global pandemic starting from March 2020, the commercial vessels that used to regularly stop in Australian ports nearly disappeared. Even when they did visit Australian ports during the pandemic, the Bureau technical staff or PMOs were often prohibited from getting onboard to repair or service the sensors. Recruiting new VOS ships during this time was also significantly hindered. Issues with coastal moorings around Australia might be slightly different. A concern, if not a real issue, is that near-surface SSTs are not measured at many mooring sites due to the lack of a surface float. For those that do have a surface float and telemetry capabilities, these NRT SSTs are not yet employed to the maximum in the SST community, as none of the Australian CMs are included in the very popular *i*Quam product or supplied to the GTS. Rectifying this situation can boost the data's uptake. Vandalism might be another problem for moorings, particularly for those near areas of intense fishing activity.

More high-quality in situ SSTs will continue to be provided around Australia in the future. The two related IMOS sub-facilities operated at the Bureau, namely, the SST sensors for Australian vessels and the RV real time air-sea flux project, have recently secured extended funding for another 4 years from July 2023 to June 2027. By then, these projects will have created an SST time series of nearly 2 decades since 2008 in the Australian coastal waters and Southern Ocean. Within this timeframe (2023−27), at least one new VOS ship is planned to be recruited and installed with the S-AWS system every year. Furthermore, at the moment, the Bureau is sending only TAC met-ocean messages to the GTS. As most meteorological agencies around the world have shifted to ingesting only BUFR format data, these data are unfortunately not fully utilized. In addition, when the BUFR moored buoy format is used for wave buoys, SSTs can then be included within the wave observation messages as well. Therefore, we will soon be sending BUFR data to the GTS as well to increase the uptake of those measurements by meteorological agencies and researchers worldwide.

Abbreviations

ANFOG	Australian National Facility for Ocean Gliders
ANMN	Australian National Mooring Network
AODN	Australian Ocean Data Network
AOML	Atlantic Oceanographic and Meteorological Laboratory
AUV	Autonomous underwater vehicle
AVHRR	Advanced Very High Resolution Radiometer
BUFR	Binary Universal Form for Representation of Meteorological Data
Bureau Australian	Bureau of Meteorology
C-MAN	Coastal Marine Automated Network
cal/val	calibration/validation

CLS	Collecte Localisation Satellites
CM	Coastal Mooring
CMC	Canadian Meteorological Center
CMEMS	Copernicus Marine Environment Monitoring Service
CODE	Coastal Ocean Dynamics Experiment
CSIRO	Commonwealth Scientific and Industrial Research Organization
CTD	Conductivity, Temperature, Depth
DBCP	Data Buoy Cooperation Panel
DV	Diurnal variation
EAC	Eastern Australian Current
ENSO	El Nino-Southern Oscillation
ERI	Engine Room Intake
ESA	European Space Agency
EUROGOOS	European Global Ocean Observing System
FNMOC	Fleet Numerical Meteorology and Oceanography Center
FRM	International SST Fiducial Reference Measurement
GAMSSA	Global Australian Multi-Sensor Sea Surface Temperature Analysis
GBR	Great Barrier Reef
GDP	Global Drifter Program
GHRSST	Group for High-Resolution Sea Surface Temperature
GLOSS	Global Sea Level Observing System
GMDSS	Inmarsat Global Maritime Distress and Safety System
GODAE	Global Ocean Data Assimilation Experiment
GOOS	Global Ocean Observing System
GOSUD	Global Ocean Surface Underway Data
GTS	Global Telecommunication System
HRSST	High-Resolution Sea Surface Temperature
ICOADS	International Comprehensive Ocean—Atmosphere Data Set
IMO	International Maritime Organization
IMOS	Integrated Marine Observing System
IOC	Intergovernmental Oceanographic Commission
iQuam	in situ Sea Surface Temperature Quality Monitor
IR	Infrared
ISAR	Infrared SST Autonomous Radiometer
L3S	Level 3 Super-located
L4	Level 4
M-AERI	Marine-Atmospheric Emitted Radiance Interferometer
MV	Merchant Vessel
NCEP	National Centers for Environmental Prediction
NDBC	National Data Buoy Center
NetCDF	Network Common Data Form
NOAA	National Oceanic and Atmospheric Administration
NOMAD	Navy Oceanographic Meteorological Automated Device
NRS	National Reference Station
NRT	Near Real Time
NSW	New South Wales
NSWIMOS	NSW IMOS node
NWP	Numerical Weather Prediction
OceanSITES	Ocean Sustained Interdisciplinary Time-series Environment Observation System

OUC	Ocean University of China
OUCFIRST	FIRST IR radiometer made by the Ocean University of China
OWS	Ocean Weather Station
PMO	Port Meteorological Officer
PSMSL	Permanent Service for Mean Sea Level
QC	Quality Control
QIMOS	Queensland IMOS node
RAMSSA	Regional Australian Multi-Sensor Sea Surface Temperature Analysis
RV	Research vessel
S-AWS	Shipborne Automatic Weather Station
SAIMOS	Southern Australia IMOS node
SAMOS	Shipboard Automated Meteorological and Oceanographic System
SBD	Short Burst Data
SEAFRAME	SEA-level Fine Resolution Acoustic Measuring Equipment
SIO	Scripps Institution of Oceanography
SISTeR	Scanning Infrared Sea Surface Temperature Radiometer
SLSTR	Sea and Land Surface Temperature Radiometer
SOLAS	Safety of Life at Sea
SOOP	Ship of Opportunity Program
SOT	Ship Observations Team
SST	Sea Surface Temperature
SVP	Surface Velocity Program
TAC	Traditional Alphanumeric Codes
TASIMOS	Tasmania IMOS node
TRUSTED	Toward fiducial Reference measUrements of Sea-Surface Temperature by European Drifters
TSG	ThermoSalinoGraph
USA	United States of America
UTC	Coordinated Universal Time
VOS	Voluntary Observing Ship
WAIMOS	Western Australia IMOS node
WCRP	World Climate Research Program
WHOI	Woods Hole Oceanographic Institute
WMO	World Meteorological Organization
WOD	World Ocean Database
XBT	eXpendable BathyThermographs

Acknowledgments

Part of the data used in this chapter were sourced from Australia's Integrated Marine Observing System (IMOS) — IMOS is enabled by the National Collaborative Research Infrastructure Strategy (NCRIS). It is operated by a consortium of institutions as an unincorporated joint venture, with the University of Tasmania as Lead Agent. The author greatly appreciates the very useful and detailed comments from Helen Beggs, Lisa Krummel, and Christopher Griffin during the writing of this chapter, which have significantly improved the quality of this study. Discussions with Diana Greenslade, Ben Hague, Ebru Kirezci, Joel Cabrie, David Hughes, and Simon Spagnol on certain topics are also helpful. Comments from Prof. Xiao Hua Wang are much appreciated as well.

References

Bailey, K., Steinberg, C., Davies, C., Galibert, G., Hidas, M., McManus, M. A., Murphy, T., Newton, J., Roughan, M., & Schaeffer, A. (2019). Coastal mooring observing networks and their data products: Recommendations for the next decade. *Frontiers in Marine Science, 6.* https://doi.org/10.3389/fmars.2019.00180

Bainbridge, S., Steinberg, C., & Furnas, M. (2008). GBROOS—An Ocean Observing system for the great barrier reef. *Proceedings of the 11th Session International Coral Reef Symposium, Ft. Lauderdale, Florida, 16,* 529—533.

Barton, I. J., Minnett, P. J., Maillet, K. A., Donlon, C. J., Hook, S. J., Jessup, A. T., & Nightingale, T. J. (2004). The Miami2001 infrared radiometer calibration and intercomparison. Part II: Shipboard results. *Journal of Atmospheric and Oceanic Technology, 21*(2), 268—283. https://doi.org/10.1175/1520-0426(2004)021<0268:tmirca>2.0.co;2

Beggs, H., Morgan, N., & Sisson, J. (2017). IMOS ship SST for satellite SST validation. In G. Corlett, & S. Bragaglia-Pike (Eds.), *18TH international GHRSST science team meeting (GHRSST XVIII)* (pp. 127—134). Qingdao China: GHRSST Project Office. https://doi.org/10.5281/zenodo.4700225, 5-6 June 2017.

Beggs, H., Verein, R., Paltoglou, G., Kippo, H., & Underwood, M. (2012). Enhancing ship of opportunity sea surface temperature observations in the Australian region. *Journal of Operational Oceanography, 5*(1), 59—73. https://doi.org/10.1080/1755876X.2012.11020132

Beggs, H., Zhong, A., Warren, G., Alves, O., Brassington, G., & Pugh, T. (2011). Ramssa - an operational, high-resolution, Regional Australian Multi-Sensor Sea surface temperature Analysis over the Australian region. *Australian Meteorological and Oceanographic Journal, 61*(1), 1—22. https://doi.org/10.22499/2.6101.001

Bell, M. J., Lefèbvre, M., le Traon, P. Y., Smith, N., & Wilmer-Becker, K. (2009). GODAE the global ocean data assimilation experiment. *Oceanography, 22*(3), 14—21. https://doi.org/10.5670/oceanog.2009.62

Capone, D. G., & Hutchins, D. A. (2013). Microbial biogeochemistry of coastal upwelling regimes in a changing ocean. *Nature Geoscience, 6*(9), 711—717. https://doi.org/10.1038/ngeo1916

Casey, K. S., Brandon, T. B., Cornillon, P., & Evans, R. (2010). The past, present, and future of the AVHRR pathfinder SST program. *Oceanography from Space: Revisited, 273*—287. https://doi.org/10.1007/978-90-481-8681-5_16

Castro, S. L., Wick, G. A., & Emery, W. J. (2012). Evaluation of the relative performance of sea surface temperature measurements from different types of drifting and moored buoys using satellite-derived reference products. *Journal of Geophysical Research: Oceans, 117*(2). https://doi.org/10.1029/2011JC007472

Centurioni, L. R., Turton, J. D., Lumpkin, R., Braasch, L., Brassington, G., Chao, Y., Charpentier, E., Chen, Z., Corlett, G., Dohan, K., Donlon, C., Gallage, C., Hormann, V., Ignatov, A., Ingleby, B., Jensen, R., Kelly-Gerreyn, B. A., Koszalka, I. M., Lin, X., ... Zhang, D. (2019). Global in-situ observations of essential climate and ocean variables at the air-sea interface. *Frontiers in Marine Science, 6.* https://doi.org/10.3389/fmars.2019.00419

Cipollini, P., Calafat, F. M., Jevrejeva, S., Melet, A., & Prandi, P. (2017). Monitoring Sea level in the coastal zone with satellite altimetry and tide gauges. *Surveys in Geophysics, 38*(1), 33—57. https://doi.org/10.1007/s10712-016-9392-0

Colacino, M., Rossi, E., & Vivona, F. M. (1970). Sea-surface temperature measurements by infrared radiometer. *Pure and Applied Geophysics PAGEOPH, 83*(1), 98–110. https:// doi.org/10.1007/BF00875103

Coppo, P., Mastrandrea, C., Stagi, M., Calamai, L., Barilli, M., & Nieke, J. (2013). The sea and land surface temperature radiometer (SLSTR) detection assembly design and performance. *Proceedings of SPIE - The International Society for Optical Engineering, 8889*. https:// doi.org/10.1117/12.2029432

Cummings, J. A. (2006). Operational multivariate ocean data assimilation. *Quarterly Journal of the Royal Meteorological Society, 131*(613), 3583–3604. https://doi.org/10.1256/ qj.05.105

Davies, C., Sommerville, E., Hidas, M., Suthers, Lara-Lopez, A., Matis, P., Kamp, Tibben, S., Abell, G., Allen, S., Berry, K., Bonham, P., Clementson, L., Coman, F., Critchley, G., Frampton, D., Latham, V., Richardson, A., Rober, S., & Tattersall, K. (2020). *National reference stations biogeochemical operations manual - version 3.3.1.* Integrated Marine Observng Ststem. https://doi.org/10.26198/5c4a56f2a8ae3

Davis, R. E. (1985). Drifter observations of coastal surface currents during CODE: The method and descriptive view. *Journal of Geophysical Research: Oceans, 90*(C3), 4741–4755. https://doi.org/10.1029/jc090ic03p04741

Deser, C., Alexander, M. A., Xie, S. P., & Phillips, A. S. (2010). Sea surface temperature variability: Patterns and mechanisms. *Annual Review of Marine Science, 2*(1), 115–143. https://doi.org/10.1146/annurev-marine-120408-151453

Domingues, C. M., Church, J. A., White, N. J., Gleckler, P. J., Wijffels, S. E., Barker, P. M., & Dunn, J. R. (2008). Improved estimates of upper-ocean warming and multi-decadal sea-level rise. *Nature, 453*(7198), 1090–1093. https://doi.org/10.1038/nature07080

Donlon, C., Minnett, P. J., Jessup, A., Barton, I., Emery, W., Hook, S., Wimmer, W., Nightingale, T. J., & Zappa, C. (2014). Ship-borne thermal infrared radiometer systems. *Experimental Methods in the Physical Sciences, 47*, 305–404. https://doi.org/10.1016/ B978-0-12-417011-7.00011-8

Donlon, C., Robinson, I., Casey, K. S., Vazquez-Cuervo, J., Armstrong, E., Arino, O., Gentemann, C., May, D., LeBorgne, P., Piolle, J., Barton, I., Beggs, H., Poulter, D. J. S., Merchant, C. J., Bingham, A., Heinz, S., Harris, A., Wick, G., Emery, B., … Rayner, N. (2007). The global Ocean Data assimilation experiment high-resolution Sea Surface temperature pilot project. *Bulletin of the American Meteorological Society, 88*(8), 1197–1213. https://doi.org/10.1175/BAMS-88-8-1197

Donlon, C., Robinson, I. S., Reynolds, M., Wimmer, W., Fisher, G., Edwards, R., & Nightingale, T. J. (2008). An infrared sea surface temperature autonomous radiometer (ISAR) for deployment aboard volunteer observing ships (VOS). *Journal of Atmospheric and Oceanic Technology, 25*(1), 93–113. https://doi.org/10.1175/2007JTECHO505.1

Durrant, T., Greenslade, D., Hemar, M., & Trenham, C. (2014). A global hindcast focussed on the central and South pacific. *CAWCR Technical Report, 40*.

Elipot, S., Sykulski, A., Lumpkin, R., Centurioni, L., & Pazos, M. (2022). A dataset of hourly sea surface temperature from drifting buoys. *Scientific Data, 9*(1). https://doi.org/10.1038/ s41597-022-01670-2

Emery, W. J., Cherkauer, K., Shannon, B., & Reynolds, R. W. (1997). Hull-mounted Sea Surface temperatures from ships of opportunity. *Journal of Atmospheric and Oceanic Technology, 14*(5), 1237–1251. https://doi.org/10.1175/1520-0426(1997)014<1237: hmsstf>2.0.co;2

Eriksen, C. C., Osse, T. J., Light, R. D., Wen, T., Lehman, T. W., Sabin, P. L., Ballard, J. W., & Chiodi, A. M. (2001). Seaglider: A long-range autonomous underwater vehicle for oceanographic research. *IEEE Journal of Oceanic Engineering, 26*(4), 424–436. https://doi.org/10.1109/48.972073

Fairall, C. W., Bradley, E. F., Godfrey, J. S., Wick, G. A., Edson, J. B., & Young, G. S. (1996). Cool-skin and warm-layer effects on sea surface temperature. *Journal of Geophysical Research: Oceans, 101*(C1), 1295–1308. https://doi.org/10.1029/95jc03190

Folland, C. K., & Parker, D. E. (1995). Correction of instrumental biases in historical sea surface temperature data. *Quarterly Journal of the Royal Meteorological Society, 121*(522), 319–367. https://doi.org/10.1002/qj.49712152206

Frankignoul, C. (1985). Sea surface temperature anomalies, planetary waves, and air-sea feedback in the middle latitudes. *Reviews of Geophysics, 23*(4), 357–390. https://doi.org/10.1029/rg023i004p00357

Freeman, E., Woodruff, S. D., Worley, S. J., Lubker, S. J., Kent, E. C., Angel, W. E., Berry, D. I., Brohan, P., Eastman, R., Gates, L., Gloeden, W., Ji, Z., Lawrimore, J., Rayner, N. A., Rosenhagen, G., & Smith, S. R. (2017). ICOADS release 3.0: A major update to the historical marine climate record. *International Journal of Climatology, 37*(5), 2211–2232. https://doi.org/10.1002/joc.4775

Gentemann, C. L. (2014). Three way validation of MODIS and AMSR-E sea surface temperatures. *Journal of Geophysical Research: Oceans, 119*(4), 2583–2598. https://doi.org/10.1002/2013JC009716

Goni, G., Roemmich, D., Molinari, R., Meyers, G., Sun, C., Boyer, T., Baringer, M., Gouretski, V., DiNezio, P., & Reseghetti, F. (2010). The ship of opportunity program. *Proceedings of OceanObs, 9*, 366–383. https://doi.org/10.5270/oceanobs09.cwp.35

Goni, G., Sprintall, J., Bringas, F., Cheng, L., Cirano, M., Dong, S., Domingues, R., Goes, M., Lopez, H., Morrow, R., Rivero, U., Rossby, T., Todd, R. E., Trinanes, J., Zilberman, N., Baringer, M., Boyer, T., Cowley, R., Domingues, C. M., … Volkov, D. (2019). More than 50 years of successful continuous temperature section measurements by the global expendable bathythermograph network, its integrability, societal benefits, and future. *Frontiers in Marine Science, 6*(JUL), 452. https://doi.org/10.3389/fmars.2019.00452

Govekar, P. D., Griffin, C., & Beggs, H. (2022). Multi-sensor Sea Surface temperature products from the Australian Bureau of Meteorology. *Remote Sensing, 14*(15). https://doi.org/10.3390/rs14153785

Greenslade, D., Hemer, M., Babanin, A., Lowe, R., Turner, I., Power, H., Young, I., Ierodiaconou, D., Hibbert, G., Williams, G., Aijaz, S., Albuquerque, J., Allen, S., Banner, M., Branson, P., Buchan, S., Burton, A., Bye, J., Cartwright, N., … Zieger, S. (2020). 15 priorities for wind-waves research: An Australian perspective. *Bulletin of the American Meteorological Society, 101*(4), E446–E461. https://doi.org/10.1175/BAMS-D-18-0262.1

Greenslade, D. J. M., Zanca, A., Zieger, S., & Hemer, M. A. (2018). Optimising the Australian wave observation network. *Journal of Southern Hemisphere Earth Systems Science, 68*(1), 184–200. https://doi.org/10.22499/3.6801.010

Hemming, M. P., Roughan, M., Malan, N., & Schaeffer, A. (2023). Observed multi-decadal trends in subsurface temperature adjacent to the East Australian Current. *Ocean Science, 19*(4), 1145–1162. https://doi.org/10.5194/os-19-1145-2023

Hidas, M. G., Proctor, R., Atkins, N., Atkinson, J., Besnard, L., Blain, P., Bohm, P., Burgess, J., Finney, K., Fruehauf, D., Galibert, G., Hoenner, X., Hope, J., Jones, C., Mancini, S., Pasquer, B., Nahodil, D., Reid, K., & Tattersall, K. (2016). Information infrastructure for Australia's integrated marine observing system. *Earth Science Informatics, 9*(4), 525–534. https://doi.org/10.1007/s12145-016-0266-2

Hill, K., Moltmann, T., Proctor, R., & Allen, S. (2010). The Australian integrated marine observing system: Delivering data streams to address national and international research priorities. *Marine Technology Society Journal, 44*(6), 65−72. https://doi.org/10.4031/MTSJ.44.6.13

Hill, K., Rintoul, S. R., Ridgway, K. R., & Oke, P. R. (2011). Decadal changes in the South Pacific western boundary current system revealed in observations and ocean state estimates. *Journal of Geophysical Research: Oceans, 116*(1). https://doi.org/10.1029/2009JC005926

Holgate, S. J., Matthews, A., Woodworth, P. L., Rickards, L. J., Tamisiea, M. E., Bradshaw, E., Foden, P. R., Gordon, K. M., Jevrejeva, S., & Pugh, J. (2013). New data systems and products at the permanent service for mean sea level. *Journal of Coastal Research, 29*(3), 493−504. https://doi.org/10.2112/JCOASTRES-D-12-00175.1

Hollmann, R., Merchant, C. J., Saunders, R., Downy, C., Buchwitz, M., Cazenave, A., Chuvieco, E., Defourny, P., de Leeuw, G., Forsberg, R., Holzer-Popp, T., Paul, F., Sandven, S., Sathyendranath, S., van Roozendael, M., & Wagner, W. (2013). The ESA climate change initiative: Satellite data records for essential climate variables. *Bulletin of the American Meteorological Society, 94*(10), 1541−1552. https://doi.org/10.1175/bams-d-11-00254.1

Hu, J., & Wang, X. H. (2016). Progress on upwelling studies in the China seas. *Reviews of Geophysics, 54*(3), 653−673. https://doi.org/10.1002/2015RG000505

Huang, Z., & Feng, M. (2021). MJO induced diurnal sea surface temperature variations off the northwest shelf of Australia observed from Himawari geostationary satellite. *Deep Sea Research Part II: Topical Studies in Oceanography, 183.* https://doi.org/10.1016/j.dsr2.2021.104925

Isern-Fontanet, J., Chapron, B., Lapeyre, G., & Klein, P. (2006). Potential use of microwave sea surface temperatures for the estimation of ocean currents. *Geophysical Research Letters, 33*(24). https://doi.org/10.1029/2006gl027801

James, R., & Fox. (1972). *Comparative sea surface temperature measurements in WMO reports on marine science affairs.*

Jessup, A. T., & Brance, R. (2008). Integrated ocean skin and bulk temperature measurements using the Calibrated Infrared in Situ Measurement System (CIRIMS) and through-hull ports. *Journal of Atmospheric and Oceanic Technology, 25*(4), 579−597. https://doi.org/10.1175/2007JTECHO479.1

Kennedy, J. J. (2014). A review of uncertainty in in situ measurements and data sets of sea surface temperature. *Reviews of Geophysics, 52*(1), 1−32. https://doi.org/10.1002/2013RG000434

Kennedy, J. J., Rayner, N. A., Smith, R. O., Parker, D. E., & Saunby, M. (2011). Reassessing biases and other uncertainties in sea surface temperature observations measured in situ since 1850: 1. Measurement and sampling uncertainties. *Journal of Geophysical Research Atmospheres, 116*(14). https://doi.org/10.1029/2010JD015218

Kent, E. C., & Challenor, P. G. (2006). Toward estimating climatic trends in SST. Part II: Random errors. *Journal of Atmospheric and Oceanic Technology, 23*(3), 476−486. https://doi.org/10.1175/JTECH1844.1

Kent, E. C., & Kaplan, A. (2006). Toward estimating climatic trends in SST. Part III: Systematic biases. *Journal of Atmospheric and Oceanic Technology, 23*(3), 487−500. https://doi.org/10.1175/JTECH1845.1

Kent, E. C., Kennedy, J. J., Berry, D. I., & Smith, R. O. (2010). Effects of instrumentation changes on sea surface temperature measured in situ. *Wiley Interdisciplinary Reviews: Climate Change, 1*(5), 718−728. https://doi.org/10.1002/wcc.55

Kent, E. C., & Kennedy, J. J. (2021). Historical estimates of surface marine temperatures. *Annual Review of Marine Science, 13*, 283–311. https://doi.org/10.1146/annurev-marine-042120-111807

Kent, E. C., Kennedy, J. J., Smith, T. M., Hirahara, S., Huang, B., Kaplan, A., Parker, D. E., Atkinson, C. P., Berry, D. I., Carella, G., Fukuda, Y., Ishii, M., Jones, P. D., Lindgren, F., Merchant, C. J., Morak-Bozzo, S., Rayner, N. A., Venema, V., Yasui, S., & Zhang, H. M. (2017). A call for new approaches to quantifying biases in observations of sea surface temperature. *Bulletin of the American Meteorological Society, 98*(8), 1601–1616. https://doi.org/10.1175/BAMS-D-15-00251.1

Kent, E. C., & Taylor, P. K. (2006). Toward estimating climatic trends in SST. Part I: Methods of measurement. *Journal of Atmospheric and Oceanic Technology, 23*(3), 464–475. https://doi.org/10.1175/JTECH1843.1

Kent, E. C., Taylor, P. K., Truscott, B. S., & Hopkins, J. S. (1993). The accuracy of voluntary observing ships' meteorological observations - results of the VSOP-NA. *Journal of Atmospheric and Oceanic Technology, 10*(4), 591–608. https://doi.org/10.1175/1520-0426(1993)010<0591:TAOVOS>2.0.CO;2

Klein, S. A., Soden, B. J., & Lau, N. C. (1999). Remote sea surface temperature variations during ENSO: Evidence for a tropical atmospheric bridge. *Journal of Climate, 12*(4), 917–932. https://doi.org/10.1175/1520-0442(1999)012<0917:RSSTVD>2.0.CO;2

Lara-Lopez, A., Moltmann, T., & Proctor, R. (2016). Australia's integrated marine observing system (IMOS): Data impacts and lessons learned. *Marine Technology Society Journal, 50*(3), 23–33. https://doi.org/10.4031/mtsj.50.3.1

Le Menn, M., Poli, P., David, A., Sagot, J., Lucas, M., O'Carroll, A., Belbeoch, M., & Herklotz, K. (2019). Development of surface drifting buoys for fiducial reference measurements of Sea-surface temperature. *Frontiers in Marine Science, 6*. https://doi.org/10.3389/fmars.2019.00578

Le Traon, P. Y., Reppucci, A., Fanjul, E. A., Aouf, L., Behrens, A., Belmonte, M., Bentamy, A., Bertino, L., Brando, V. E., Kreiner, M. B., Benkiran, M., Carval, T., Ciliberti, S. A., Claustre, H., Clementi, E., Coppini, G., Cossarini, G., De Alfonso Alonso-Muñoyerro, M., Delamarche, A., … Zacharioudaki, A. (2019). From observation to information and users: The Copernicus marine service perspective. *Frontiers in Marine Science, 6*(May). https://doi.org/10.3389/fmars.2019.00234

Legler, D. M., Freeland, H. J., Lumpkin, R., Ball, G., McPhaden, M. J., North, S., Crowley, R., Goni, G. J., Send, U., & Merrifield, M. A. (2015). The current status of the real-time in situ Global Ocean Observing System for operational oceanography. *Journal of Operational Oceanography, 8*(Suppl. 2), s189–s200. https://doi.org/10.1080/1755876x.2015.1049883

Lehmann, A., & Myrberg, K. (2008). Upwelling in the baltic sea - a review. *Journal of Marine Systems, 74*. https://doi.org/10.1016/j.jmarsys.2008.02.010

Liu, C., Freeman, E., Kent, E. C., Berry, D. I., Worley, S. J., Smith, S. R., Huang, B., Zhang, H. M., Cram, T., Ji, Z., Ouellet, M., Gaboury, I., Oliva, F., Andersson, A., Angel, W. E., Sallis, A. R., & Adeyeye, A. (2022). Blending TAC and BUFR marine in situ data for ICOADS near-real-time release 3.0.2. *Journal of Atmospheric and Oceanic Technology, 39*(12), 1943–1959. https://doi.org/10.1175/JTECH-D-21-0182.1

Lumpkin, R., Grodsky, S. A., Centurioni, L., Rio, M. H., Carton, J. A., & Lee, D. (2013). Removing spurious low-frequency variability in drifter velocities. *Journal of Atmospheric and Oceanic Technology, 30*(2), 353–360. https://doi.org/10.1175/JTECH-D-12-00139.1

Lumpkin, R., & Pazos, M. (2007). Measuring surface currents with surface velocity program drifters: The instrument, its data, and some recent results. In *Lagrangian analysis and prediction of coastal and ocean Dynamics* (pp. 39−67). https://doi.org/10.1017/cbo9780511535901.003

Luo, B., & Minnett, P. J. (2020). Evaluation of the ERA5 sea surface skin temperature with remotely-sensed shipborne marine-atmospheric emitted radiance interferometer data. *Remote Sensing, 12*(11). https://doi.org/10.3390/rs12111873

Lynch, T. P., Morello, E. B., Evans, K., Richardson, A. J., Rochester, W., Steinberg, C. R., Roughan, M., Thompson, P., Middleton, J. F., Feng, M., Sherrington, R., Brando, V., Tilbrook, B., Ridgway, K., Allen, S., Doherty, P., Hill, K., & Moltmann, T. C. (2014). IMOS national reference stations: A continental-wide physical, chemical and biological coastal observing system. *PLoS One, 9*(12). https://doi.org/10.1371/journal.pone.0113652

Masson, S., Terray, P., Madec, G., Luo, J. J., Yamagata, T., & Takahashi, K. (2012). Impact of intra-daily SST variability on ENSO characteristics in a coupled model. *Climate Dynamics, 39*(3), 681−707. https://doi.org/10.1007/s00382-011-1247-2

Matthews, J. B. R. (2013). Comparing historical and modern methods of sea surface temperature measurement − Part 1: Review of methods, field comparisons and dataset adjustments. *Ocean Science, 9*(4), 683−694. https://doi.org/10.5194/os-9-683-2013

McPhaden, M. J., Busalacchi, A. J., Cheney, R., Donguy, J. R., Gage, K. S., Halpern, D., Ji, M., Julian, P., Meyers, G., Mitchum, G. T., Niiler, P. P., Picaut, J., Reynolds, R. W., Smith, N., & Takeuchi, K. (1998). The tropical ocean-global atmosphere observing system: A decade of progress. *Journal of Geophysical Research: Oceans, 103*(7), 14169−14240. https://doi.org/10.1029/97jc02906

McPhaden, M. J., Zhang, X., Hendon, H. H., & Wheeler, M. C. (2006). Large scale dynamics and MJO forcing of ENSO variability. *Geophysical Research Letters, 33*(16). https://doi.org/10.1029/2006GL026786

Meldrum, D., Charpentier, E., Fedak, M., Lee, B., Lumpkin, R., Niiler, P., & Viola, H. (2010). *Data buoy observations: The status quo and anticipated developments over the next decade.* https://doi.org/10.5270/oceanobs09.cwp.62

Merchant, C. J., Embury, O., Rayner, N. A., Berry, D. I., Corlett, G. K., Lean, K., Veal, K. L., Kent, E. C., Llewellyn-Jones, D. T., Remedios, J. J., & Saunders, R. (2012). A 20year independent record of sea surface temperature for climate from Along-Track Scanning Radiometers. *Journal of Geophysical Research: Oceans, 117*(12). https://doi.org/10.1029/2012JC008400

Minnett, P. J., Alvera-Azcárate, A., Chin, T. M., Corlett, G. K., Gentemann, C. L., Karagali, I., Li, X., Marsouin, A., Marullo, S., Maturi, E., Santoleri, R., Saux Picart, S., Steele, M., & Vazquez-Cuervo, J. (2019). Half a century of satellite remote sensing of sea-surface temperature. *Remote Sensing of Environment, 233.* https://doi.org/10.1016/j.rse.2019.111366

Minnett, P. J., Knuteson, R. O., Best, F. A., Osborne, B. J., Hanafin, J. A., & Brown, O. B. (2001). The marine-atmospheric emitted radiance interferometer: A high-accuracy, seagoing infrared spectroradiometer. *Journal of Atmospheric and Oceanic Technology, 18*(6), 994−1013. https://doi.org/10.1175/1520-0426(2001)018<0994:tmaeri>2.0.co;2

Minnett, P. J., Smith, M., & Ward, B. (2011). Measurements of the oceanic thermal skin effect. *Deep-Sea Research Part II Topical Studies in Oceanography, 58*(6), 861−868. https://doi.org/10.1016/j.dsr2.2010.10.024

Moltmann, T., Turton, J., Zhang, H. M., Nolan, G., Gouldman, C., Griesbauer, L., Willis, Z., Piniella, á. M., Barrell, S., Andersson, E., Gallage, C., Charpentier, E., Belbeoch, M., Poli, P., Rea, A., Burger, E. F., Legler, D. M., Lumpkin, R., Meinig, C., … Zhang, Y. (2019). A Global Ocean Observing System (GOOS), delivered through enhanced collaboration across regions, communities, and new technologies. *Frontiers in Marine Science, 6.* https://doi.org/10.3389/fmars.2019.00291

Nightingale, T. J. (2006). The scanning infrared sea surface temperature radiometer (SISTeR). In *Proceedings of the second working meeting on MERIS and AATSR calibration and geophysical validation (MAVT-2006.*

O'Carroll, A. G., Armstrong, E. M., Beggs, H., Bouali, M., Casey, K. S., Corlett, G. K., Dash, P., Donlon, C., Gentemann, C. L., Høyer, J. L., Ignatov, A., Kabobah, K., Kachi, M., Kurihara, Y., Karagali, I., Maturi, E., Merchant, C. J., Marullo, S., Minnett, P., ... Wimmer, W. (2019). Observational needs of sea surface temperature. *Frontiers in Marine Science, 6.* https://doi.org/10.3389/fmars.2019.00420

O'Carroll, A. G., Eyre, J. R., & Saunders, R. W. (2008). Three-way error analysis between AATSR, AMSR,E, and in situ sea surface temperature observations. *Journal of Atmospheric and Oceanic Technology, 25*(7), 1197−1207. https://doi.org/10.1175/2007JTECHO542.1

O'neill, L. W., Esbensen, S. K., Thum, N., Samelson, R. M., & Chelton, D. B. (2010). Dynamical analysis of the boundary layer and surface wind responses to mesoscale SST perturbations. *Journal of Climate, 23*(3), 559−581. https://doi.org/10.1175/2009JCLI2662.1

Poli, P., Lucas, M., O'Carroll, A., Le Menn, M., David, A., Corlett, G. K., Blouch, P., Meldrum, D., Merchant, C. J., Belbeoch, M., & Herklotz, K. (2019). The Copernicus surface velocity platform drifter with barometer and reference sensor for temperature (SVP-brst): Genesis, design, and initial results. *Ocean Science, 15*(1), 199−214. https://doi.org/10.5194/os-15-199-2019

Proctor, R., Roberts, K., & Ward, B. J. (2010). A data delivery system for IMOS, the Australian integrated marine observing system. *Advances in Geosciences, 28,* 11−16. https://doi.org/10.5194/adgeo-28-11-2010

Ravichandran, M. (2011). In-situ Ocean Observing system. In *Operational Oceanography in the 21st century* (pp. 55−90). Dordrecht: Springer. https://doi.org/10.1007/978-94-007-0332-2_3

Rayner, N. A., Brohan, P., Parker, D. E., Folland, C. K., Kennedy, J. J., Vanicek, M., Ansell, T. J., & Tett, S. F. B. (2006). Improved analyses of changes and uncertainties in Sea Surface temperature measured in situ since the mid-nineteenth century: The HadSST2 dataset. *Journal of Climate, 19*(3), 446−469. https://doi.org/10.1175/jcli3637.1

Ridgway, K. R., & Dunn, J. R. (2003). Mesoscale structure of the mean East Australian Current System and its relationship with topography. *Progress in Oceanography, 56*(2), 189−222. https://doi.org/10.1016/s0079-6611(03)00004-1

Ridgway, K. R., Coleman, R. C., Bailey, R. J., & Sutton, P. (2008). Decadal variability of East Australian Current transport inferred from repeated high-density XBT transects, a CTD survey and satellite altimetry. *Journal of Geophysical Research: Oceans, 113*(8). https://doi.org/10.1029/2007JC004664

Ridgway, K. R., & Ling, S. D. (2023). Three decades of variability and warming of nearshore waters around Tasmania. *Progress in Oceanography, 215.* https://doi.org/10.1016/j.pocean.2023.103046

Roughan, M., Hemming, M., Schaeffer, A., Austin, T., Beggs, H., Chen, M., Feng, M., Galibert, G., Holden, C., Hughes, D., Ingleton, T., Milburn, S., & Ridgway, K. (2022). Multi-decadal ocean temperature time-series and climatologies from Australia's long-term National Reference Stations. *Scientific Data, 9*(1). https://doi.org/10.1038/s41597-022-01224-6

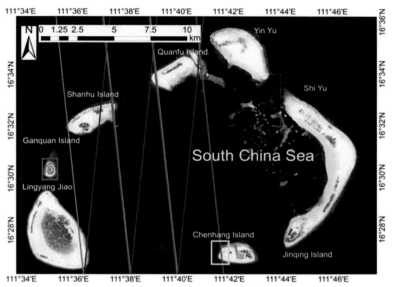

Figure 3.3 Location of the Yongle Atoll in the south China Sea. The basemap is a 10-m Sentinel-2A image acquired on April 12, 2019. Three red lines, three green lines, and three purple lines correspond to the ICESat-2's laser tracks of three strong beams on October 22, 2018, February 22, 2019, and April 12, 2019, respectively. The blue box corresponds to Ganquan Island where the in-situ airborne lidar data were captured to validate our bathymetric result. The yellow box corresponds to Chenhang Island where a sampled ICESat-2 along-track dataset was illustrated.

$$L_{deep} = \frac{gT^2}{2\pi} \tag{3.9}$$

A harmonic wave retains its period T when propagating over seafloors with gentle slopes (i.e., slope$\leq$0.2), but the dispersion relationship remains. In offshore areas with deep waters, L_{deep} is nearly stable for gravity waves. ICESat-2 can obtain water wave-numbers (k) or wavelengths in near-shore areas (i.e., $L_{shallow}$) as shown in Fig. 3.3d and offshore areas (i.e., L_{deep}) that have been validated in a previous study. With an observed L_{deep}, the period T can be retrieved in offshore areas using Eq. (3.9). After the wave period T in deep waters is determined, we can further retrieve the depths h in near-shore areas with observed wavelengths $L_{shallow}$ using Eq. (3.8), that is, the red curve in Fig. 3.3d.

4. Coastal bathymetry from satellite multispectral imagery

4.1 Introduction of satellite multispectral imagery

The term, "remote sensing" encompasses a series of well-established procedures for bathymetric surveys of the seabed (Kenny et al., 2003), but these procedures are not without limitations in terms of both the sensors and those imposed by the environment (Hedley et al., 2012). Remote sensing for bathymetry can be subdivided into active and passive techniques. Now, passive remote sensing for bathymetry is predominantly performed using multispectral satellite imagery. Multispectral imagery is preferable because electromagnetic radiation at different wavelengths can penetrate the water column to different depths and the land—water interface can be clearly defined. Bathymetry is calculated using analytical (Benny & Dawson, 1983; Spitzer & Dirks, 1986) or empirical methods (Ji et al., 1992; Lafon et al., 2002), based on the statistical relationships between image pixel values and ground truth depth measurements and is applicable in shallow coastal waters. Accuracy is high using these methods for depths of water penetration not exceeding 20 m but becomes less accurate in deeper waters (Gao, 2009). Turbidity is an issue for imagery derived bathymetry in all water types. Hyperspectral satellite data can provide an important insight into the water column properties when deriving satellite bathymetry (e.g., turbidity, oceanographic) (Lee et al., 1999) due to the greater number and narrower bands; however, the spatial resolution is generally much larger than with multispectral (1 or 2 orders of magnitude).

Optical multispectral satellite-derived bathymetry (SDB) that implements analytical or empirical methods, based on the statistical relationships between image pixel values and field measured water depth measurements, apply the general physical principle that sea water transmittances at near-visible wavelengths are functions of a general optical equation depending on the intrinsic optical properties of sea water. A variety of empirical models have been proposed, from linear functions (Lyzenga, 1978), band ratios to log transformed regression models (Stumpf et al., 2003), and nonlinear inverse models (Su et al., 2008) with varying degrees of corrections applied (atmospheric, sun glint, seafloor). The evolution of empirical models has been largely linked to the chronology of satellite platforms: coarse spatial resolution using Landsat TM (Lyzenga, 1981; Spitzer & Dirks, 1986; Vanderstraete et al., 2003); medium spatial resolution SPOT images in shallow waters (Lafon et al., 2002; Liu et al., 2010) or RapidEye in lakes (Giardino et al., 2014); and high spatial resolution with the use of commercial satellites such as WorldView (Doxani et al., 2012; Kanno & Tanaka, 2012), QuickBird (Conger et al., 2006; Lyons et al., 2011), and IKONOS (Ohlendorf et al., 2011).

4.2   Methods and accuracy of obtaining coastal bathymetry from satellite multispectral imagery

The theoretical analytical model was first established in 1978 by Lyzenga (1978) based on the classical radiative transfer theory of double-layer flow, which used analysis of water reflectivity to retrieve water depth. The analytical expression of radiance received by the optical remote sensor as a function of water depth and bottom reflection is established by neglecting the reflection effect inside the water body (Ma et al., 2018). In the absence of measured water depth data, the water depth can be retrieved from the image (Liu et al., 2019). However, as the parameters involved are complex, it is necessary to determine the optical model parameters of the water body in the study area and iteratively calculate them to obtain the estimated water depth (Liu et al., 2019).

The semi-theoretical and semi-empirical model (also called empirical model) is based on the theory of light radiation attenuation in the water and uses the combination of a theoretical model and empirical parameters to determine the water depth using passive optical remote sensing (Wang et al., 2022). The radiance received by the optical remote sensor is expressed as the sum of deep-water area radiance and bottom reflected radiance. In 1973, the log-linear model was proposed by Polcyn and Lyzenga (Polcyn & Lyzenga, 1973) and became the more widely used model, and it was later simplified by Tanis and Byrnes (Tanis & Byrnes, 1985) to derive a single-band model in 1986. As shown by Paredes and Spero (Paredes & Spero, 1983), assuming that the reflectance ratio of each band on different substrate types is constant, a dual-band model is derived, which can also be developed into a multiband model. In order to avoid a situation where the difference between the radiance received by the optical remote sensor and the radiance in the deep-water area is negative in the log-linear model, Stumpf et al. (2003) proposed a log-transformed ratio model in 2003. In 2015, Hamylton et al. (2015) compared the empirical and optimization methods on the Great Barrier Reef: Lizard Island (a continental island fringing reef) and Sykes Reef (a planar platform reef). Many Chinese scholars have also studied the semi-theoretical and semi-empirical models. For example, in 2007, Tian et al. (2007) used Landsat-TM data to derive the remote sensing bathymetry equation, constructed a waveband ratio model to remove the deep bathymetric reflectivity, and completed the remote sensing mapping of the near-coastal bathymetry in Jiangsu. In 2003, Dang and Ding (2003) later used Landsat-TM data to build a bathymetric inversion model of the sea area near Yongshu Reef in the South China Sea by combining wavebands, and they also improved the accuracy of the estimated inversion bathymetry by adopting the method of dividing the substrate types. In 2019, Lu et al. (2019) established the band ratio model by using measured points and three multispectral images of Landsat-8, SPOT-6, and WorldView-2. The semi-theoretical and semi-empirical model is widely used due to its simple parameters and optimizability to different study areas using only measured bathymetry data (Casal et al., 2019).

The statistical model directly establishes the statistical relationship between the radiance of the remote sensing image and the measured water depth without considering the physical mechanism of water depth remote sensing. The traditional statistical model expressions include the linear function, logarithmic function, and power function, although their accuracy is not high. In 1998, Zhang et al. (1998) showed that, using linear regression, the water depth can be inversed by adding a nonlinear correction term in the form of a power function. In 2009, the water depths of the radial sand ridges in the South Yellow Sea off the coast of Jiangsu Province were retrieved by Zhang et al. (2009) (reference) (using linear, exponential, reciprocal, and quadratic regression), as well as other types of water depth remote sensing models, and the average relative errors based on the traditional statistical models were mostly between 30% and 45%. Machine learning is a statistical method that uses self-learning, self-organization, self-adaptation, and nonlinear dynamic processing (Ao et al., 2019; Cheng & Zhu, 2007). The accuracy of machine learning is affected by the sample data. When there is a certain number of measured water depth data, the fitting effect is significantly better than with the two models described above. In 2005, Wang and Zhang (2005) used a BP neural network to invert the bathymetry of the Nangang section of the Yangtze River Estuary. In 2018, Wang et al. (2018) used support vector machines to perform bathymetric inversions of the North Island in the Xisha Islands. In 2019, Qiu et al. (2019) used the random forest algorithm to invert the bathymetry of Ganquan Dao in Xisha Islands and found that the machine learning—based bathymetry inversion is more accurate than that obtained using the traditional statistical model, with an average relative error of 10%—20%. In 2020, Ai et al. (2020) proposed a model based on a convolutional neural network, which used different remote sensing images in four spectral bands, red, green, blue, and near-infrared, to retrieve the water depth. The accuracy of the convolutional neural network model was better than traditional neural network (Ai et al., 2020). Moreover, when a sufficient number of measured water depth data from different areas can be obtained, the robustness of the machine learning model can be improved, making this method suitable for different water bodies and areas with highly transparent water.

5. Case studies of satellite-based coastal bathymetry inversion

5.1 Case study 1

The Yongle Atoll in the South China Sea was selected as the study area to demonstrate the performance of our method of deriving highly accurate shallow water bathymetry. The Yongle Atoll, a remote near-continuous annular coral atoll, is located in the Xisha Islands of the South China Sea, which is 300 km southeast to Hainan Island and 80 km southwest to Yongxing Island (Yang et al., 2015).

The bathymetric map around Yongle Atoll in the South China Sea, which is illustrated in Fig. 3.3. Across the study area, a total of 86.10 km^2 shallow water was separated from land and deep water using the Sentinel-2 imagery. The elevation of shallow water in the study area varied from -19.42 to 3.39 m. Across our study area, the elevation range is 22.81 m, which means that the maximum depth below water we can extract using our method is more than 22 m (Fig. 3.4).

Fig. 3.5 demonstrates the comparison between the bathymetric maps from the Sentinel-2 time-series and a traditional single image around Ganquan Island, Yongle Atoll. Fig. 3.5 indicates that different bathymetric maps show a similar spatial pattern but exhibit some differences. Specifically, the bathymetric map derived from a single Sentinel-2 image exhibits some obviously spatial variations, while the bathymetric map from multiple images exhibits much smoother. In addition, clouds and shadows in a single Sentinel-2 image can introduce bathymetric gaps in the derived bathymetric map in Fig. 3.5c—e. The percentages of clear pixels of three single Sentinel-2 images in Fig. 3.5c—e are 93.19%, 99.86%, and 98.64%, respectively. Even though three single images with very good quality were selected, the elevation accuracies of the three derived bathymetric maps are lower, and some bathymetric gaps exist. In summary,

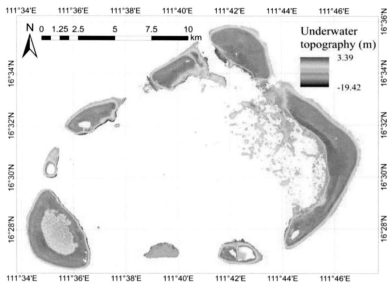

Figure 3.4 Derived shallow water bathymetric map around Yongle Atoll in the South China Sea using the method by combining Sentinel-2 time-series and ICESat-2 data. In this bathymetric map, the elevation of underwater topography that varies from -19.42 m (blue) to 3.39 m (red) was illustrated using a colormap. Note that the derived bathymetric result was based on the WGS84 datum in units of meters.

averaging from Sentinel-2 time-series not only reduces the impact of inevitable noise but also solves the bathymetric gaps caused by clouds and shadows in single Sentinel-2 images.

5.2 Case study 2

The case area in this study is a typical tidal flat along the Jiangsu coast, China (Fig. 3.6), which is characterized by a series of Radial Submarine Sand Ridges located on the Yellow Sea seafloor. Accordingly, the largest tidal flat is distributed on the Chinese continental shelf, with a length of approximately 200 km in the north-south direction and a width of approximately 90 km in the east-west direction (Fan Xu et al., 2016). Due to the active tidal processes at various spatial scales and the large sediment supply from both the old Yangtze River and old Yellow River, tidal flats along the Jiangsu coast, are well developed (Gong et al., 2012).

Based on the Sentinel-2-derived inundation frequency map in Fig. 3.2a, we can derive the tidal flat topography with an area of 1323.72 km^2 based on the cylindrical equal area projection and height range of 2.91 m based on orthometric heights, which is illustrated in Fig. 3.2b. In Fig. 3.2b, red pixels with higher elevations (i.e., lower inundation frequencies) are mainly located in the core area of tidal flats, and blue pixels with lower elevations (i.e., higher inundation frequencies) mainly located in

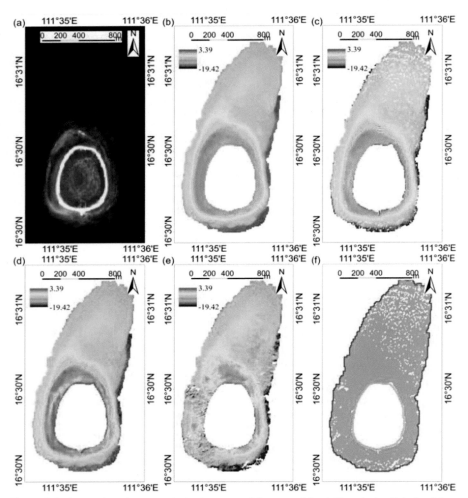

Figure 3.5 Comparison between the bathymetry of Ganquan Island, Yongle Atoll derived from different methods. (a) Spatial distribution of the Sentinel-2 imagery captured on February 24, 2019. (b) Bathymetric map derived from multiple images and (c)−(e) bathymetric maps from traditional single Sentinel-2 images near Ganquan Island, Yongle Atoll. In (f), yellow curve represents the boundary between the land and water, blue curve represents the boundary between shallow water and deep water, and white area represents the bathymetric gaps using traditional single Sentinel-2 image. Note that the same color map (i.e., with the same elevation range) was used in (b)−(e).

the marginal area of tidal flats. In addition, some holes within the tidal flat in the three purple boxes in Fig. 3.2a exist because some pixels correspond to the permanent land (with the highest elevations). In the derived tidal flat topography in Fig. 3.2b, these holes were interpolated and filled using the natural neighbor interpolation method

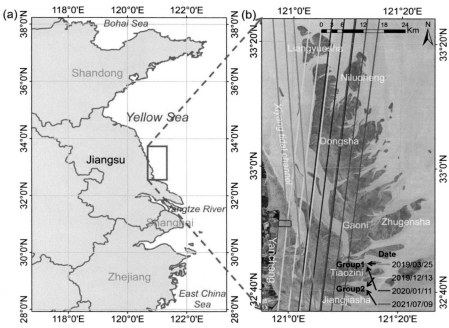

Figure 3.6 Overview of the tidal flat along the Jiangsu coast, China. (a) Location of the study area, and (b) the basemap from the Sentinel-2 false color composite acquired on 2021/02/02. In (a), the red rectangle corresponds to the study area in (b). In (b), 12 lines with four colors represent laser tracks of Ice, Cloud, and Land Elevation Satellite-2 strong beams on four dates (i.e., 2019/03/25; 2019/12/13; 2020/01/11; and 2021/07/09) across the area, and the blue rectangle indicates the location with airborne lidar data for the accuracy assessment.

(Beutel et al., 2010), which is widely used in Digital Elevation Model (DEM) construction. Only approximately 1% of the total pixels within the derived topography were produced by this spatial interpolation. Fig. 3.2c shows the self-comparison between elevations from our derived tidal flat results and ICESat-2-derived surface points (for modeling) with an RMSE of 0.43 m. The validations in Fig. 3.2d—e show RMSEs of 0.49 m using ICESat-2 points (not for modeling) and 0.23 m using airborne lidar data, respectively. From Figs. 3.2f and 3.7h, tidal creeks are easy to identify due to the obvious height difference, for example, four arrows from Figs. 3.2g and 3.7h illustrate four tidal creeks in the cross section. In addition, to directly evaluate the elevation accuracy of ICESat-2 ground points on tidal flats, two laser tracks of ICESat-2 flew over the area with high-resolution airborne lidar data in the blue box in Fig. 3.6b. One of the two along-track surface profiles is compared with the airborne LiDAR data in Fig. 3.2h, which corresponds to the red line in Fig. 3.2g and the left track on 2019/12/13 in Fig. 3.6b. The ICESat-2 tidal flat surface points of the two tracks agree well with the airborne LiDAR data with an RMSE of 0.21 m (not shown) (Fig. 3.7).

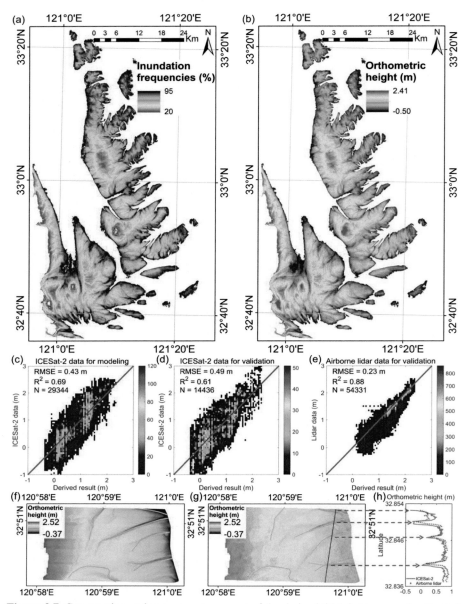

Figure 3.7 Construction and accuracy assessment of the derived tidal flat topography. Spatial distribution of inundation frequencies (a) and orthometric heights (b) for tidal flats in the study area. (c) Self-comparison between elevations from our derived tidal flat results and Ice, Cloud, and Land Elevation Satellite-2 (ICESat-2) data in Group 1. (d) Validation between elevations from our results and ICESat-2 data in Group 2. (e) Validation between elevations from our results and airborne lidar data near Yancheng city, that is, the blue rectangle in Fig. 3.6b. (f) Derived tidal flat topography and (g) airborne lidar data within the 5.43 km² case area near Yancheng city. A case along-track surface profile is presented in (h) that corresponds to the red line in (g) and the left track on 2019/12/13 in Fig. 3.6b. Gray lines delineate tidal creeks from our results and high-resolution airborne lidar data by visual interpretation.

5.3   Case study 3

In the two study areas, the water depths derived by the waves are validated in different ways. Specifically, near the Virgin Islands with very clear waters, the derived water depths can be directly validated by the ICESat-2 bathymetric points (green points), as shown in Fig. 3.3d. After performing a bathymetric error correction, the accuracy of ICESat-2 bathymetric points is generally 0.5 m at depths within 30 m (Parrish et al., 2019; Yue Ma et al., 2020). In turbid waters near Bar Harbor, ICESat-2 cannot consistently obtain a signal from the seafloor due to strong absorption and scattering (Zhang et al., 2021). Alternatively, a local digital elevation model (DEM) of the seafloor is used, which is referenced to the vertical heights of NAVD88 and horizontal coordinates of WGS84 (Friday et al., 2011). This DEM is produced by multibeam sonars and provided by NOAA's National Geophysical Data Center (NGDC). The NGDC data were collected in 2009, and the temporal separation between the NGDC data and ICESat-2 data has a limited impact on the validation because Bar Harbor has a rocky coastline (Friday et al., 2011). The grid spacing for this DEM is ~ 10 m, and the accuracy is between 0.1 m and 5% of the water depth. In addition, to compensate for the tidal effect, orthometric heights rather than water depths are used in the validation process. As shown in Fig. 3.3d, the WGS84 ellipsoid heights of the ICESat-2 points are converted to NAVD88 orthometric heights via the Vdatum tool (Wang et al., 2013) (Fig. 3.8).

Using the wave-based method, the water depths of the two tracks near the Virgin Islands and the other two tracks near Bar Harbor are calculated and illustrated in Fig. 3.3 Near the Virgin Islands in Figs. 3.3a and 3.9b, the seafloor points of ICESat-2 after bathymetric error correction are used to validate the derived water depths. In Figs. 3.3c and 3.9d, near Bar Harbor with turbid harbor waters, the in situ seafloor topographies (green solid curve) measured by multibeam sonars are used as the true values. It is notable that the NAVD88 orthometric heights rather than water depths are used in the validation process to compensate for the tidal effect. The validation results of the four tracks have correlation coefficients R^2 ranging from 0.80 to 0.95, root mean square errors (RMSEs) ranging from 1.3 to 2.0 m, and mean absolute percent errors (MAPEs) ranging from 9% to 22%. Generally, the derived water depths are consistent with the true values. In particular, Figs. 3.3c and 3.9d demonstrate that when the attenuation precludes deep penetration of the green laser of ICESat-2, this method can alternatively estimate water depths by combining wave physics with the actual surface measurements. As mentioned in Section 3.1, this method is more suitable for underwater topography with a gentle slope in principle, but it shows the potential to estimate water depths in highly variable underwater topography, for example, in Fig. 3.3c and 3.9.

5.4   Case study 4

Two study areas with multidate ICESat-2 and Sentinel2 data are involved. As shown in Fig. 3.10, the first study area is Yongle Atoll (latitude: 16.25°-16.40°N, longitude:

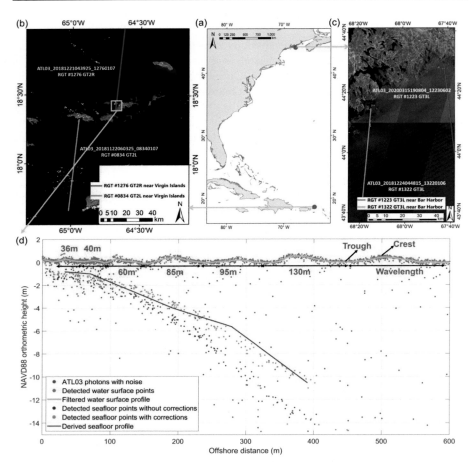

Figure 3.8 Overview of the study areas and sampled ICESat-2 ATL03 data. (a) Location of two study areas; (b) two ICESat-2 ground tracks near the Virgin Islands; (c) two ICESat-2 tracks near Bar Harbor; and (d) geolocated photons from the ICESat-2 ATL03 product of one track near the Virgin Islands corresponding to the orange box in (a). In (b) and (c), the ATL03 datasets used are listed near the corresponding track, and the basemaps are from the Sentinel-2 false color composite. Notably, in (b) and (c), only the ground track of the ICESat-2 strong beam is illustrated because the strong track is very close to the weak beam (90 m perpendicular distance), and they overlap in the base map. In (d), the *x*-axis denotes the relative offshore distance, and the *y*-axis is the NAVD88 orthometric height.

111.30°-111.50°E) that locates in the southwestern part of Xisha archipelago in the South China Sea. Yongle Atoll is composed of several islands as well as some reefs and sandbanks, and the total area that is out of the water in Yongle Atoll is several km^2 (Xu et al., 2011). It has abundant aquatic resources and a convenient waterway in shallow lagoon inside the atoll.

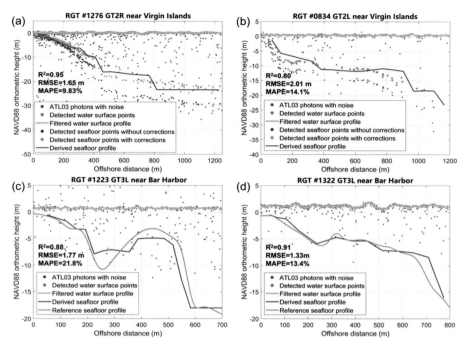

Figure 3.9 Validations of derived water depths for four tracks of ICESat-2 in two study areas. (a) And (b) two tracks near the Virgin Islands with clear waters where the true values are the bathymetric points of ICESat-2 after error corrections. (c) and (d) two tracks near bar harbor with turbid waters where the true values are the in situ seafloor topographies measured by multibeam sonars. Notably, in (a)–(d), the points and derived depths correspond to strong beams of ICESat-2 laser tracks.

The second study area is the lagoon between the Acklins Island (the southeastern island in Fig. 3.11) and Long Cay (the northwestern island) and the surrounding waters outside these islands. The geographic location of this area has a range of Latitude: 22.10°-22.60°N and Longitude 73.90°-74.40°W. This study area locates to the southeast of Bahama. Inside and outside of the lagoon, the total shallow water area (within 25 m in depth) is approximately 1500 km^2 and mainly with reefs, sands, and rocks in bottom.

Fig. 3.12a–d correspond to the bathymetric maps derived from the linear band models on February 24, 2019, March 10, 2020, March 20, 2020, and March 25, 2020, respectively. Fig. 3.12e–h correspond to the bathymetric maps derived from the band ratio models on four dates. Then, the bathymetric maps derived from the two empirical models near Ganquan Island (in the purple box in Fig. 3.10) were validated by the in-situ data captured by an airborne Optech Aquarius lidar system. The topography variation between the time when the airborne lidar surveyed and the

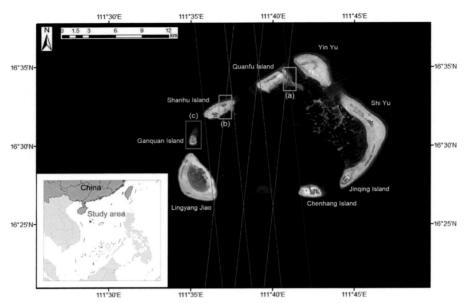

Figure 3.10 ICESat-2 laser beam trajectories near Yongle Atoll, in the South China Sea. The satellite image used as a basemap is from the Sentinel-2 imagery on February 24, 2019. Three red trajectories, three green trajectories, and three blue trajectories correspond to the flight routes of ICESat-2 on October 22, 2018, February 22, 2019, and April 12, 2019, respectively. The purple box marked by (c) near Ganquan island illustrates the in-situ bathymetric measurements (that were used for validation) from an airborne bathymetric lidar. In Fig. 3.13, the SDB maps from linear band models and band ratio models were generally well estimated (with four-date mean R^2 of 0.91 and 0.85, and four-date mean RMSE of 1.44 and 1.85 m, respectively).

time when the ICESat-2 and Sentinel-2 flew over this area was not considered because few constructions were made around this island in recent years.

In Fig. 3.13, spatial distributions of the Sentinel-2 imagery (captured on 24/02/2019 in Fig. 3.10a), the in-situ Optech Aquarius lidar points (in Fig. 3.10f), and derived bathymetric maps near Ganquan Island were illustrated in sequence. Fig. 3.10b—e correspond to the SDB maps derived from the linear band models on February 24, 2019, March 10, 2020, 20/03/2020, and March 25, 2020, respectively, and Fig. 3.10g—j correspond to SDB maps derived from the band ratio models on four dates. It should be noted that on different dates, due to some rocks out of the water surface, clouds, and shadows, the derived bathymetric maps may have blank areas in Fig. 3.10 (the entire maps near Yongle Atoll) and Fig. 3.10 (the enlarged maps near Ganquan Island). To further verify the accuracy of the estimation results, the error scatter diagram, R^2, RMSE, the regression line, and the regression equation between the retrieved water

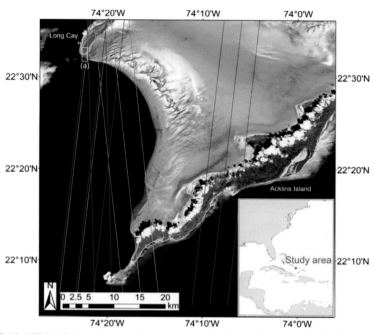

Figure 3.11 ICESat-2 laser beam trajectories near the lagoon between the Acklins Island (the southeastern island) and Long Cay (the northwestern island), to the southeast of Bahama. The satellite image used as a basemap is from the Sentinel-2 imagery on January 27, 2020. Eight red lines correspond to the laser trajectories of ICESat-2 on February 11, 2019, March 12, 2019, and September 02, 2019, respectively, and they were used to train the empirical models. Five green lines correspond to the laser trajectories of ICESat-2 on November 12, 2018 and June 03, 2019, respectively, and they were used to validate the derived bathymetric results.

depths and in-situ depths near Ganquan Island. For the linear band models, the error scatter diagrams correspond to the maps in Fig. 3.10b—e. The SDB maps from linear band models and band ratio models were generally well estimated (with four-date mean R^2 of 0.91 and 0.85, and four-date mean RMSE of 1.44 and 1.85 m, respectively).

The shallow water bathymetric maps near Acklins Island and Long Cay were generated and drawn in Fig. 3.14, where Fig. 3.14a—d correspond to maps derived from linear band models on four dates and Fig. 3.14e—h correspond to the maps from band ratio models. Generally, near Acklins Island and Long Cay, the SDB maps from linear band models and band ratio models were well estimated (with four-date mean R^2 of 0.90 and 0.89, and four-date mean RMSE of 1.18 and 1.24 m, respectively).

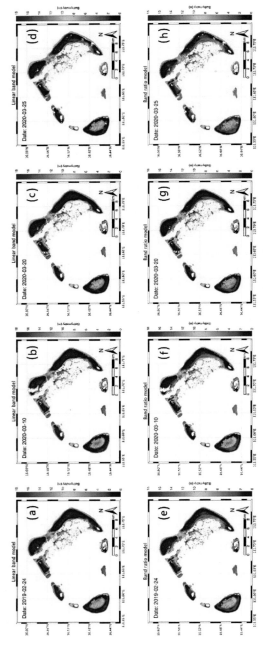

Figure 3.12 Shallow water bathymetric maps derived from the two models for Yongle Atoll using multidate satellite images. (a) The linear band models on February 24, 2019, (b) on March 10, 2020, (c) on March 20, 2020, and (d) on March 25, 2020, respectively; (e) the band ratio models on February 24, 2019, (f) on March 10, 2020, (g) on March 20, 2020, and (h) on March 25, 2020, respectively.

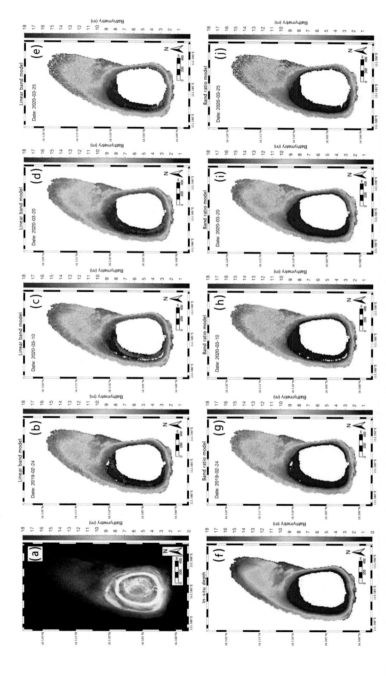

Figure 3.13 Spatial distributions of the Sentinel-2 imagery, Optech Aquarius lidar datasets, and derived bathymetric maps near Ganquan Island, Yongle Atoll. (a) Sentinel-2 imagery on February 24, 2019 image of Ganquan Island; (b)–(e) SDB maps derived from the linear band model on February 24, 2019, March 10, 2020, March 20, 2020, and March 25, 2020, respectively. (f) In-situ truth data from the airborne Optech Aquarius bathymetric LiDAR. (g)–(j) SDB maps derived from the band ratio model on 24/02/2019, March 10, 2020, March 20, 2020, and March 25, 2020, respectively.

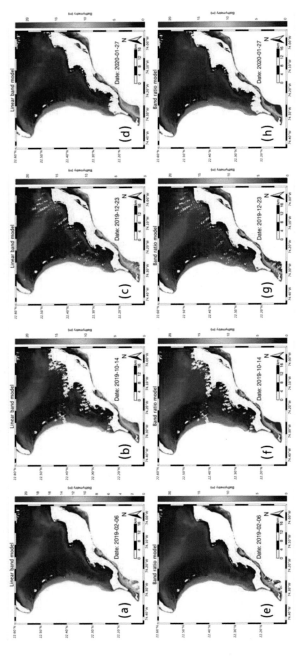

Figure 3.14 Shallow water bathymetric maps derived from the two models for the lagoon of Acklins Island and long Cay using multidate satellite images. (a) The linear band models on February 06, 2019, (b) on October 14, 2019, (c) on December 23, 2019, and (d) on January 27, 2020, respectively; (e) the band ratio models on February 06, 2019, (f) on October 14, 2019, (g) on December 23, 2019, and (h) on January 27, 2020, respectively. Note that land areas (over the water level), clouds covered areas, and water depths deeper than 22 m were subtracted in maps.

6. Conclusions

The space borne ICESat-2 LiDAR can provide high accurate bathymetric points, which can be used in place of the in-situ auxiliary bathymetric information to train the classical empirical models (i.e., the linear model and the band ratio model). Then, the bathymetric map in shallow waters can be produced with only satellite remotely sensed data via empirical models (based on the novel ICESat-2 LiDAR data and the Sentinel-2 multispectral imagery). First, an improved DBSCAN method was proposed to detect the bathymetric signal points from the noisy raw ICESat-2 data photons. Compared to the official results of the bathymetric signal photons, the DBSCAN method performs much better in daytime and at night. Second, for the first time, the bathymetric errors caused by the refraction effect in water column, the refraction effect on the water surface, and the fluctuation effect on the water surface were analyzed and corrected for the ICESat-2 datasets. Third, in our selected study areas, the ICESat-2 bathymetric points were used as the priori measurement to train the local linear band models and the band ratio models with the preprocessed Sentinel-2 images.

In addition, we can adopt ICESat-2 data to extract the wave parameters of sea surfaces and then successfully estimates the bathymetry regardless of the water clarity based on classical wave theory. These two procedures are seemingly similar to SAR-derived bathymetry. However, in terms of the methodology, the observations of SARs are the sea surface roughness caused by internal waves or swells with large wave periods, whereas the observations of lidars are the direct measurement of the surface profile of the wave field, that is, the dynamic characteristics of surface waves. As a wave is shoaling to shallow water zones, the wave height increases, which leads to an increase in the orbital velocity of water particles within the wave. The velocity of water particles probably becomes very high in shallow water areas near the coastline (depths <5 m), and they can no longer be tracked by SAR as smearing of the SAR images occurs. In Fig. 3.3, the surface profiles and depths even in very shallow waters (<2 m) can still be precisely observed and retrieved by ICESat-2. We further explore the novelty of combining wave physics with sea surface measurements and creates a method for "indirectly" extracting depth information in those cases, in which direct seafloor measurements cannot be achieved. Currently, there is still a lack of space-based bathymetric datasets for turbid water areas, and this method aims to provide an improvement. This study also indicates that ICESat-2 can obtain high-resolution and reliable wave characteristics in both offshore and near-shore areas. The combination of wave theory and other auxiliary in-situ data will play an important role in understanding the sediment movement caused by waves and predicting the effect of waves on marine engineering construction in the future.

In the future, the ICESat-2 can provide accurate along-track bathymetric information wherever it flies over, and along with multispectral satellite, we are able to obtain bathymetric maps in shallow water of coastal areas, surroundings of islands and reefs, and inland waters (e.g., rivers, lakes and reservoirs) at a global scale, especially for the remote and sensitive areas without priori data. High-resolution bathymetric data is a fundamental reference for a wide range of coastal applications, e.g., the near-shore geomorphology, sediment process, fishing, hydrology, mineral exploration, cable routing, coastal planning, and hydrodynamic model developing. Meanwhile, since ancient times, coastal areas provide space for large and growing concentrations of human population, settlements, and economic activities.

References

Ai, B., Wen, Z., Wang, Z., Wang, R., Su, D., Li, C., & Yang, F. (2020). Convolutional neural network to retrieve water depth in marine shallow water area from remote sensing images. *Ieee Journal of Selected Topics in Applied Earth Observations and Remote Sensing, 13*, 2888−2898. https://doi.org/10.1109/jstars.2020.2993731

Ao, Y., Li, H., Zhu, L., Ali, S., & Yang, Z. (2019). The linear random forest algorithm and its advantages in machine learning assisted logging regression modeling. *Journal of Petroleum Science and Engineering, 174*, 776−789. https://doi.org/10.1016/j.petrol.2018.11.067

Benny, A. H., & Dawson, G. J. (1983). Satellite imagery as an aid to bathymetric charting in the red sea. *The Cartographic Journal, 20*(1), 5−16. https://doi.org/10.1179/caj.1983.20.1.5

Beutel, A., Mølhave, T., & Agarwal, P. K. (2010). Natural neighbor interpolation based grid DEM construction using a GPU. *Geo Informations Systeme: Proceedings of the ACM International Symposium on Advances in Geographic Information Systems*, 172−181. https://doi.org/10.1145/1869790.1869817

Casal, G., Monteys, X., Hedley, J., Harris, P., Cahalane, C., & McCarthy, T. (2019). Assessment of empirical algorithms for bathymetry extraction using Sentinel-2 data. *International Journal of Remote Sensing, 40*(8), 2855−2879. https://doi.org/10.1080/01431161.2018.1533660

Cheng, K., & Zhu, Y. A. (2007). Summary of machine learning and related algorithms. *Statistics Information Forum, 22*(5), 105−112.

Conger, C. L., Hochberg, E. J., Fletcher, C. H., & Atkinson, M. J. (2006). Decorrelating remote sensing color bands from bathymetry in optically shallow waters. *IEEE Transactions on Geoscience and Remote Sensing, 44*(6), 1655−1660. https://doi.org/10.1109/tgrs.2006.870405

Dang, F., & Ding, Q. (2003). A technique for extracting water depth information from multispectral scanner data in the South China sea. *Marine Science Bulletin, 22*(3), 55−60.

Doxani, G., Papadopoulou, M., Lafazani, P., Pikridas, C., & Tsakiri-Strati, M. (2012). Shallow-water bathymetry over variable bottom types using multispectral worldview-2 image. *The International Archives of the Photogrammetry, Remote Sensing and Spatial Information Sciences, XXXIX-B8*, 159−164. https://doi.org/10.5194/isprsarchives-xxxix-b8-159-2012

Friday, D. Z., Taylor, L. A., Eakins, B. W., Carignan, K. S., Grothe, P. R., Lim, E., & Love, M. R. (2011). *Procedures, data sources and analysis*. NOAA National Geophysical Data Center.

Gao, J. (2009). Bathymetric mapping by means of remote sensing: Methods, accuracy and limitations. *Progress in Physical Geography: Earth and Environment, 33*(1), 103−116. https://doi.org/10.1177/0309133309105657

Giardino, C., Bresciani, M., Cazzaniga, I., Schenk, K., Rieger, P., Braga, F., Matta, E., & Brando, V. (2014). Evaluation of multi-resolution satellite sensors for assessing water quality and bottom depth of lake garda. *Sensors, 14*(12), 24116−24131. https://doi.org/10.3390/s141224116

Gong, Z., Wang, Z., Stive, M. J. F., Zhang, C., & Chu, A. (2012). Process-based morphodynamic modeling of a schematized mudflat dominated by a long-shore tidal current at the Central Jiangsu Coast, China. *Journal of Coastal Research, 28*(6), 1381–1392. https://doi.org/10.2112/JCOASTRES-D-12-00001.1

Hamylton, S. M., Hedley, J. D., & Beaman, R. J. (2015). Derivation of high-resolution bathymetry from multispectral satellite imagery: A comparison of empirical and optimisation methods through geographical error analysis. *Remote Sensing, 7*(12), 16257–16273. https://doi.org/10.3390/rs71215829

Hedley, J. D., Roelfsema, C. M., Phinn, S. R., & Mumby, P. J. (2012). Environmental and sensor limitations in optical remote sensing of coral reefs: Implications for monitoring and sensor design. *Remote Sensing, 4*(1), 271–302. https://doi.org/10.3390/rs4010271

Ji, W., Civco, D. L., & Kennard, W. C. (1992). Satellite remote bathymetry: A new mechanisms for modeling. *Photogrammetric Engineering & Remote Sensing, 58*(5), 545–549.

Kanno, A., & Tanaka, Y. (2012). Modified Lyzenga's method for estimating generalized coefficients of satellite-based predictor of shallow water depth. *IEEE Geoscience and Remote Sensing Letters, 9*(4), 715–719. https://doi.org/10.1109/lgrs.2011.2179517

Kenny, A. J., Cato, I., Desprez, M., Fader, G., Schüttenhelm, R. T. E., & Side, J. (2003). An overview of seabed-mapping technologies in the context of marine habitat classification. *ICES Journal of Marine Science, 60*(2), 411–418. https://doi.org/10.1016/S1054-3139(03)00006-7

Lafon, V., Froidefond, J. M., Lahet, F., & Castaing, P. (2002). SPOT shallow water bathymetry of a moderately turbid tidal inlet based on field measurements. *Remote Sensing of Environment, 81*(1), 136–148. https://doi.org/10.1016/s0034-4257(01)00340-6

Lee, Z., Carder, K. L., Mobley, C. D., Steward, R. G., & Patch, J. S. (1999). Hyperspectral remote sensing for shallow waters: 2. Deriving bottom depths and water properties by optimization. *Applied Optics, 38*(18), 3831–3843. https://doi.org/10.1364/AO.38.003831

Liu, S., Zhang, J., & Ma, Y. (2010). Bathymetric ability of SPOT-5 multi-spectral image in shallow coastal water. *2010 18th International Conference on Geoinformatics, Geoinformatics.* https://doi.org/10.1109/GEOINFORMATICS.2010.5567951

Liu, Y., Deng, R., Qin, Y., Cao, B., Liang, Y., Liu, Y., Tian, J., & Wang, S. (2019). Rapid estimation of bathymetry from multispectral imagery without in situ bathymetry data. *Applied Optics, 58*(27), 7538–7551. https://doi.org/10.1364/AO.58.007538

Lu, T., Chen, S., Tu, Y., Yu, Y., Cao, Y., & Jiang, D. (2019). Comparative study on coastal depth inversion based on multi-source remote sensing data. *Chinese Geographical Science, 29*(2), 192–201. https://doi.org/10.1007/s11769-018-1013-z

Lyons, M., Phinn, S., & Roelfsema, C. (2011). Integrating quickbird multi-spectral satellite and field data: Mapping bathymetry, seagrass cover, seagrass species and change in moreton bay, Australia in 2004 and 2007. *Remote Sensing, 3*(1), 42–64. https://doi.org/10.3390/rs3010042

Lyzenga, D. R. (1978). Passive remote sensing techniques for mapping water depth and bottom features. *Applied Optics, 17*(3), 379–383. https://doi.org/10.1364/AO.17.000379

Lyzenga, D. R. (1981). Remote sensing of bottom reflectance and water attenuation parameters in shallow water using aircraft and landsat data. *International Journal of Remote Sensing, 2*(1), 71–82. https://doi.org/10.1080/01431168108948342

Ma, Y., Xu, N., Liu, Z., Yang, B., Yang, F., Wang, X. H., & Li, S. (2020). Satellite-derived bathymetry using the ICESat-2 lidar and Sentinel-2 imagery datasets. *Remote Sensing of Environment, 250.* https://doi.org/10.1016/j.rse.2020.112047

Ma, Y., Xu, N., Sun, J., Wang, X. H., Yang, F., & Li, S. (2019). Estimating water levels and volumes of lakes dated back to the 1980s using Landsat imagery and photon-counting lidar datasets. *Remote Sensing of Environment, 232.* https://doi.org/10.1016/j.rse.2019.111287

Ma, Y., Zhang, J., Zhang, J., Zhang, Z., & Wang, J. (2018). Progress in shallow water depth mapping from optical remote sensing. *Advances in Marine Science, 36,* 331–351. https://doi.org/10.3969/j.issn.1671-6647.2018.03.001

Markus, T., Neumann, T., Martino, A., Abdalati, W., Brunt, K., Csatho, B., Farrell, S., Fricker, H., Gardner, A., Harding, D., Jasinski, M., Kwok, R., Magruder, L., Lubin, D., Luthcke, S., Morison, J., Nelson, R., Neuenschwander, A., Palm, S., … Zwally, J. (2017). The ice, cloud, and land elevation satellite-2 (ICESat-2): Science requirements, concept, and implementation. *Remote Sensing of Environment, 190,* 260–273. https://doi.org/10.1016/j.rse.2016.12.029

Ohlendorf, S., Müller, A., Heege, T., Cerdeira-Estrada, S., & Kobryn, H. T. (2011). Bathymetry mapping and sea floor classification using multispectral satellite data and standardized physics-based data processing. *Proceedings of SPIE - The International Society for Optical Engineering, 8175.* https://doi.org/10.1117/12.898652

Paredes, J. M., & Spero, R. E. (1983). Water depth mapping from passive remote sensing data under a generalized ratio assumption. *Applied Optics, 22*(8), 1134–1135. https://doi.org/10.1364/AO.22.001134

Parrish, C. E., Magruder, L. A., Neuenschwander, A. L., Forfinski-Sarkozi, N., Alonzo, M., & Jasinski, M. (2019). Validation of ICESat-2 ATLAS bathymetry and analysis of ATLAS's bathymetric mapping performance. *Remote Sensing, 11*(14). https://doi.org/10.3390/rs11141634

Polcyn, F. C., & Lyzenga, D. R. (1973). *Calculations of water depth from ERTS-MSS data* (Vol. 1). Ann Arbor, MI, USA: Environmental Research Institute of Michigan.

Qiu, Y., Shen, W., Hui, X., & Zhang, H. (2019). Satellite-derived bathymetry using random forest model. *The Journal of Ocean Technology, 38*(5), 98–103. https://doi.org/10.3969/j.issn.1003-2029.2019.05.018

Spitzer, D., & Dirks, R. W. J. (1986). Classification of bottom composition and bathymetry of shallow waters by passive remote sensing. *Remote sensing for resources development and environmental management. Proc. 7th ISPRS Commission VII symposium, Enschede* (Vol. 2, 775–777).

Stumpf, R. P., Holderied, K., & Sinclair, M. (2003). Determination of water depth with high-resolution satellite imagery over variable bottom types. *Limnology & Oceanography, 48*(1), 547–556. https://doi.org/10.4319/lo.2003.48.1_part_2.0547

Su, H., Liu, H., & Heyman, W. (2008). Automated derivation of bathymetric information from multi-spectral satellite imagery using a non-linear inversion model. *Marine Geodesy, 31*(4), 281–298. https://doi.org/10.1080/01490410802466652

Tanis, F. J., & Byrnes, H. J. (1985). Optimization of multispectral sensors for bathymetry applications. *Proceedings of the 19th International Symposium on Remote Sensing of Environment,* 865–874.

Tian, J., Wang, J., & Du, X. (2007). Study on water depth extraction from remote sensing imagery in Jiangsu coastal zone. *National Journal of Remote Sensing, 11*(3), 373–379. https://doi.org/10.11834/jrs.20070351

Vanderstraete, T., Goossens, R., & Ghabour, T. K. (2003). Remote sensing as a tool for bathymetric mapping of coral reefs in the Red Sea (Hurghada – Egypt). *BelGéo,* (3), 257–268. https://doi.org/10.4000/belgeo.16652

Wang, H., Huang, W., Wu, D., & Cheng, Y. (2022). Bathymetry inversion method based on adaptive empricial semi- analytical model without in situ data-A case study in south

Coastal and near-coastal eddies are a subtype of ocean eddies that occur close to shorelines and coastal regions. These eddies are often influenced by local factors such as coastal geometry, seabed topography, and the interaction between ocean currents and land. Coastal eddies can vary in size and intensity, and they play a significant role in shaping coastal ecosystems and processes. They can transport nutrients and plankton, affect sediment transport, and impact local fishing and navigation activities. Understanding the dynamics of coastal and near-coastal eddies is crucial for managing coastal resources, predicting harmful algal blooms, and ensuring safe maritime operations.

1.2 Importance in coastal systems

Eddies play a pivotal role in coastal systems, orchestrating a symphony of intricate interactions that profoundly influence the dynamics and health of these fragile environments.

Back in the 1990s, several studies recognized some of the mechanisms by which offshore eddies transport nourishment to the coastal region. In the South Atlantic Bight, cyclonic frontal eddies spin off from the western (coastal) edge of the main current and move into coastal waters (Lee et al., 1991). East Australian Current eddies have been shown to contribute to coastal nutrient enrichment (Tranter et al., 1986).

In the coastal context, eddies emerge as dynamic vortices of immense significance, shaping a plethora of physical, chemical, and biological processes that govern the delicate balance between land and sea.

One of the primary roles of eddies in coastal systems is nutrient transport. These swirling currents serve as nature's nutrient mixers, upwelling nutrient-rich waters from the depths and transporting them to the surface. This upward movement of nutrients fuels the growth of phytoplankton, the base of the marine food chain. As eddies transport these nutrients along the coastline, they create zones of elevated biological productivity, which in turn attract a diverse array of marine species and support thriving ecosystems.

Eddies also play a crucial role in sediment transport along coastal regions. Their circular motion can influence the movement of sediments, shaping the contours of beaches, estuaries, and underwater canyons. By redistributing sediments, eddies contribute to the maintenance of coastal morphology, influencing erosion and deposition patterns that directly impact coastal communities and infrastructure.

In addition to their physical impacts, eddies influence coastal weather patterns and ocean circulation. They can modulate the exchange of heat and moisture between the ocean and the atmosphere, thereby influencing local climate conditions. Moreover, eddies interact with larger-scale currents and fronts, affecting coastal circulation patterns and potentially influencing the dispersal of pollutants or the spread of harmful algal blooms.

Eddies can also have significant economic implications for coastal regions. Their effects on nutrient distribution, fisheries, and sediment transport directly impact industries such as fishing, tourism, and navigation.

An illustrative example is found along the north coasts of Morocco, Algeria, and Tunisia, where the African Current features a series of intricate anticyclonic eddies propagating eastward. These eddies incorporate substantial amounts of coastal and modified Atlantic waters, contributing to an overall eastward transport. The turbulent shear resulting from the interaction of the current with the African coast leads to the advection of coastal waters into the existing eddy field. This mixing process induces biological activity and chemical reactions, observable through variations in ocean color depicted in satellite imagery (Arnone & La Violette, 1986).

In regions like the South China Sea and the Bay of Bengal, the convergence of mesoscale eddies, coastal upwelling, and river discharges introduces abundant nutrients to coastal waters, fostering the growth of phytoplankton (Ye et al., 2023). Differing in physical structure from adjacent waters, these eddies play a crucial role in regulating nutrient availability and influencing primary production in these seas.

Off the coast of Chile in the Eastern Pacific Ocean, the presence of mesoscale eddies and striations, quasi-zonal mesoscale jet-like features, facilitates the advection and mixing of physical properties. This process holds the potential to impact biogeochemistry and marine ecosystems (Auger et al., 2020).

In the context of climate change, the dynamics of mesoscale eddies can drive the severity of marine heatwaves (MHWs) and contribute to coral bleaching across depths by altering thermocline depths and modulating the cooling effect of internal waves (Wyatt et al., 2023).

Understanding the behavior of eddies in coastal systems is therefore essential for sustainable management of these resources and for mitigating potential hazards they might pose to coastal communities.

In essence, the importance of eddies in coastal systems is multifaceted and profound. These swirling phenomena are intricate architects of life, shaping the very fabric of coastal ecosystems, and influencing a cascade of processes that impact both the natural environment and human societies in ways that are both complex and deeply interconnected.

2. Physical mechanisms underlying formation and evolution

The mesmerizing choreography of coastal and near-coastal eddies unfolds through intricate interactions between oceanic forces and localized environmental factors. These eddies, born from the convergence of complex dynamics, embark on a captivating journey from inception to dissipation, sculpting coastal waters and influencing thriving ecosystems.

Eddies can emerge in the wake of coastal headlands or capes, influenced by topographic effects and rotating flows (Davies et al., 1995). The interaction between coastal currents and eddies near coastal topography also contributes to their formation (An & McDonald, 2005). Numerical studies have delved into the nonlinear evolution of coastal currents and eddies near topography to unravel their behavior (An & McDonald, 2005).

The genesis of coastal eddies often stems from the convergence of diverse forces. Coastal geometry, ocean currents, wind patterns, and the interplay between different water masses collectively initiate these swirling currents. As adjacent currents diverge due to varying speeds or directions, a rotating column of water emerges, giving rise to an eddy. Research highlights the Black Sea's abundance of small-scale eddies formed by the interaction of water masses with distinct temperature and salinity properties (Gunduz et al., 2020). The meeting of warm and cold waters induces density gradients, resulting in pressure differences and the initiation of rotational currents. These currents, spiraling inward or outward along density interfaces, lead to the formation of small-scale eddies.

Underwater topography and coastal irregularities further guide and amplify these eddy motions. The coastline's shape influences eddy initiation, where diverging adjacent currents create a rotating column of water, birthing the eddy (Davies et al., 1995). Ongoing studies explore the detailed dynamics associated with the interaction between eddies, coastal currents, and sea level (Shinoda et al., 2023). Orographic wind jets normal to a coast can also generate eddies (Wang et al., 2008). The distribution, lifespan, intensity, and frequency of eddies remain least explored along the southeast coast of India (Arunraj et al., 2018).

Eddies constitute a crucial component of the ocean's mesoscale field, comprising coherent vortices alongside a diverse array of structures such as filaments, squirts, and spirals. More than half of the kinetic energy in ocean circulation resides in the mesoscale eddy field, while the remainder is predominantly found in the large-scale circulation.

Once initiated, coastal eddies embark on a dynamic journey of evolution, shaped by internal and external influences. These eddies, far from static entities, continuously interact with surrounding waters, incorporating elements such as temperature and salinity through their swirling and rotating motions. In the Gulf of Mexico, eddies persist as notable features, their properties influenced by the Loop Current system and coastal low-salinity, highly biologically productive waters (Brokaw et al., 2020). However, the intricate interaction among eddies, coastal currents, and sea level remains complex and not yet fully understood (Shinoda et al., 2023).

The fate of coastal eddies over time is intricately linked to underlying physical processes and the surrounding environment (Shinoda et al., 2023). Gradually losing energy, these eddies dissipate, releasing entrained waters back into surrounding currents. Alternatively, they may merge with larger-scale features or interact with other eddies, resulting in complex dynamics with amplified or dampened effects (Shinoda et al., 2023). The destiny of coastal and near-coastal eddies is shaped by various factors, and their interactions with other features contribute to the complexity of their dynamics.

Seasonal changes, weather patterns, and oceanic conditions further influence the evolution of coastal eddies. These factors can alter the rotational velocities and sizes of eddies, impacting their influence on coastal waters. Li et al. (2022) revealed seasonal variations in the number and size of transient coastal eddies near Fraser Island, with more generated in summer and fewer in winter. Coherent eddies, smaller in size, were more prevalent in summer. Seasonal variations in the Rim Current were also linked to the formation of anticyclonic eddies near the coast (Korotenko, 2017). Weather patterns, such as wind, play a role in eddy generation and propagation (Manso-Narvarte et al., 2021). Additionally, coastal eddies can have feedback effects on the larger oceanic circulation as they interact with broader currents and fronts in coastal and offshore regions.

The formation and evolution of coastal and near-coastal eddies weave captivating tales of nature's fluid intricacy. Sculpted by a myriad of interacting forces, these eddies shape coastal ecosystems, influence marine resources and offer glimpses into the complex interactions defining the interface between land and sea. Studying their formation, behavior, and evolution provides valuable insights into the delicate balance of coastal dynamics, enhancing our understanding of broader oceanic processes that govern our planet.

2.1 Physical mechanisms

The mesmerizing choreography of coastal and near-coastal eddies is orchestrated by a symphony of intricate physical mechanisms, each playing its part in shaping the swirling currents that grace nearshore waters. These mechanisms, arising from the interplay of diverse forces, form the foundation of the eddy dance that shapes coastal environments.

The intricate geometry of coastlines and the submerged topography beneath the ocean surface significantly contribute to the formation of coastal eddies and the dynamics of ocean currents. When ocean currents encounter irregularities along the shoreline, they are compelled to adjust their paths, creating regions of convergence and divergence that set the stage for the initiation of rotating vortices. Submerged features like underwater canyons or seamounts can also guide and enhance the spinning motion of eddies as water masses interact with these topographical variations. The impact of coastline and underwater topography on ocean currents and eddies has been extensively explored through numerical model simulations and observational analyses, underscoring the pivotal role of these factors in shaping coastal ocean dynamics (Gruetzner et al., 2013; Signorini et al., 1992).

For instance, a study focusing on the Texas—Louisiana shelf utilized numerical model simulations to evaluate along-shelf and cross-shelf surge and current variability as influenced by shoreline geometry and bottom topography (Signorini et al., 1992). Another study demonstrated how ocean bathymetry and associated currents influence the weakening of sea ice and the occurrence of sea-ice leads in the Arctic (Willmes et al., 2023). These findings underscore the substantial influence of coastline geometry and underwater topography in shaping the intricate patterns of ocean currents and eddies.

The Coriolis effect, arising from the Earth's rotation, imparts a twisting motion to vortices, ultimately giving rise to coastal and near-coastal eddies. This phenomenon in fluid dynamics significantly influences the behavior of vortices and waves in diverse natural systems. For example, research has demonstrated the impact of the Coriolis effect on traveling wave solutions in the generalized Rotation-Camassa-Holm equation, a model describing the propagation of shallow-water waves in the presence of Earth's rotation (Xu & Yang, 2020). Additionally, the Coriolis effect's role in tidal residual eddies and their influence on water exchange in coastal areas has been studied, underscoring its significance in the dynamics of eddies in specific marine environments (Chao, 1990; Yang & Wang, 2013). Furthermore, investigations into mesoscale eddies in the Black Sea have highlighted the Coriolis effect's involvement in the generation and evolution of coastal eddies, affecting the transport of water and impurities along the coast (Korotenko et al., 2022). These collective studies underscore the substantial impact of the Coriolis effect on the dynamics of coastal and near-coastal eddies across various natural systems.

Lagrangian tracks show that surface particles can be temporarily trapped in eddies and frontal convergent zones, limiting their transport (Mantovanelli et al., 2017). Coastal eddies have been observed in the Southern California Bight, where they are characterized as small-scale surface features (DiGiacomo & Holt, 2001). The theory of underwater acoustic propagation and inversion has complexities due to the nonlinear ocean dynamics occurring at multiple scales, the coupling of which is responsible for the generation of eddies.

The interaction between different water masses with distinct temperature and salinity properties is another crucial factor. As warm and cold waters meet, gradients in density can arise. These density gradients give rise to pressure differences, setting the stage for the initiation of rotational currents. These currents, in turn, lead to the formation of eddies as water spirals inward or outward along the density interfaces (Gorman et al., 2018).

As mentioned previously, wind patterns exert a significant influence on coastal eddy formation. Winds generate friction and drag on the ocean's surface, leading to the accumulation of waters along the coast. This accumulation can trigger the development of coastal eddies as the excess water seeks to balance out these wind-induced imbalances. For example, strong onshore winds can pile up water along the coast, creating a pressure gradient that initiates the formation of eddies. The interaction between winds and existing ocean currents can amplify or dampen the rotational motion of the eddies, shaping their intensity and longevity. When winds align with the direction of ocean currents, they can enhance the rotational motion of eddies, making them stronger and more persistent. Conversely, when winds oppose the direction of the currents, they can weaken or dissipate the eddies. Wind patterns can also lead to tidal flow separation and the generation of tidal eddies. For example, protruding beach can impact tidal currents, potentially leading to tidal flow separation and the formation of eddies (Radermacher et al., 2017).

Coastal eddies are also influenced by larger-scale oceanic features such as coastal currents and fronts. The presence of coastal currents and fronts can guide and steer eddy formation, shaping their trajectories and determining the locations where they are more likely to develop. Coastal eddies often form in regions where there are strong gradients in ocean currents or where different water masses meet. These gradients and interactions between water masses can initiate rotational currents and lead to the formation of eddies. The interaction between coastal eddies and larger-scale ocean currents can lead to a complex interplay of forces, impacting the eddies' evolution and dispersal. For example, coastal eddies can interact with boundary currents, such as the Gulf Stream, and be either advected or entrained by these larger currents. This interaction can influence the size, shape, and lifespan of the eddies.

The physical mechanisms underlying the formation of coastal eddies are a testament to the intricate dance between forces in the oceanic realm. The convergence of coastal geometry, wind patterns, temperature gradients, and the Coriolis effect creates a dynamic stage on which swirling vortices emerge. Understanding these mechanisms not only unravels the secrets of coastal eddies but also deepens our appreciation for the captivating complexity of our oceans' behavior.

2.2   Additional influencing factors

Another pivotal factor influencing coastal eddy formation is the seasonal variability in oceanic conditions. Changes in temperature, salinity, and currents due to seasonal shifts can alter the gradients and dynamics of coastal waters. Warmer or colder water masses can interact with the coastline differently during different times of the year, impacting the convergence and divergence patterns that initiate eddy formation. These seasonal fluctuations provide a backdrop against which coastal eddies emerge, with their intensity and prevalence fluctuating accordingly.

Gunduz et al. (2020) showed that the Rim Current plays a crucial role in the formation of coastal eddies. The Rim Current exhibits seasonal fluctuations, with a stronger flow during the winter months and a weaker flow during the summer months. These seasonal variations can lead to the formation of anticyclonic eddies near the coast. The presence of the East Australian Current is also essential for the generation of Capricorn eddies in the southern Great Barrier Reef region (Li et al., 2022).

The interaction between eddies and these features can also lead to alterations in their paths and intensities. The intricate interplay between these different scales of oceanic behavior adds a layer of complexity to coastal eddy dynamics.

The factors influencing the formation and evolution of coastal eddies are as varied and dynamic as the oceans themselves. From the seasonal rhythms of oceanic conditions to the intricate interplay with local currents and the sculpted terrain of the coastline, each factor weaves its thread into the intricate tapestry of coastal eddy behavior. The study of these factors not only enriches our understanding of coastal dynamics but also unveils the fascinating ways in which nature orchestrates the movement of our planet's waters.

3.   Observations and measurements

Unveiling the secrets of coastal and near-coastal eddies demands a mix of observation and measurement techniques. Through an array of innovative tools and technologies, scientists endeavor to capture the essence of these dynamic phenomena, unraveling their behavior and shedding light on their profound impact on coastal ecosystems and processes.

Satellite remote sensing stands as a vanguard in the realm of coastal eddy observation. Advanced satellites equipped with altimeters and radiometers offer a bird's-eye view of the oceans, allowing researchers to detect sea surface height anomalies, temperature variations, and other crucial parameters associated with eddies. These observations provide insights into eddy formation, evolution, and movement over vast areas.

In-situ measurements, conducted directly in the waters where eddies are present, provide a more detailed understanding of their physical properties. Instrumentation such as buoys, profiling floats, and autonomous underwater vehicles venture into the heart of eddies, collecting data on temperature, salinity, currents, and more. These measurements offer a dynamic portrait of eddy characteristics, revealing their internal structures and interactions with surrounding waters.

High-resolution ocean models also play a pivotal role in the study of coastal eddies. These computational simulations, fueled by extensive data and complex equations, allow scientists to replicate and analyze the behavior of eddies under various conditions. Models provide a platform on which to explore the effects of different factors, from wind patterns to oceanic gradients, shedding light on the underlying mechanisms governing eddy formation, propagation, and dissipation.

Novel technologies, such as underwater gliders and remote-operated vehicles, venture into the depths of the ocean to capture the three-dimensional complexities of eddies. These technologies provide unprecedented access to eddy behavior below the sea surface, allowing scientists to examine their vertical structures and interactions with subsurface currents.

Incorporating data from diverse sources, scientists can paint a comprehensive portrait of coastal eddies, revealing their intricate dynamics and multifaceted roles in nearshore ecosystems. The combination of satellite observations, in-situ measurements, high-resolution models, and cutting-edge technologies offers a holistic understanding of these swirling phenomena.

The observations and measurements of coastal and near-coastal eddies form a range of scientific endeavors, combining remote sensing, in-situ exploration, modeling, and innovation. With these techniques, researchers have unveiled the intricacies of eddy behavior, allowing us to comprehend their significance in shaping coastal environments, driving oceanic processes, and influencing the delicate equilibrium of our planet's dynamic waters.

3.1 Observation techniques

Peering into the intricate world of coastal eddies demands a sophisticated array of observation techniques that traverse the realms of both cutting-edge technology and strategic deployment. These techniques, carefully thought out and meticulously executed, allow scientists to unravel the mysteries of eddy behavior, capturing their nuances and dynamics with unprecedented clarity.

Satellite-based remote sensing constitutes a cornerstone of coastal eddy observation. Earth-observing satellites armed with altimeters and radiometers enable researchers to detect the sea-surface height anomalies and temperature gradients associated with eddies. By monitoring alterations in sea level and thermal patterns, scientists can discern the presence, movement, and evolution of eddies across extensive coastal regions, offering a macroscopic perspective on their impact. However, owing to the constraints of data resolution, most observations are not directly employed for exploring coastal and near-coastal eddies but are often integrated with other observational and numerical data.

One study, for instance, observed an anticlockwise eddy in Toyama Bay, Japan, utilizing ship-mounted Acoustic Doppler Current Profiler (ADCP) and compared the data with numerical simulation results and satellite remote sensing data for chlorophyll-a concentration (Chiba et al., 2015). Another study focused on evaluating the impact of wet troposphere path delays on altimeter sea-level measurements, addressing the correction of these delays using onboard microwave radiometers and atmospheric reanalyses (Legeais et al., 2014).

In-situ measurements, gathered directly from the eddies, provide a more detailed and localized perspective. These measurements are acquired through diverse methods, including buoy networks, profiling floats, and autonomous underwater vehicles (AUVs). Buoy networks, strategically anchored, provide real-time data on sea surface temperature, salinity, and currents. Profiling floats, equipped with sensors, delve into various depths, unraveling the three-dimensional structures of eddies by measuring physical properties. AUVs navigate through eddying currents, collecting data on their characteristics and interactions with surrounding waters. For instance, research has concentrated on path planning for AUVs influenced by ocean currents, as well as the development of sensors for 2D estimation of velocity relative to water and tidal currents. These in-situ measurements and technologies play a pivotal role in advancing our comprehension of coastal and near-coastal eddies and their impact on the marine environment (Kim, 2020; Meurer et al., 2020; Wen et al., 2021).

High-resolution ocean models complement these observations by creating virtual laboratories in which researchers can manipulate various factors to simulate eddy behavior. By inputting data from satellite observations and in-situ measurements, these models generate dynamic visualizations that replicate eddy formation, propagation, and dissipation under different conditions. This approach allows scientists to dissect the underlying mechanisms that govern eddy dynamics, enhancing our understanding of their intricate behavior. Studies have revealed that increasing horizontal resolution from eddy-resolving to submesoscale-enabled, coupled with incorporating high-resolution bathymetry and tides, significantly enhances the replication of eddy kinetic energy and the representation of features such as Gulf Stream penetration (Chassignet & Xu, 2021). Moreover, the impact of wind and hydrodynamic processes on the dispersion of turbid waters has been investigated through a combination of satellite observations, in-situ measurements, and numerical modeling (Lavrova et al., 2015). These examples underscore how high-resolution ocean models, informed by real-world data, serve as virtual laboratories for scrutinizing the intricate behavior of ocean eddies.

Emerging technologies have further expanded the observational toolkit for coastal eddies. Underwater gliders, equipped with sensors and communication systems, glide through eddying waters, capturing real-time data on their physical properties. Remote-operated vehicles dive into the depths, capturing high-resolution imagery and measurements that illuminate the vertical structures of eddies.

In unison, these observation techniques create a multidimensional portrait of coastal eddies, unveiling their complex behaviors and implications. By combining satellite

5. Coastal eddies and climate change

5.1 Potential changes in coastal eddy characteristics

1. *Climate change and oceanic conditions*: Ongoing research delves into the impact of climate change on coastal eddies, anticipating shifts in oceanic conditions that could influence these eddies' characteristics. Alterations in temperature, salinity, and currents may reshape the gradients and dynamics underlying eddy formation. For instance, rising ocean temperatures could affect the density-driven forces initiating eddies, potentially modifying their sizes and strengths (Melentyev et al., 2005; Yun et al., 2024).

    Studies suggest that under greenhouse warming conditions, mesoscale eddies are likely to become more frequent with larger amplitudes and radii, particularly in regions characterized by strong ocean currents like the Antarctic Circumpolar Current and Western Boundary Currents. However, the impact of climate change on eddy dynamics may vary across different oceanic regions and their associated currents. The Gulf Stream, for example, exhibited increased eddy occurrence and radius concurrent with a significant decrease in amplitude, attributed to the influence of eddy lifespans (Yun et al., 2024).

    While the specific changes in coastal eddy characteristics due to climate change are still under investigation, it is evident that these small-scale features play a crucial role in the broader coastal environment. They are instrumental in the exchange of heat, salt, volume, and biogeochemical properties in the ocean (Yun et al., 2024).

2. *Sea-level rise and coastal geometry*: Elevated sea levels have the potential to alter coastal geometry, thereby influencing the processes that guide eddy formation. As coastlines undergo changes and underwater topography transforms, the dynamics of interaction between ocean currents and coastal features may shift. These alterations in geometry have the potential to influence the locations and patterns of eddy initiation, subsequently impacting their trajectories and the regions they influence. However, given the nascent stage of studies on coastal and near-coastal eddies, there is currently insufficient long-term data to make meaningful comparisons. Nevertheless, investigating the influence of rising sea levels on coastal eddy dynamics is a promising avenue for future research.

3. *Human activities and anthropogenic factors*: Human activities, such as coastal development, pollution, and fishing practices, can introduce additional variables that influence coastal eddies. Changes in coastal morphology due to human intervention can alter the convergence and divergence patterns that drive eddy formation. Nutrient inputs from agriculture and urban runoff can impact nutrient availability, potentially affecting the nutrient-driven productivity within eddies (Van Sebille et al., 2020). These factors highlight the complex relationship between human activities and coastal eddies, emphasizing the need for sustainable coastal management practices.

4. *Ocean-circulation changes and current patterns*: Large-scale changes in ocean circulation patterns, such as shifts in major currents and fronts, can have downstream effects on coastal and near-coastal eddies. Alterations in the transport of water masses and the interactions between different currents can influence the pathways and intensities of eddies. These changes can cascade through the coastal environment, impacting ecosystems and resource availability.

    Numerous investigations have documented notable changes in oceanic phenomena, exemplified by the decadal shifts in the Kuroshio Extension (KE) jet, the influence of mesoscale eddies on the Leeuwin Current, and alterations in Atlantic Water circulation patterns north of Svalbard spanning the past 12 years (Athanase et al., 2008). For instance, a study examining the changes in Atlantic Water circulation patterns north of Svalbard over the last 12 years revealed a shift in circulation dynamics. During this period, the Atlantic Water advanced farther north along novel pathways, and the boundary current exhibited increased instability with heightened mesoscale activity (Athanase et al., 2008). Another study focused on the mesoscale eddies associated with the Leeuwin Current in the Eastern Indian Ocean, underscoring the vital role of eddies in nutrient input and the overall productivity of the region (Beckley et al., 2005). Additionally, research on the decadal shifts of the Kuroshio Extension (KE) jet emphasized the impact of meridional shifts of the jet on decadal time scales, with downstream effects on coastal and near-coastal eddies (Sasaki & Schneider, 2011).

5. *Feedback effects and complex interactions*: Changes in the characteristics of coastal and near-coastal eddies can trigger feedback effects that reverberate through the broader oceanic system. Altered nutrient transport, mixing patterns, and water exchange can influence the distribution of marine species, nutrient cycling, and ecosystem health. These complex interactions can have cascading effects on coastal processes and marine resources.

The potential changes in the characteristics of coastal and near-coastal eddies represent a chapter of uncertainty and adaptation in the story of our oceans. As climate dynamics and human influence continue to evolve, these currents may undergo transformations that ripple through coastal ecosystems, impacting marine life, fisheries, and the intricate balance between land and sea. Monitoring and understanding these potential changes are crucial for anticipating and managing the implications of this evolving narrative within our planet's dynamic aquatic realm.

5.2   Future research directions

As the curtain rises on a new era marked by the intersection of coastal eddies and climate change, the quest for understanding and managing these dynamic currents gains heightened urgency. The interplay between eddies and the evolving climate landscape presents a captivating realm for exploration, calling on researchers to delve deeper into the intricate connections, potential impacts, and adaptation strategies that define the future of coastal environments. Future research directions emerge as a compass guiding us through uncharted waters, offering insights that can illuminate the path forward.

5.2.1   Quantifying eddy-climate interactions

A fundamental avenue of research lies in deciphering the intricate interactions between coastal eddies and climate change. Understanding how changes in sea-surface temperature, oceanic currents, and atmospheric conditions influence the initiation, propagation, and dissipation of eddies is crucial. Advanced modeling and observational efforts can shed light on the feedback loops between eddies and the broader climate system, allowing us to discern the roles these currents play in shaping regional and global climate patterns.

5.2.2   Eddy-driven nutrient transport and biological responses

Exploring the impacts of eddies on nutrient transport and biological responses in the context of climate change is a burgeoning field. Investigating how altered nutrient availability within eddies influences phytoplankton growth, marine food webs, and fisheries productivity is paramount. Such research can provide insights into the potential shifts in marine ecosystems and the adaptive strategies that can be employed to mitigate the consequences of changing nutrient dynamics.

5.2.3   Coastal resilience and ecosystem-based adaptation

Understanding how coastal ecosystems, including wetlands, mangroves, and seagrass beds, can act as buffers against the impacts of climate change is pivotal. Research should be focused on quantifying the protective potential of these natural features, exploring their ability to mitigate storm surges, erosion, and sea-level rise. Integrating these insights into comprehensive ecosystem-based adaptation strategies can offer solutions for enhancing coastal resilience.

5.2.4   Integrated observational networks

Creating integrated observational networks that capture the nuances of coastal eddies' behavior and their responses to climate change is a pressing research direction. Combining data from satellite remote sensing, in-situ measurements, ocean models, and emerging technologies such as autonomous vehicles can provide a comprehensive understanding of eddy dynamics. These networks can facilitate real-time monitoring, enabling timely responses to changing conditions.

5.2.5   Community engagement and socio-economic implications

The influence of climate change on coastal eddies extends beyond ecological realms, impacting human communities and economies. Future research should delve into the socio-economic implications of changing eddy characteristics, considering the effects on fisheries, tourism, and coastal infrastructure. Engaging local communities and stakeholders in research efforts can inform strategies that balance ecological conservation with human well-being.

5.2.6   Global collaboration and knowledge sharing

Addressing the intricate nexus of coastal eddies and climate change necessitates global collaboration and knowledge sharing. Researchers, policymakers, and stakeholders across disciplines must come together to pool resources, share findings, and formulate strategies for adaptive management. International partnerships can enhance data collection, modeling accuracy, and the translation of research into actionable policies.

In conclusion, the horizon of research on coastal and near-coastal eddies and climate change stretches wide, beckoning us to explore the uncharted territory that defines the future of our coastal landscapes. By unraveling the complex relationships, assessing potential impacts, and fostering resilience through innovative strategies, we have the power to navigate this evolving landscape and ensure the harmony between land, sea, and humanity endures in the face of a changing climate.

6.   Summary and conclusions

6.1   Coastal and near-coastal eddies: dominant players on the global stage

In the grand choreography of oceanic currents, coastal and near-coastal eddies emerge as dominant protagonists, commanding attention with their intricate movements and profound influence. These eddies, born of the dynamic interplay between coastal topography and oceanic forces, hold a special place within the vast expanse of the oceans. Their significance transcends mere local phenomena, as they stand as key contributors to a substantial portion of the global eddy population.

6.2   Navigating the eddy landscape

The world's oceans are a mosaic of swirling currents known as eddies, each a circular motion of water that contrasts with the prevailing flow. Within this intricate tapestry, coastal and near-coastal eddies take center stage, arising in proximity to coastlines and driven by the complex interactions of tides, currents, and coastal features. Their proximity to land distinguishes them as coastal, while their behavior and impact set them apart as dynamic entities that shape coastal ecosystems and beyond.

6.3   Drivers of connectivity and resilience

The prevalence of coastal and near-coastal eddies is more than a numerical statistic; it underscores their pivotal role in global oceanic dynamics. These eddies act as conduits that bridge coastal and open ocean realms, facilitating the exchange of water masses, nutrients, and marine life. This connectivity enhances the resilience of marine ecosystems, driving genetic diversity, supporting fisheries, and bolstering the intricate web of life that relies on the oceans' bounty.

6.4   Agents of local and global change

Coastal and near-coastal eddies are not confined to the role of localized actors; they wield influence on regional and global scales. Their interactions with coastal habitats, currents, and climate patterns resonate far beyond their birthplaces, shaping weather systems, influencing oceanic circulation, and contributing to the intricate puzzle of Earth's environmental dynamics.

6.5   A story of balance and complexity

In the grand narrative of Earth's oceans, coastal and near-coastal eddies tell a tale of balance and complexity. Their prevalence within the global eddy tapestry highlights their essential role in shaping coastal ecosystems, maintaining the delicate equilibrium of marine life, and contributing to the global-scale choreography of oceanic currents.

As we delve deeper into their intricacies, we unlock insights that reveal the profound interconnections that define our planet's watery realms.

6.6 Unveiling complexity and ecological significance

Our journey into the realm of coastal eddies has illuminated their multifaceted nature and ecological significance. These swirling currents, driven by a delicate interplay of forces, serve as engines of nutrient transport, habitat creation, and larval dispersal. Their influence ripples through the entire coastal ecosystem, fostering biodiversity, bolstering fisheries, and shaping the delicate balance between species and resources.

6.7 Impact on human society and adaptation

The impacts of coastal eddies stretch beyond ecological realms, resonating with human societies and coastal communities. Eddies influence fisheries, impact coastal erosion, and play a role in shaping weather patterns. Recognizing their role in these processes prompts us to consider adaptive strategies that balance human needs with ecological integrity, fostering resilient coexistence with these dynamic currents.

6.8 Climate-change nexus and uncharted waters

As the tides of climate change sweep across the globe, the interaction between coastal eddies and shifting climate patterns emerges as a fascinating frontier of study. Future research holds the promise of uncovering the intricate links between eddy behavior and climate change, exploring potential impacts on ecosystems, weather patterns, and the delicate balance of the oceans. The uncharted waters of this nexus present a canvas for innovation and insight.

6.9 Integrated research and multidisciplinary collaboration

The study of coastal eddies demands an integrated approach that transcends disciplinary boundaries. Collaborations between oceanographers, climatologists, ecologists, and social scientists are imperative to unravel the complex interplay of forces and processes that govern these currents. Shared data, modeling efforts, and insights from diverse perspectives enrich our understanding and inform holistic management strategies.

6.10 Stewardship and sustainability

As we chart future directions for research on coastal eddies, the call for stewardship and sustainability rings clear. Our expanding knowledge must translate into actionable policies and practices that safeguard these dynamic currents and the ecosystems they shape. The preservation of coastal environments, the vitality of marine resources, and the resilience of communities depend on our commitment to responsible management.

In the grand tapestry of the oceans, coastal eddies etch their unique patterns, adding to the intricate weave of life, forces, and interactions that define our planet's aquatic realms. As we bid adieu to this exploration, we stand on the threshold of endless possibilities, poised to navigate the complexities of coastal eddies with curiosity, compassion, and a commitment to the delicate balance of our world's coastal waters. With every current that swirls, there is an opportunity to deepen our connection to the oceans and weave a narrative of coexistence that reverberates through generations to come.

AI disclosure

During the preparation of this work, the author(s) used Chatgpt in order to enhance the quality of the language. After using this tool/service, the author(s) reviewed and edited the content as needed and take(s) full responsibility for the content of the publication.

References

An, B. W., & McDonald, N. R. (2005). Coastal currents and eddies and their interaction with topography. *Dynamics of Atmospheres and Oceans, 40*(4), 237−253. https://doi.org/10.1016/j.dynatmoce.2005.04.002

Arnone, R. A., & La Violette, P. E. (1986). Satellite definition of the bio- optical and thermal variation of coastal eddies associated with the African Current. *Journal of Geophysical Research, 91*(2), 2351−2364. https://doi.org/10.1029/JC091iC02p02351

Arunraj, K. S., Jena, B. K., Suseentharan, V., & Rajkumar, J. (2018). Variability in eddy distribution associated with East India coastal current from high-frequency radar observations along southeast coast of India. *Journal of Geophysical Research: Oceans, 123*(12), 9101−9118. https://doi.org/10.1029/2018jc014041

Athanase, M., Provost, C., Artana, C., Pérez-Hernández, M. D., Sennéchael, N., Bertosio, C., Garric, G., Lellouche, J. M., & Prandi, P. (2008). Changes in Atlantic water circulation patterns and volume transports north of Svalbard over the last 12 years. *Journal of Geophysical Research: Oceans, 126.*

Auger, P., Villegas, V., Belmadani, A., Donoso, D., & Hormazabal, S. (2020). Evaluating the signature of oceanic striations on the distribution of biogeochemical properties in the Eastern Pacific Ocean off Chile. *EGU General Assembly. 2020-22115.* https://doi.org/10.5194/egusphere-egu2020-22115

Beckley, L. E., Muhling, B., & Waite, A. M. (2005). Ichthyoplankton assemblages and primary production in meso-scale eddies associated with the Leeuwin Current, Eastern Indian Ocean. *American Fisheries Society 29th Annual Larval Fish Conference, 54*, 1113−1128.

Brokaw, R. J., Subrahmanyam, B., Trott, C. B., & Chaigneau, A. (2020). Eddy surface characteristics and vertical structure in the Gulf of Mexico from satellite observations and model simulations. *Journal of Geophysical Research: Oceans, 125*(2). https://doi.org/10.1029/2019JC015538

Chao, S.-Y. (1990). Tidal modulation of estuarine plumes. *Journal of Physical Oceanography, 20*(7), 1115−1123. https://doi.org/10.1175/1520-0485(1990)020<1115:tmoep>2.0.co;2

Chassignet, E. P., & Xu, X. (2021). On the importance of high-resolution in large-scale ocean models. *Advances in Atmospheric Sciences, 38*(10), 1621−1634. https://doi.org/10.1007/s00376-021-0385-7

Chelton, D. B., Schlax, M. G., Samelson, R. M., & de Szoeke, R. A. (2007). Global observations of large oceanic eddies. *Geophysical Research Letters, 34*(15). https://doi.org/10.1029/2007GL030812

Chen, G., & Han, G. (2019). Contrasting short-lived with long-lived mesoscale eddies in the global ocean. *Journal of Geophysical Research: Oceans, 124*(5), 3149−3167. https://doi.org/10.1029/2019jc014983

Chiba, H., Hamada, K., Michida, Y., & Hashimoto, S. (2015). Oceanographic observations by onboard CTD and ADCP in Toyama Bay. *Journal of Japan Institute of Navigation, 132*(0), 86−96. https://doi.org/10.9749/jin.132.86

Cipollone, A., Masina, S., Storto, A., & Iovino, D. (2017). Benchmarking the mesoscale variability in global ocean eddy-permitting numerical systems. *Ocean Dynamics, 67*(10), 1313−1333. https://doi.org/10.1007/s10236-017-1089-5

Crawford, W. R., Brickley, P. J., & Thomas, A. C. (2007). Mesoscale eddies dominate surface phytoplankton in northern Gulf of Alaska. *Progress in Oceanography, 75*(2), 287−303. https://doi.org/10.1016/j.pocean.2007.08.016

Davies, P. A., Dakin, J. M., & Falconer, R. A. (1995). Eddy formation behind a coastal headland. *Journal of Coastal Research, 11*(1), 154−167.

Deppeler, S. L., & Davidson, A. T. (2017). Southern Ocean phytoplankton in a changing climate. *Frontiers in Marine Science, 4*. https://doi.org/10.3389/fmars.2017.00040

DiGiacomo, P. M., & Holt, B. (2001). Satellite observations of small coastal ocean eddies in the Southern California Bight. *Journal of Geophysical Research: Oceans, 106*(C10), 22521−22543. https://doi.org/10.1029/2000jc000728

Duncombe, J. (2021). Eddy killing in the ocean. *Eos, 102*. https://doi.org/10.1029/2021eo161292

Gorman, A. R., Smillie, M. W., Cooper, J. K., Bowman, M. H., Vennell, R., Holbrook, W. S., & Frew, R. (2018). Seismic characterization of oceanic water masses, water mass boundaries, and mesoscale eddies SE of New Zealand. *Journal of Geophysical Research: Oceans, 123*(2), 1519−1532. https://doi.org/10.1002/2017JC013459

Gruetzner, J., Uenzelmann-Neben, G., & Franke, D. (2013). The influence of deep ocean currents in shaping the Argentine continental slope during the Neogene. *Jahrestagung der Deutschen Geophysikalischen Gesellschaft (DGG), 73*.

Gunduz, M., Özsoy, E., & Hordoir, R. (2020). A model of Black Sea circulation with strait exchange (2008−2018). *Geoscientific Model Development, 13*(1), 121−138. https://doi.org/10.5194/gmd-13-121-2020

Hou, Y., Jin, F.-F., Gao, S., Zhao, J., Liu, K., Qu, T., & Wang, F. (2022). An "Eddy β-Spiral" mechanism for vertical velocity dipole patterns of isolated oceanic mesoscale eddies. *Frontiers in Marine Science, 9*. https://doi.org/10.3389/fmars.2022.1036783

Justić, D., Kourafalou, V., Mariotti, G., He, S., Weisberg, R., Androulidakis, Y., Barker, C., Bracco, A., Dzwonkowski, B., Hu, C., Huang, H., Jacobs, G., Le Hénaff, M., Liu, Y., Morey, S., Nittrouer, J., Overton, E., Paris, C. B., Roberts, B. J., … Wiggert, J. (2022). Transport processes in the Gulf of Mexico along the river-estuary-shelf-ocean continuum: A review of research from the Gulf of Mexico research initiative. *Estuaries and Coasts, 45*(3), 621−657. https://doi.org/10.1007/s12237-021-01005-1

Kim, J. (2020). Cooperative localization and unknown currents estimation using multiple autonomous underwater vehicles. *IEEE Robotics and Automation Letters, 5*(2), 2365−2371. https://doi.org/10.1109/lra.2020.2972889

Korotenko, K. A. (2017). Modeling processes of the protrusion of near-coastal anticyclonic eddies through the Rim Current in the Black Sea. *Oceanology, 57*(3), 394–401. https://doi.org/10.1134/S0001437017020114

Korotenko, K., Osadchiev, A., & Melnikov, V. (2022). Mesoscale eddies in the Black Sea and their impact on river plumes: Numerical modeling and satellite observations. *Remote Sensing, 14*(17). https://doi.org/10.3390/rs14174149

Lavrova, O., Krayushkin, E., Golenko, M., & Golenko, N. (2015). Propagation of the Vistula Lagoon outflow plume into the Baltic Sea: Satellite observations, in-situ measurements and numerical modeling. *International Geoscience and Remote Sensing Symposium (IGARSS), 2015*, 2299–2302. https://doi.org/10.1109/IGARSS.2015.7326267

Le Vu, B., Stegner, A., & Arsouze, T. (2018). Angular momentum eddy detection and tracking algorithm (AMEDA) and its application to coastal eddy formation. *Journal of Atmospheric and Oceanic Technology, 35*(4), 739–762. https://doi.org/10.1175/jtech-d-17-0010.1

Lee, T. N., Yoder, J. A., & Atkinson, L. P. (1991). Gulf Stream frontal eddy influence on productivity of the southeast U.S. continental shelf. *Journal of Geophysical Research: Oceans, 96*(C12), 22191–22205. https://doi.org/10.1029/91JC02450

Legeais, J.-F., Ablain, M., & Thao, S. (2014). Evaluation of wet troposphere path delays from atmospheric reanalyses and radiometers and their impact on the altimeter sea level. *Ocean Science, 10*(6), 893–905. https://doi.org/10.5194/os-10-893-2014

Li, G., He, Y., Liu, G., Zhang, Y., Hu, C., & Perrie, W. (2020). Multi-sensor observations of submesoscale eddies in coastal regions. *Remote Sensing, 12*(4). https://doi.org/10.3390/rs12040711

Li, Z., Cai, Z., Liu, Z., Xiaohua, W., & Jianyu, H. (2022). A novel identification method for unrevealed mesoscale eddies with transient and weak features-Capricorn Eddies as an example. *Remote Sensing of Environment, 274*, 112981. https://doi.org/10.1016/j.rse.2022.112981

Lu, X., Chen, Y., & Wang, C. (2022). A method for detection and parameter inversion of ocean eddies in SAR images. *2022 3rd International Conference on Geology, Mapping and Remote Sensing, ICGMRS 2022*, 874–877. https://doi.org/10.1109/ICGMRS55602.2022.9849288

Ma, C., Li, S., Wang, A., Yang, J., & Chen, G. (2019). Altimeter observation-based eddy nowcasting using an improved conv-LSTM network. *Remote Sensing, 11*(7). https://doi.org/10.3390/rs11070783

Manso-Narvarte, I., Rubio, A., Jordà, G., Carpenter, J., Merckelbach, L., & Caballero, A. (2021). Three-dimensional characterization of a coastal mode-water eddy from multiplatform observations and a data reconstruction method. *Remote Sensing, 13*(4). https://doi.org/10.3390/rs13040674

Mantovanelli, A., Keating, S., Wyatt, L. R., Roughan, M., & Schaeffer, A. (2017). Lagrangian and Eulerian characterization of two counter-rotating submesoscale eddies in a western boundary current. *Journal of Geophysical Research: Oceans, 122*(6), 4902–4921. https://doi.org/10.1002/2016JC011968

Melentyev, V., Chernook, V., & Melentyev, K. (2005). Stationary spiraling eddies and self-cleaning processes in the white sea in presence of climate change and their relationship with ecology of the Greenland seal: Results of airborne-satellite-in situ study. *Mitigation and Adaptation Strategies for Global Change, 10*(1), 115–126. https://doi.org/10.1007/s11027-005-7834-y

Meurer, C., Fuentes-Perez, J. F., Schwarzwalder, K., Ludvigsen, M., Sorensen, A. J., & Kruusmaa, M. (2020). 2D estimation of velocity relative to water and tidal currents based

The primary freshwater sources for the estuary are the Parramatta, Lane Cove, and Duck Rivers (Lee & Birch, 2012) (see Fig. 5.1a). Estuarine water undergoes effective tidal mixing during extended periods of low rainfall (<5 mm rainfall/day) and transitions into a stratified state during intermittent periods of high rainfall (>50 mm rainfall/day) lasting a few days (Birch & Rochford, 2010; Lee et al., 2011). While recent studies have primarily examined plume behavior and hydrodynamic characteristics, the investigation of the dominant physical processes accountable for temporal variations in stratification and vertical mixing during periods of high river flow in the Sydney estuary remains unexplored.

Residual circulation, as an important indicator of estuarine health, may be created by two forcing mechanisms, that is, nonlinear interactions between tidal oscillations and irregular bathymetry (Zimmerman, 1978) and buoyancy inflow with a horizontal salinity gradient (Pritchard, 1956). Recent studies identify a range of different processes related to residual flow formation, including subtidal runoff, a longitudinal salinity gradient, asymmetric vertical mixing, wind straining, and lateral processes (Cheng et al., 2010). Extensive studies have been conducted on tide-induced residual circulation in estuaries, bays, and lagoons, which result from the nonlinear interactions between tidal oscillatory flow and the boundary geometry and bottom topography (Robinson, 1981). Tide-induced residual circulation in Sydney Estuary was investigated through numerical experiments (Das et al., 2000); they illustrated how nonlinear terms, irregular bathymetry, bottom friction coefficients, and homogenous wind forcing influence the generation of residual flows and alterations in residual circulation patterns.

In this study, a horizontal salinity gradient-induced residual circulation has been incorporated for Sydney Estuary. Similar to other estuaries (Hansen & Rattray, 1965; Pritchard, 1956), this horizontal salinity gradient has been identified as the primary component of estuarine residual flows. Furthermore, the residual flows, characterized by the same vertical structure as the density-driven circulation, will arise due to the asymmetric tidal mixing resulting from strain-induced periodic stratification and the horizontal salinity gradient (Cheng et al., 2010). This vertical structure of residual flows will be evaluated in detail using a numerical model.

The first objective of this study was to develop and calibrate a three-dimensional hydrodynamic model against field survey data to reveal the hydrodynamic characteristics with reasonable accuracy. In order to predict the effects of different forcing fields on stratification/mixing and hydrodynamic circulation, a set of numerical experiments was designed by adding tides, winds, and river discharge successively to the system. In addition, we use the hydrodynamic model to describe the structure of residual flow in the horizontal and vertical and then identify the dominant forcing mechanisms affecting the residual flows within the estuary.

2. Field observation

Surface water elevations were acquired from the FD tidal gauge with recordings taken at a 60-minute sampling interval. A bottom-mounted (upward looking) acoustic

Doppler current profile (ADCP) was deployed in 32 m of water in the main channel (Fig. 5.1a), by the Sydney Harbor Ports Authority. The 300 kHz ADCP had a vertical bin size of 1 m from 0.5 to 27.5 m of water depth. Measurement range was limited in the surface and bottom blank area. Current speed and direction data were extracted for the month of January 2013 at 6-minute time intervals. Data were then averaged to hourly.

The CTD profiling system near BH was configured with three vertically suspended TRDI Citadel CTDs, supported by mid-water floats, and a Seabird SBE37 mounted at the bottom. The sensors collected samples at depths of 1.3 m, 7.3 m, 10.6 m and 13.7 m at 5-minute sampling intervals (Fig. 5.1c).

Two 600 kHz ADCPs were operated in bottom-tracking mode to conduct transect profiling along four estuary cross-sections, on December 4, 2013, over a full flood-ebb cycle during a spring-tide cycle. The time lapse between consecutive measurements at each cross-section was approximately 15 min, and one round trip for the four sections was finished within 1 hour (section locations are shown in Fig. 5.1b).

A CTD probe (YSI Model with a YSI 6136 optical turbidity probe) was lowered manually from the surface to the bottom at 2s recording intervals. Monthly CTD profiles measured across the whole estuary were used as input to generate initial condition for salinity in the model and to compare the predicted spatial salinity variations (see Appendix for details).

3. Numerical model

3.1 Model description

An unstructured-grid, finite-volume, three-dimensional primitive equation Coastal Ocean Model (FVCOM) was used to reproduce coastal oceanic and estuarine circulation. The FVCOM simulates water surface elevation, velocity, temperature and salinity by resolving equations of momentum, continuity, temperature, salinity, and density in an integrated form to preserve mass conservation.

$$\frac{\partial u}{\partial t}+u\frac{\partial u}{\partial x}+v\frac{\partial u}{\partial y}+w\frac{\partial u}{\partial z}-fv=-\frac{1}{\rho_o}\frac{\partial(P_{H+P_a})}{\partial x}-\frac{1}{\rho_o}\frac{\partial q}{\partial x}+\frac{\partial}{\partial z}\left(K_m\frac{\partial u}{\partial z}\right)+F_u$$

$$(5.1)$$

$$\frac{\partial v}{\partial t}+u\frac{\partial v}{\partial x}+v\frac{\partial v}{\partial y}+w\frac{\partial v}{\partial z}+fu=-\frac{1}{\rho_o}\frac{\partial(P_{H+P_a})}{\partial y}-\frac{1}{\rho_o}\frac{\partial q}{\partial y}+\frac{\partial}{\partial z}\left(K_m\frac{\partial v}{\partial z}\right)+F_v$$

$$(5.2)$$

$$\frac{\partial w}{\partial t}+u\frac{\partial w}{\partial x}+v\frac{\partial w}{\partial y}+w\frac{\partial w}{\partial z}=-\frac{1}{\rho_o}\frac{\partial q}{\partial z}+\frac{\partial}{\partial z}\left(K_m\frac{\partial w}{\partial z}\right)+F_w \qquad (5.3)$$

$$\frac{\partial u}{\partial x} + \frac{\partial v}{\partial y} + \frac{\partial w}{\partial z} = 0 \tag{5.4}$$

$$\frac{\partial T}{\partial t} + u\frac{\partial T}{\partial x} + v\frac{\partial T}{\partial y} + w\frac{\partial T}{\partial z} = \frac{\partial}{\partial z}\left(K_h\frac{\partial T}{\partial z}\right) + F_T \tag{5.5}$$

$$\frac{\partial S}{\partial t} + u\frac{\partial S}{\partial x} + v\frac{\partial S}{\partial y} + w\frac{\partial S}{\partial z} = \frac{\partial}{\partial z}\left(K_h\frac{\partial S}{\partial z}\right) + F_S \tag{5.6}$$

$$\rho = \rho(T, S, p) \tag{5.7}$$

Where x, y, and z are the east, north, and vertical axes in the Cartesian coordinate system; u, v, and w the velocity components along the x, y, and z axes; T, temperature; S, salinity; ρ, density; P_a, air pressure at the sea surface; P_H, hydrostatic pressure; q, nonhydrostatic pressure; f, Coriolis parameter; g, gravitational acceleration; K_m, vertical eddy viscosity coefficient; K_h, thermal vertical eddy diffusion coefficient; F_u and F_v, horizontal momentum diffusion terms; F_T and F_S, thermal and salt diffusion terms (Chen et al., 2003).

In addition, specific modules can be added to simulate sediment dynamics and water quality. The use of an unstructured triangular grid in the horizontal plane gives greater flexibility for areas with a complex coastline and allows for greater refinement of local grids in coastal areas, a suitable approach for the SHE. A sigma vertical coordinate is used to better represent the irregular bottom topography in the three-dimensional domain. The model uses the Smagorinsky (Smagorinsky, 1963) and the modified Mellor-Yamada Level 2.5 (Mellor & Yamada, 1982) turbulence closure schemes for horizontal and vertical mixing, respectively. The empirical constants and simplified higher level equations in MY-2.5 (Mellor & Yamada, 1982) are based on neutral boundary layer and pipe data, also known as the k-kl model (where k is the turbulent kinetic energy and l is the turbulent macroscale). To ensure accuracy for the MY-2.5 model, modifications were made on parameters involving velocity, temperature gradients, turbulent kinetic energy, and mixing length scales (Mellor & Yamada, 1982). A Richardson-number-dependent dissipation function was introduced to reduce turbulent dissipation in stratified flow, which generates over-mixing at the bottom boundary layer (Wang, 2002). FVCOM features the modified MY-2.5 model (Chen et al., 2003), which includes (a) the upper and lower bounds limits of the stability function (Galperin et al., 1988); (b) the wind-driven surface wave breaking-induced turbulent energy input at the surface and internal wave parameterization by (Mellor & Blumberg, 2004); and (c) the improved parameterization of pressure-strain covariance and shear instability-induced mixing in the strongly stratified region (Kantha & Clayson, 2000). An alternative turbulence scheme in FVCOM is the k-ε model (where ε is the turbulence dissipation rate), which is similar in dynamics to the k-kl model, but also includes the buoyancy parameterization (Canuto et al., 2001). The choice of a particular turbulence scheme depends on the hydrodynamic model configuration; the turbulence parameters should be determined via local conditions. The wetting/drying algorithm as a point treatment incorporating a bottom viscous layer of specified thickness ($D_{min} = 5$ cm in this case) was applied to take into account the impact of tidal flats. Water depths at nodes and cells were checked against D_{min}; if less than D_{min}, the velocity in the triangle cell and the flux at the cell boundaries were set to zero in calculating of volume flux to ensure mass conservation (Chen et al., 2003).

3.2   Model configuration

The computational domain encompassed approximately 23 km offshore, incorporating a portion of the continental shelf, and extended approximately 30 km upstream westward to include freshwater discharge points (Fig. 5.2). The grid consisted of 79,278 elements (triangles) and 43,584 nodes (nodes of triangles) forming a mesh of triangles with variable cell width ranging from 2000 m at the open ocean boundary to 30 m within the estuary. Over 50% of the cells were less than 50 m wide. The vertical resolution consisted of 15 sigma layers, with a uniform thickness in the middle (11% of the total depth), and higher resolution near the surface and bottom (1% of the total depth). The bathymetry data were interpolated linearly to a spatial resolution of 30 m.

The model was simulated for a duration of 31 days, running from January 01, 2013 to January 31, 2013, with consideration of tidal forcing, surface winds, and freshwater inflow (river discharge). The model was initialized with a zero velocity field and constant temperature of 25°C and an initial horizontal salinity gradient estimated based on the CTD survey conducted on January 10, 2013.

3.3   Forcing fields and limitation

3.3.1   External forcing

Open boundary conditions were specified by tidal elevations simulated using the TPXO7.2 global ocean tide model (http://volkov.oce.orst.edu/tides/global.html). The data set consisted of four diurnal constituents Q_1, O_1, P_1, K_1, four semi-diurnal constituents M_2, S_2, N_2, K_2, two long-period components MF, MM, and three shallow-water constituents M_4, MS_4, MN_4. The amplitude and phase for each constituent were derived at each open boundary node from the coarser global model by linear interpolation.

Surface wind fields were obtained from the global atmospheric reanalysis model ECMWF (European Centre for Medium-Range Weather Forecasts http://www.ecmwf.int/) product ERA-interim. The spatial resolution was 0.125 degrees × 0.125 degrees with output every 6 h. The ECMWF wind data were compared with observed wind fields of two land-based meteorological stations (at Fort Denison and southern end of model domain) provided by Bureau of Meteorology (BOM). For a 1-year period (2013), the ECMWF wind data showed a correlation coefficient of 0.68 and 0.59, respectively in magnitude, 0.58 and 0.53 in direction with observations.

3.4   Catchment—Estuary continuum modeling

Modeling the catchment-to-estuary continuum is crucial for a more comprehensive understanding of the dynamic interactions and processes occurring within coastal region interconnected aquatic systems. A continuum model provides an integrated framework that considers the entire catchment-to-estuary system by integrating the complex

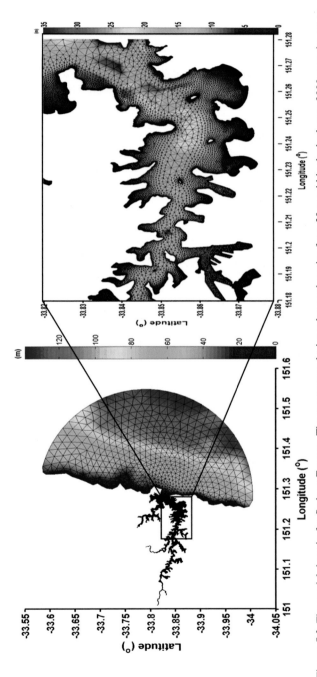

Figure 5.2 The model domain for Sydney Estuary. The unstructured triangular mesh varies from 30 m within the harbor to 2000 m at the ocean boundary. The model depth is indicated by the colored shading.
From Xiao, Z., Wang, X. H., Roughan, M. & Harrison, D. (2019). Estuarine, Coastal and Shelf Science, 217, 132–147. https://doi.org/10.1016/j.ecss. 2018.11.004

interactions and feedback mechanisms between riverine and estuarine processes, including the transport of water, sediment, and nutrients from rivers to estuaries.

Catchment rainfall-runoff processes were simulated and calibrated with a purpose built GIS-based stormwater model called Watershed Basin Network Model (WBNM Fig. 5.3). WBNM employs a runoff routing method to compute the flood hydrograph arising from storm rainfall. This involves dividing the catchment into subcatchments using the stream network, with each subcatchment assigned a lag time determined by its size. These lag times are determined based on investigations into the nonlinear variation of lag time observed in actual catchments (Boyd et al., 1996). It distinguishes overland flow routing from channel routing, thereby permitting modifications to either or both of these processes, such as in catchments undergoing urbanization. Twelve major inflow locations were identified and are labeled as Q1-Q12 in Fig. 5.3a. The calibration for rainfall-runoff was conducted at station Q1 (Fig. 5.3b), situated in the primary river that flows into the SHE. The model captures the flow peaks with a minor underestimation. Freshwater discharge rate, water temperature, and salinity were introduced at the periphery of the model boundary elements. The coupling between catchment and estuary is currently one-way coupled, with no feedback from the estuary to the rivers.

WBNM simplifies certain hydrological and nutrient processes, potentially leading to a loss of detail and accuracy in representing complex catchment dynamics. The model assumes the statistical properties of the catchment (e.g., rainfall patterns, land use) remain constant over time which cannot represent the dynamic environment. The model's spatial and temporal resolution is too coarse to capture small-scale variations or short duration events that can be significant in some catchments. In addition, accurate parameterization is crucial for model performance. Difficulties in obtaining precise parameter values or uncertainties in model inputs can affect the reliability of the results.

For future catchment-to-estuary modeling, utilization of integrated modeling approaches that consider hydrodynamics, water quality, sediment transport, and ecological components will provide a more holistic understanding of the system and how different processes interact. A catchment model which assesses the appropriate spatial and temporal scales of SHE is required. Considering the size of the catchment and the level of detail and precision needed for the model study is desired.

3.5   Bottom friction sensitivity test

To model bottom boundary conditions, a nonlinear bottom friction law was applied in the model to represent the turbulent frictional process as

$$\tau_b = \rho C_d |u_b| u_b \tag{5.8}$$

where τ_b is the bottom stress, C_d is the bottom drag coefficient, and u_b is the bottom current velocity at depth of z_b. When z_b falls within the constant stress bottom layer where water is neutrally stratified, C_d can be defined as (Wang, 2002)

$$C_d = \max \left[\frac{k^2}{\ln\left(\dfrac{Z_{ab}}{z_O}\right)^2}, 0.0025 \right] \tag{5.9}$$

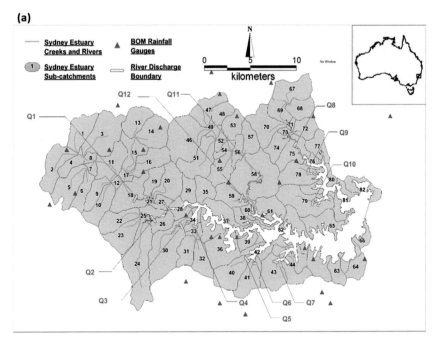

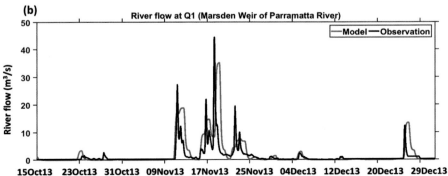

Figure 5.3 Hydrology model Watershed Bounded Network Model (WBNM) used for rainfall-runoff estimation. (a) The Sydney catchment was divided into 82 subcatchments (as labeled 1−82). The *red lines* (numbered Q1-Q12) showed the location of the freshwater input at the model boundary as edge sources. A total of 38 rainfall stations were used for rainfall pattern generation. (b) River inflow calibration at the Marsden Weir of Parramatta river (Q1) during October 15, 2013−December 31, 2013.

From Xiao, Z., Wang, X. H., Roughan, M. & Harrison, D. (2019). Estuarine, Coastal and Shelf Science, 217, 132−147. https://doi.org/10.1016/j.ecss.2018.11.004

where $k = 0.4$ is the von Karman's constant, z_{ab} is the logarithmic bottom layer height, and z_o is the bottom roughness. Since z_{ab} varies with location and z_o is determined by sediment type and vegetation, C_d varies through the domain to keep a constant bottom stress layer. Limited document shows the spatial distribution of bed roughness within the Sydney Estuary, which is used to determine the bottom drag coefficient. Uniformly distributed bed roughness (2 cm) was used within the estuary to account for the mounds from biotic activity on the bed (Lee et al., 2011). The sensitivity of residual flows to the bottom drag coefficient was tested by applying Eq. (5.9) and constant C_d of 0.25 and 0.325 through the whole domain.

3.6 Design of numerical experiments

Several numerical experiments were conducted to investigate hydrodynamic processes under various forcing conditions (Table 5.1). The first model run was forced by tidal elevations only at the open boundary (Exp. 1). To evaluate the wind forcing and fresh-water buoyancy impact, wind fields, and freshwater discharge with a horizontal salinity gradient were applied to the model successively (Exp. 2, Exp. 3). A sensitivity test of the bottom drag coefficient on the residual flows was conducted in Exp. 4.

4. Model validation

4.1 Error statistics

The model skill score (SS), used to quantify model errors, is defined as the ratio of the RMSE to the standard deviation of the observation:

$$SS = 1 - \frac{\sum (X_{mod} - X_{obs})^2}{\sum (X_{obs} - \overline{X}_{obs})^2} \tag{5.10}$$

where X is the variable and $\overline{X}$ is the temporal average. The correlation coefficient (CC) between model and observation is also used to evaluate model performance:

$$CC = \frac{\sum (X_{mod} - \overline{X}_{mod})(X_{obs} - \overline{X}_{obs})}{\left[\sum (X_{mod} - \overline{X}_{mod})^2 \sum (X_{obs} - \overline{X}_{obs})^2\right]^{1/2}} \tag{5.11}$$

Table 5.1 Setup of numerical experiments.

	Tidal forcing	Wind forcing	Freshwater buoyancy	Salinity gradient	Bottom drag coefficient
Exp. 1	√				Eq. (5.9)
Exp. 2	√	√			Eq. (5.9)
Exp. 3	√	√	√	√	Eq. (5.9)
Exp. 4	√				0.25; 0.325

SS was calculated as 0.99 at FD and 0.94 at BH; CC was estimated as 0.99 (FD) and 0.97 at BH (Table 5.2). A harmonic analysis was conducted using the T-Tide program in Matlab (Pawlowicz et al., 2002). The errors of amplitudes and phases of the six major tidal constituents (M_2, S_2, N_2, Q_1, K_1, O_1) are minor (Table 5.3). The amplitude and phase errors for M_2 were within 1 cm and 3 degrees for the two stations.

4.2 Tidal elevations

Hourly water elevation was obtained at FD and BH and was compared with model elevations in Fig. 5.4a; Fig. 5.5a. The root mean square error was 0.04 and 0.09 m (less than 0.5% of the water column depth). Harmonic analysis was conducted using the T-Tide program in Matlab (Pawlowicz et al., 2002). The amplitude and phase of 30 tidal constituents were resolved. The harmonic constants of six major tidal constituents (M_2, S_2, N_2, Q_1, K_1, O_1) were in good agreement with the FD tidal gauge harmonic analysis

Table 5.2 Model errors in the depth-averaged current velocities (m/s: along-estuary, cross-estuary) and bottom salinity (psu) at the mooring station.

	Mean error	RMS error	Skill score	Correlation coefficient
FD water level	−0.01	0.04	0.99	0.99
BH water level	−0.02	0.09	0.94	0.97
BH u	−0.03	0.05	0.84	0.95
BH v	0.01	0.03	0.52	0.71
BH salt	−0.21	0.42	0.20	0.68
BH sediment	0.15	0.30	0.45	0.70

Modified from Xiao, Z., Wang, X. H., Roughan, M. & Harrison, D. (2019). Estuarine, Coastal and Shelf Science, 217, 132−147. https://doi.org/10.1016/j.ecss.2018.11.004

Table 5.3 Comparison of modeled and observed harmonic parameters for the main tidal constituents at the Fort Denison tide gauge.

Tidal constituent	Amplitude (m)			Phase (degree)		
	Observed	Model	Amplitude deviation (%)	Observed	Model	Phase deviation (deg)
Q_1	0.02	0.02	−4.2	253.7	250.8	−2.8
K_1	0.17	0.20	13.9	326.6	321.3	−5.3
O_1	0.10	0.10	−1.1	286.8	286.2	−0.6
N_2	0.12	0.11	−4.5	260.4	269.4	9.1
M_2	0.51	0.50	−2.2	278.2	277.6	−0.6
S_2	0.13	0.12	−7.1	307.2	301.6	−5.6

Data from Xiao, Z., Wang, X. H., Roughan, M. & Harrison, D. (2019). Estuarine, Coastal and Shelf Science, 217, 132−147. https://doi.org/10.1016/j.ecss.2018.11.004

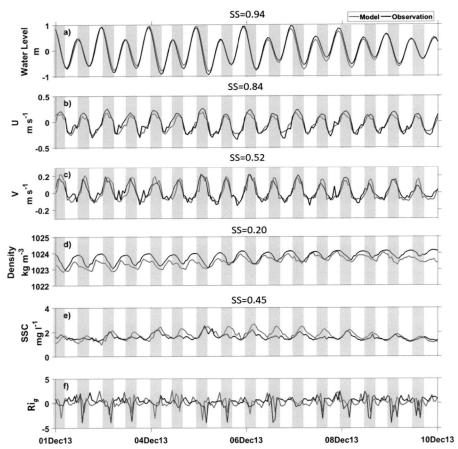

Figure 5.4 Observed (*black line*) and modeled (*red line*) data from Balls head **(a)** surface water level (m); **(b)** *bottom* along-estuary velocity (m/s) (positive indicates ebbing); **(c)** *bottom* cross-estuary velocity (m/s) (positive indicates ebbing); **(d)** *bottom* density (kg/m^3); **(e)** *bottom* SSC (mg/L); **(f)** *bottom* normalized gradient Richardson number $\log_{10}(Ri_g/0.25)$ (positive indicates stratification, negative mixing). Ebb tides are indicated by the *shaded background*. *SS*, Skill score.

From Xiao, Z. Y., Wang, X. H., Song, D., Jalón-Rojas, I. & Harrison, D. (2020). Estuarine, Coastal and Shelf Science, 236, 106605. https://doi.org/10.1016/j.ecss.2020.106605

(Table 5.3). In shallow water, the progression of tides was distorted due to bottom friction and channel storage changes, which were often expressed by the nonlinear growth of compound constituents and harmonics of the principal astronomical tidal components (Friedrichs & Aubrey, 1988). The nonlinear distortion parameters derived from the dominant semi-diurnal constituent M_2 and its first harmonic M_4 (amplitude ratio $M_4/M_2 = 0.007$ and relative phase $2M_2-M_4 = 358$ degrees) suggested a weak tidal asymmetry (amplitude).

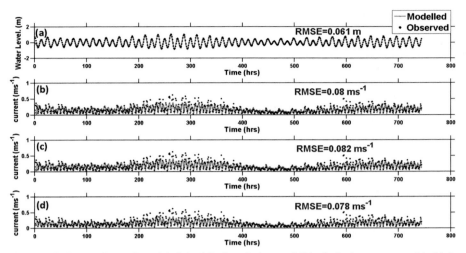

Figure 5.5 Comparison of modeled (*red line*) and observed (*black dots*) parameters (a) tidal elevations for Fort Denison; (b, c, d) depth-averaged current speed at the ADCP station in Exp. 1, (RMSE = 0.08 m/second), Exp. 2 (RMSE = 0.082 m/second) and Exp. 3 (RMSE = 0.078 m/second), respectively.

4.3 Bottom current and salinity

Bottom velocities over the spring and neap tides showed overall a good level of agreement between model and observation for both the magnitudes and phases. The magnitude of the velocities u and v showed flood-ebb asymmetries with intensified strengths of up to 0.3 m/second for u and, up to 0.2 m/second for v during spring ebb tides, and maxima of 0.2 m/second for u and 0.05 m/second for v during neap ebb tides. The RMSE of u was 0.05 m/second, v, 0.03 m/second (Table 5.2). The mean error showed that u was underestimated by 0.03 m/second and v overestimated by 0.01 m/second (Table 5.2). The SS values for the u and v velocity components were 0.84 and 0.52, respectively, and the CC values 0.95 and 0.71 (Table 5.2). The errors are likely partly due to the coarse grid resolution in the complex curving flow channel, in which it is difficult to represent the bottom stress layer and the bathymetry variability accurately. Differences in the measurements averaged over a grid cell compared to the ADCP point measurement will further contribute separation between observation and predictions.

The bottom density values at a depth of 13.7 m ($\sim$2 m above the bottom) are compared in Fig. 5.4d. The salinity fluctuated within a narrow range of 2 psu following the tidal cycles (not shown here). The modeled surface salinities were higher than the observations by up to 1 psu during storm, due to lack of sufficient monitoring following rainfall to address infrequent high precipitation conditions. The RMSE of the salinity was calculated as 0.42 psu (Table 5.2). The SS and CC for the salinity were relatively low, at 0.20 and 0.68, respectively (Table 5.2). The salinity fluctuation trend was reproduced by the model, but the fluctuation range was reduced. The model reduced the top-to-bottom salinity difference but showed lower salinity at the bottom and higher salinity at the surface than was observed, thus underestimating the stratification peaks during storm events (not shown here). This discrepancy is also partly due to overestimation of vertical mixing by the model turbulence scheme (noting the salinity dominated density variations in this system), as indicated by normalized Ri_g in Fig. 5.4f. Apart from the flood-ebb asymmetries in current velocity, Ri_g also indicated flood-ebb asymmetries in turbulent mixing, with intensified mixing during spring ebb, which was reasonably reproduced by the model (Fig. 5.4f).

4.4 Model limitations

The setup of the numerical model involves certain assumptions, and calibration through a direct comparison between observation and model results provides verification and justification of the assumptions made in developing the model. The calibration of the model used here has been extensive and thorough and will continue to be refined. The fixed mooring station near BH provides the salinity and current variations in a time series from water surface to bottom. Observations of the temporal variations in the lateral processes in the SHE are limited; these are claimed to be as important as the along-estuary circulation (Lerczak & Geyer, 2004). The longitudinal salinity pressure gradient is greatly influenced by the freshwater inflow. In the SHE, the main river discharge rate is monitored and used for hydrology model calibration, but there is a lack of data for the other tributaries. Also limited are field data to conduct model sensitivity tests to determine the effects of the MY-2.5 turbulent coefficients on mixing behaviors. For simplicity, a constant water temperature of 25°C in the SHE during the simulation period was assumed. The observed horizontal and vertical temperature ranges near BH are up to 0.5 and 1.5°C during quiescent conditions and up to 1.5 and 2.0°C during stratification. Such variations are considered too small to have significant impact on estuarine circulations (Lee & Birch, 2012) but can be important for life in the marine environment. A structured observing system on temperature is therefore needed in the SHE that heating and cooling effects can be conducted in the model.

5.   Results and discussion

5.1 M2 and K1 amplitudes

Tidal-wave damping occurred in the central estuary for both M_2 and K_1, as shown in co-tidal charts (Fig. 5.6). The tidal amplitudes of M_2 and K_2 were preserved in the Middle Harbour reach as there is no sharp change in bathymetry gradient and no damping caused by the channel orientation. When it reached the shallow waters in the upper estuary, the M_2 amplitude had increased by 5 cm compared with that at the estuary mouth. Tidal-wave damping or amplification is caused by an imbalance between topographic convergence and frictional distortion (Friedrichs & Aubrey, 1988). With a strong convergence and weak friction, the energy is concentrated in the channel, and the amplitude increases with distance. With a weak convergence, friction causes the amplitude to decrease (Friedrichs & Aubrey, 1988).

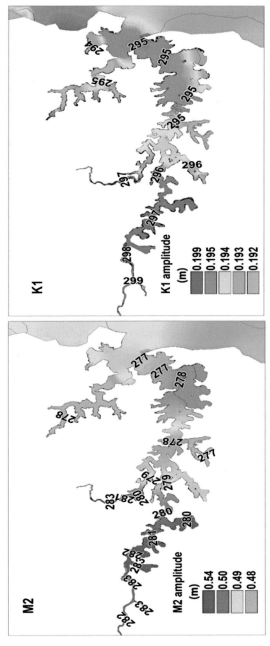

Figure 5.6 Co-tidal charts of the M_2 and K_1 constituents for the basic case (Exp. 1). Height (*color*) is shown in meters (note that the scales differ), phase in degrees by the *solid contour lines*.

The headlands and island in the SHE interact with tidal currents to cause stronger friction, plus the large area of the side embayments contains a great amount of tidal energy; therefore, the tidal-wave amplitude is reduced in the middle estuary. The channel in the upper estuary becomes narrow and converges the tidal wave, thus increasing its amplitude. The maximum phase lag in M_2 and K_1 between the estuary mouth and inner estuary was 12 min.

5.2   Current velocity

A comparison between the observed and modeled depth averaged current speed (Fig. 5.5b and c and d) showed good agreement. The root mean square errors were 0.08, 0.082, and 0.078 m/second in Exp. 1, Exp. 2, and Exp. 3, respectively. The model underestimated the current magnitude by approximately 20% during spring tides; however, it performed well during neap tides. Exp. 3, which included tides, winds, and river discharge forcing working together showed the best correlation with the observation data.

The current ellipse, which consists of major and minor axis as the maximum and minimum current velocity along the inclination of the orbits and phase, is more conventional for understanding the characteristics of tidal current. The M_2 current ellipse showed a well-defined orientation because the major axis was nearly an order of magnitude larger than the minor axis (Fig. 5.7). The observed and modeled (Exp. 1) ellipse axes of M_2 were rotated in opposite direction from surface to bottom, while it showed the same changing trend between Exp. 3 and observation. In Exp3, the addition of freshwater inflow with a salinity gradient can alter the vertical current structure through the density gradient, influencing current patterns and their inclination. Inaccurate bathymetry and coastline information will impact on the predictions of current ellipse inclination since the current ellipses were mainly orientated along the coastline and local isobaths.

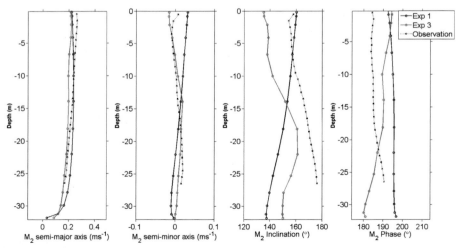

Figure 5.7 Tidal current ellipse major axis, minor axis, inclination, and orientation comparison between modeled and observed data. Note that the *black lines* represent observed velocities (ADCP), *blue lines* represent Exp. 1, and *red lines* represent Exp. 3.

Vertical profiles of current speed in Exp. 1 and Exp. 3 were compared with observed velocity profiles at the ADCP station (mid-estuary) over the spring and neap tides (Fig. 5.8). Generally, the magnitude of the current speed fluctuated

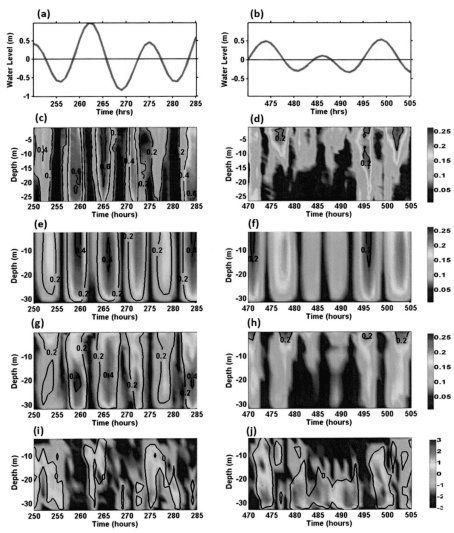

Figure 5.8 Current profiles with and without salinity gradients. Vertical and temporal variations of model predicted sea surface elevation (a, b) and current speed (c, h) during spring (*left*) and neap (*right*) tidal cycles. Velocities are as follows: Observed (c, d), Exp. 1 (e, f), and Exp. 3 (g, h). (i, j) gradient Richardson number $\log_{10}$ (R_{ig}/0.25) in Exp. 3. The *solid black line* indicates $\log_{10}(R_{ig}/0.25) = 0$; positive value (*red*) indicates stratification and negative value (*blue*) indicates mixing.

following the tidal cycles and spring tides were more than twice as strong as neap tides. Current speeds were small (≤ 0.1 m s^{-1}) and uniform from surface to bottom during the slack waters (Fig. 5.8a and b). During the mid-tides, surface-middle layer currents were intensified and decreased slowly at the bottom. However, the observed current speed during neap tides (Fig. 5.8d) was reduced considerably, and the fluctuations following the tidal cycles were weakened. This phenomenon was possibly due to the occurrence of stratification during neap tides when tidal mixing was weak. This feature was not reproduced in Exp. 1 (Fig. 5.8f) forced by tides only, while the phenomenon was reproduced in Exp. 3 (Fig. 5.8h) where the model included freshwater discharge forcing and a horizontal salinity gradient. This indicated that the stratification observed during neap tides was caused by the salinity gradient and river inflow. Salinity-induced stratification was suppressed by the strong tidal mixing during spring tides; however, stratification was preserved during neap tides, thus forming asymmetric stratification during the spring-neap tides.

5.3 Stratification asymmetry

Tidal mixing and the steady buoyancy flux provided by freshwater input compete in influencing the stratification's stability. The gradient Richardson number (R_{ig}), a parameter used to assess the relative strength of a density gradient in shear flows with stable stratification, was utilized to evaluate the spring-neap and flood-ebb variations of stratification (Fig. 5.8i and j). The ADCP in 32 m of water showed most evident variations in stratification since the greater tidal straining effects were experienced at this site. R_{ig} was estimated based on modeled velocity and the density at the ADCP station using the equation below:

$$R_{ig} = \frac{N^2}{\left[(\partial u/\partial z)^2 + (\partial v/\partial z)^2\right]} \tag{5.12}$$

where $N^2 = -(g/\rho_o)(\partial \rho/\partial z)$ called the buoyancy frequency, g is the gravitational acceleration, and ρ is the seawater density (1025 kg/m^3). For stably stratified shear flow, $R_{ig} > 0.25$ or normalized value $\log_{10}(R_{ig}/0.25) > 0$; for well-mixed flow, $R_{ig} < 0.25$ or $\log_{10}(R_{ig}/0.25) < 0$. During spring tides (Fig. 5.8i), the first peak value of R_{ig} at the beginning of the ebb followed by a decrease in stratification until a well-mixed water column was formed during the flood. During the ebb phase, the second peak of R_{ig} during ebb indicated that the water column was stratified by the tidal straining of the density fields. In contrast, the continuous high values of R_{ig}, observed in the lower half of the water column during neap tides (Fig. 5.8j), resulted from reduced tidal mixing further suppressed by stratification in the presence of a horizontal salinity gradient. The vertical profile of the eddy viscosity (K_m), which serves as an indicator of turbulent mixing intensity, demonstrated that the asymmetric turbulent mixing was attributed to strain-induced periodic stratification throughout the flood-ebb cycle and asymmetric tidal mixing during the spring-neap periods at the ADCP site (Fig. 5.9). In Exp. 1, when mixing was strong and the tidal current remained unstratified, the

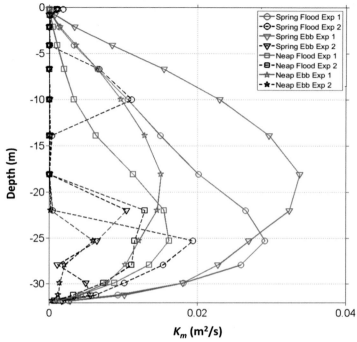

Figure 5.9 Eddy viscosity with and without salinity gradients. Vertical profiles of eddy viscosity coefficient K_m (m²/second) in Exp. 1 (*red solid lines*) and Exp. 3 (*black dash lines*) during spring flood, spring ebb, neap flood, and neap ebb at the ADCP site.

maximum K_m reached approximately ~ 0.03 m²/second. However, in Exp. 3, where stratification was present, the tidal mixing weakened, resulting in a lower K_m compared to Exp. 1 across all tidal phases. Tidal straining effects decreased K_m on ebbs by a factor of two in comparison with the flood values in Exp. 3 during both spring and neap tides. At neap ebb tides, the minimum K_m value was approximately 0.008 m²s⁻¹, mainly attributed to the tidal straining effect and turbulent mixing being at their lowest levels.

To gain insight into the influence of tidal straining mechanisms in generating periodic stratification in the presence of a horizontal salinity gradient, the equation for conservation of salt (S) was applied by vertically differentiating the advection and diffusion of salts:

$$\frac{\partial}{\partial t}\frac{\partial S}{\partial z} + u\frac{\partial}{\partial x}\left(\frac{\partial S}{\partial z}\right) + v\frac{\partial}{\partial y}\left(\frac{\partial S}{\partial z}\right) + w\frac{\partial}{\partial z}\left(\frac{\partial S}{\partial z}\right) + \frac{\partial u}{\partial z}\left(\frac{\partial S}{\partial x}\right) + \frac{\partial v}{\partial z}\left(\frac{\partial S}{\partial y}\right)$$
$$+ \frac{\partial w}{\partial z}\left(\frac{\partial S}{\partial z}\right) = \frac{\partial^2}{\partial z^2}\left(K_h\frac{\partial S}{\partial z}\right) + \frac{\partial F_S}{\partial z}$$

$$(5.13)$$

where x is along-estuary axis (positive seaward); y is cross-estuary axis (positive northward); z is vertical axis (positive upward); terms on the left are the unsteady term, advection term (along-estuary, lateral, and vertical) and the tidal straining term (interaction between vertical velocity shear and horizontal/vertical salinity gradients); and terms on the right are vertical and horizontal diffusion of turbulence. The traditional tidal straining model is to assume lateral and vertical flow is negligible and horizontal diffusion is much weaker than the vertical diffusion process, then the simplified equation becomes (Nepf & Geyer, 1996)

$$\frac{\partial}{\partial t}\frac{\partial S}{\partial z} = -\frac{\partial u}{\partial z}\frac{\partial S}{\partial x} + \frac{\partial^2}{\partial z^2}\left(K_h\frac{\partial S}{\partial z}\right) \tag{5.14}$$

where on the right the first term is tidal straining and the second term is vertical diffusion. It has been demonstrated that lateral circulation also plays an important role in controlling stratification through lateral density gradients (Giddings et al., 2011; Scully & Geyer, 2012). This study primarily focused on the impact of tidal straining on stratification at tidal time scales and did not take into account the influence of lateral circulation. As the river plume flows seaward, the salinity gradually increases along the estuary until it reaches the ocean's salinity, resulting in a positive horizontal salinity gradient ($\frac{\partial S}{\partial x} > 0$). An increased stratification with time is given by $\frac{\partial}{\partial t}\frac{\partial S}{\partial z} < 0$, while $\frac{\partial}{\partial t}\frac{\partial S}{\partial z} > 0$ indicates the weakened stratification (Nepf & Geyer, 1996). During ebb tides, waters flowed seaward (positive x direction) and vertical velocity shear became positive $\frac{\partial u}{\partial z} > 0$, which leads to $\frac{\partial}{\partial t}\frac{\partial S}{\partial z} < 0$. During flood tides, the reversed flow directions produced a negative velocity shear $\frac{\partial u}{\partial z} < 0$, thus the stratification was weakened as indicated by $\frac{\partial}{\partial t}\frac{\partial S}{\partial z} > 0$. The modeled depth-averaged stratification stability and the tidal straining term were calculated at the mooring location during different tidal phases (Table 5.4). Tidal straining as the product of vertical velocity shear and longitudinal salinity gradient enhanced stratification ($\frac{\partial}{\partial t}\frac{\partial S}{\partial z} < 0$) during spring/neap ebb and diminished the stratification ($\frac{\partial}{\partial t}\frac{\partial S}{\partial z} > 0$) during the spring flood. However, the tidal straining appears to be too weak to diminish stratification during neap flood.

5.4 Residual currents

Residual flow, influenced by nonlinear tidal forcing, wind stress, and horizontal density gradients, plays a significant role in long-term material transport, even though its magnitude is much smaller than that of tidal currents (Das et al., 2000). Exp. 1, which was solely forced by tides, was utilized to evaluate tide-induced residual currents. Subsequently, Exp. 3 was employed to assess the effects of stratification on residual flow, resulting from freshwater inflow and lateral density gradients. A time series of model results was extracted over a tidal cycle (29.6 days). The residual flow (U_r) was calculated as:

$$Ur = \left(\sum_{t=1}^{t}(U_i(t) - U_{i,pout}(t))\right)/t \tag{5.15}$$

Table 5.4 Spring-neap and flood-ebb variability of stratification at the ADCP station calculated from the model output.

TimeStep	DateTime	$\frac{\partial S}{\partial z}$ ($\times 10^{-2}$)	$\frac{\partial u}{\partial z}$ ($\times 10^{-3}$)	$\frac{\partial S}{\partial x}$ ($\times 10^{-4}$)	$-\frac{\partial u}{\partial z}\frac{\partial S}{\partial x}$ ($\times 10^{-5}$)	$\frac{\partial}{\partial t}\frac{\partial S}{\partial z}$ ($\times 10^{-5}$)	TidalPhase	Stratification
257	January 12, 1303:00	−3.9	−1	7.8	1.05	8.0	Spring flood	Weakened
268	January 12, 1314:00	−4.0	75	21	−0.6	−1.1	Spring ebb	Enhanced
470	January 21, 1301:00	−3.4	6	3.2	−0.002	−1.7	Neap flood	Enhanced
478	January 21, 1308:00	−3.3	64	7.6	−0.3	−1.8	Neap ebb	Enhanced

is the magnitude of modeled depth-averaged velocity at grid point i and time t; is the component of overall harmonic tidal constituents induced current velocity at grid point i and time t.

5.4.1   Tide-induced residual circulation

In order to comprehend the influence of bottom friction on residual currents, a sensitivity test (Exp. 4) was conducted. This test involved applying a range of bottom friction coefficients (C_d), calculated using Eq. (5.9), with values up to a maximum of 0.325, across the entire model domain under tidal forcing only. The results (Figure not shown here) revealed that the magnitude of residual currents generally decreased with higher C_d values during both spring and neap tides. However, the residual circulation pattern remained unchanged, consistent with observations made by (Das et al., 2000). The tide-induced residual flows at the surface layer were averaged separately over spring and neap tidal cycles. Comparing neap tides to spring tides, the magnitude of residual currents within the estuary was found to be reduced by a factor of two (Fig. 5.10a and b). Extensive development of residual eddies occurred adjacent to headlands, and during spring tides, a clockwise circulation at the estuary mouth was intensified due to the stronger tidal currents passing through the mouth. The mechanism behind residual eddy formation has been attributed to the transformation of vorticity from the tidal velocity field to the mean (residual) field, which occurs as a result of nonlinear bottom frictional forcing and boundary geometrical effects (Das et al., 2000).

5.4.2   Salinity gradients and freshwater inflow induced residual circulation

The spatial distribution of surface residual currents induced by tides, winds, and river inflow in Exp. 3 was significantly different (Fig. 5.10c and d) compared with tide-induced residual currents. The magnitude of residual currents in Exp. 3 was overall enhanced, and the greatest impact occurred in the shallow embayments and the estuary mouth. Opposite flow directions were observed at the estuary mouth between spring (inflow) and neap (outflow) possibly due to asymmetric mixing conditions and the maximum magnitude was concentrated at the estuary mouth where maximum currents were found (Cheng et al., 2010). Current velocity results (Fig. 5.8) revealed that the variations in the magnitude of tidal currents over the spring-neap tidal cycles were due to horizontal salinity gradients and freshwater inflow with current amplitudes at spring tides much higher than at neap tides. The asymmetries in the amplitude of tidal currents altered the turbulent mixing conditions in the estuary, thereby influencing the strength of the residual flows during the tidal cycles.

Vertical profiles of along-estuary residual current U in Exp. 1 and 3 were compared with observed profiles at the ADCP station over spring and neap tides (Fig. 5.11). In Exp. 1, the tide-induced residual currents (Fig. 5.11c and d) did not replicate the fluctuation trends observed in the actual profiles (Fig. 5.11a and b), exhibiting a significantly reduced magnitude. In Exp. 3, when incorporating the effects of buoyancy

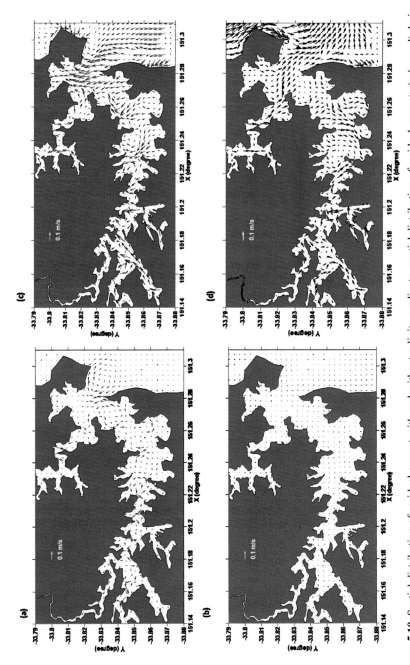

Figure 5.10 Spatial distribution of residual current with and without salinity gradients zspatial distribution of residual currents (m/second) during spring and neap tides from Exp. 1 (a, b) and Exp. 3 (c, d).

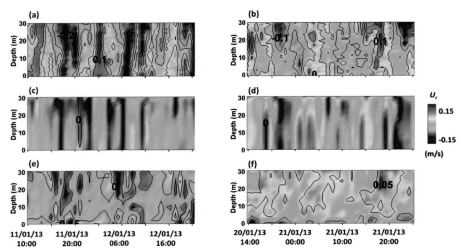

Figure 5.11 Residual current profile with and without salinity gradients. Vertical profile of along-estuary component of residual current U (m/second) over spring (*left*) and neap (*right*) tides as observed (*top*), and in Exp. 1 (*middle*) and Exp. 3 (*bottom*).

and salinity gradients (Fig. 5.11e and f), the residual currents were considerably reproduced, exhibiting a similar changing trend as the observed profiles during neap tides. The comparison of residual currents between Exp. 1 and Exp. 3 indicated the significance of river inflow and a horizontal salinity gradient in shaping the vertical structure of residual flows (refer to Fig. 5.11, further discussed later). Additionally, the effects of tidal mixing asymmetry were observed to intensify the strength of the residual flows at the estuary mouth. Despite the modeling period occurring during a dry month with a limited river discharge, the presence of persistent salinity gradients and river inflow played a crucial role in inducing stratification and generating the residual flows.

5.4.3   Residual flow mechanisms

Salinity gradient pressure resulting from river inflow generating estuarine residual flows was investigated. This mechanism produced shear flow, along with asymmetric turbulent mixing. The presence of asymmetric stratification within the tidal cycle may lead to turbulent mixing asymmetry, which, in turn, results in a net flow of residual currents (Jay & Smith, 1990). The magnitude and timing of stratification play a crucial role in influencing the level of turbulent mixing, which, in turn, determines the behavior of the estuarine residual flow. The along-estuary component of the residual flow was primarily generated during neap tides, characterized by weaker tidal mixing (Fig. 5.11f). Strain-induced stratification reached its peak at the end of the neap ebb tide (indicated by peak R_{ig} in Fig. 5.11j at 480 hours or 492 hours), and a stronger along-estuary residual flow was subsequently developed following the maximum stratification (Fig. 5.11f). As the transition to spring tides occurred, the strength of residual flows significantly weakened in comparison to neap tides (Fig. 5.11e). The growth of residual flow resulting from the pulse of stratification during spring ebb tides (indicated by peak R_{ig} in Fig. 5.5i at 265 hours or 275 hours) was evident in the observed residual current profile (Fig. 5.11a). However, the model did not successfully replicate these observations. The bias between modeled and observed residual currents may be attributed to the underperformance of the model's turbulent scheme (e.g., over mixing), or the model interpreted bathymetry spatial gradients being not accurate enough. The temporal fluctuations in the along-estuary residual flow illustrate how asymmetric turbulent mixing affects the shear flow, contributing significantly to the formation of estuarine residual flow (Stacey et al., 2001). Additional investigations are necessary to decompose estuarine residual flows and discern the contributions made by different forcing mechanisms in their development.

6. Conclusions

Sydney Estuary, a microtidal estuary mainly dominated by the semi-diurnal tidal component (M_2), showed weak tidal asymmetry (the amplitude ratio $M_4/M_2 = 0.007$). The three-dimensional model provided valuable insights into the temporal and spatial variations of water level, current velocity, and residual current during dry periods. Several model forcings were tested, and the model driven by tides, wind, river inflow, and a horizontal salinity gradient showed the best agreement with the observed data. During dry periods, the combined effect of tidal mixing and (low) freshwater input controlled turbulent mixing and stratification in the Sydney Estuary. River inflow and a horizontal salinity gradient-induced stratification were preserved during neap tides when tidal mixing was diminished. The evidence of strain-induced periodic stratification during ebbs was observed through the product of the vertical gradient of velocity and the horizontal gradient of salinity. The model successfully captured the spring-neap and intratidal asymmetries in stratification.

In contrast to most previous studies that focused on high precipitation-induced stratification, this study revealed asymmetric stratification within a tidal cycle during dry conditions using the gradient Richardson number (R_{ig}), rather than relying on the distribution of the freshwater plume. Overall, enhanced stratification was associated with $R_{ig} > 0.25$ and reduced eddy viscosity (K_m); however, a weakened stratification was associated with $R_{ig} < 0.25$ and increased eddy viscosity (K_m). By comparing these numerical experiments, we discovered that the presence of a horizontal salinity gradient, achieved by introducing freshwater at the estuary's head, and served as the primary driving force for estuarine circulation. This circulation, in turn, played a crucial role in maintaining salinity stratification.

The two primary mechanisms responsible for creating residual flows were: (i) the nonlinear interactions of tidal oscillatory flow with the land boundary geometry and bottom topography and (ii) the along-estuary salinity gradient pressure resulting from river inflow, combined with asymmetric turbulent mixing. The distribution of R_{ig} illustrated the magnitude and timing of stratification, which, in turn, influenced the levels of turbulent mixing—an essential factor in the development of estuarine residual flow.

Sydney Estuary provides a good example of a complex estuary, featuring multiple tributaries and shallow embayments connecting with the main meandering channel. To achieve model calibration, a comprehensive river and estuary observation network is essential, providing real-time river discharge and continuously monitored estuarine data. Currently, the modeled salinity data rely on limited observation data for calibration, while it has been shown that the horizontal salinity gradient plays a crucial role in estuarine circulation. Within the numerical models, we can explore the mechanisms and forcing on the water mass transport in this estuarine system; however, the fate of the sediment will be determined by much more complex mechanisms such as wind waves, interactions between continental shelf and estuarine processes, etc. Further numerical modeling on the sediment dynamics is yet to be conducted, where real-time monitored suspended sediment concentration data will be collected.

7. Appendix—Additional survey data

7.1 Cross-channel transects and monthly discrete samplings

To accurately compare observed and modeled depth averaged currents across the channel, for each ADCP transect, velocity observations were averaged over a distance corresponding to model grid cell widths, with the comparisons shown in Figs. 5.12 and Fig. 5.13. Observed current velocities were typically larger than model velocities at all four transects (0.10 m/second < RMSE <0.25 m/second). The way to calculate ADCP depth-averaged velocities was using snapshot values over the distance and comparing to the hourly averaged velocities from model at the corresponding grid cell, thus observed depth-averaged velocities were artificially high. The difference in the current direction between model and observation could be caused by the horizontal and vertical resolution not being high enough to capture small features of the bathymetric data.

The spatial precision of salinity dynamics was examined by the monthly measured CTD profiles taken on 20th—November 21, 2013 under high-runoff conditions and 18th —19th December 2013 under low-runoff conditions (Fig. 5.14; Fig. 5.15; Fig. 5.16). Fig. 5.17 a shows along-estuary salinity gradients with a vertically sharp distribution during spring tide, whereas during neap tide, the salinity gradients were piling from surface to bottom due to stratification (Fig. 5.17c). According to the vertical structure of the water-column salinity, the estuarine system was relatively well mixed in the lower estuary near the mouth but reasonably stratified upstream. Overall, the validation indicated a reasonable calibrated hydrodynamic model in simulating tides and salinity dynamics in SHE.

7.2 Sediment model setup and compared to observations

The UNSW-Sed module was two-way coupled to the SHE hydrodynamic model. This allowed the sediment concentration to affect the seawater density, bottom drag coefficient and the flux Richardson number, and thus modulate the estuarine circulation

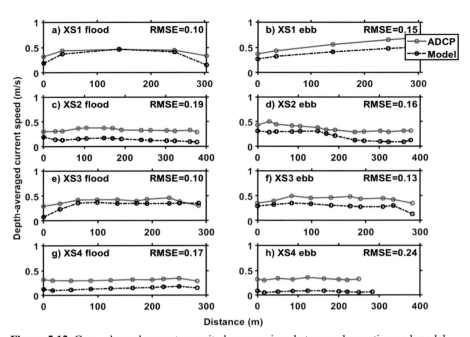

Figure 5.12 Cross-channel current magnitude comparison between observation and model comparison between measured and modeled depth-averaged current speeds along XS1−XS4 during spring flood and ebb. ADCP transect locations are shown in Fig. 5.1.

From Xiao, Z., Wang, X. H., Roughan, M. & Harrison, D. (2019). Estuarine, Coastal and Shelf Science, 217, 132−147. https://doi.org/10.1016/j.ecss.2018.11.004

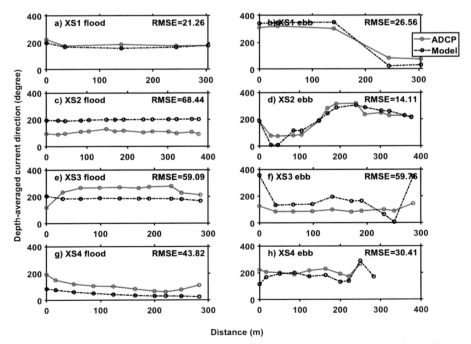

Figure 5.13 Cross-channel current direction comparison between observation and model comparison between measured and modeled depth-averaged current direction along XS1−XS4 during spring flood and ebb. ADCP transect locations are shown in Fig. 5.1.

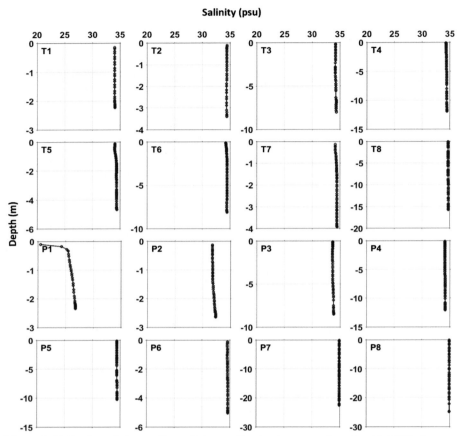

Figure 5.14 In-situ salinity profiles salinity profiles (psu) measured on Oct 15, 2013, for estuarine initial salinity field setup.

(Song & Wang, 2013; Wang, 2002; Wang et al., 2005; Wang & Pinardi, 2002). The UNSW-Sed model is based on the concept of critical stresses for deposition and erosion of noncohesive sediments, using the equation of advection-diffusion for a passive tracer in an incompressible fluid. The sediment model in ROMS takes a different and more complex approach, which accounts for the sediment properties and the seabed as multilayered for bed roughness and bottom stress calculations. The model takes account of the morphological evolution of the sea bottom, which is not the focus of this study.

Based on the assumption of a constant settling velocity w_s for suspended sediments and the continuity equation for salinity and temperature, the sediment transport equation can be written as (Wang, 2002):

$$\frac{\partial C}{\partial t} + \frac{\partial}{\partial x}(uC) + \frac{\partial}{\partial y}(vC) + \frac{\partial}{\partial z}[(w + w_s)C] = \frac{\partial}{\partial z}\left(K_h\frac{\partial C}{\partial z}\right) + F_c \qquad (A.1)$$

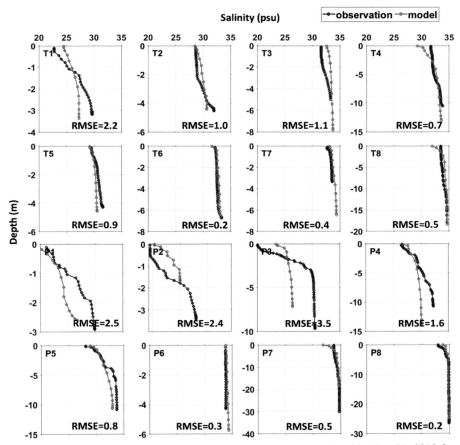

Figure 5.15 In-situ salinity profiles salinity profiles (psu) measured on 20–Nov 21, 2013 for salinity calibration during high river flow conditions.

where C is the suspended sediment concentration; the vertical eddy diffusivity for C is set equal to K_h; a first-order iterative upstream scheme is used for the horizontal diffusion term F_c to reduce implicit diffusion with an antidiffusive velocity (Smolarkiewicz, 1984).

The density of clear seawater (without sediment contribution) was determined from the equation of state (Mellor, 1998). When the impact of suspended sediment concentration on the seawater density is considered, the seawater density is calculated as:

$$\rho = \rho_w + \left(1 - \frac{\rho_w}{\rho_s}\right) C \tag{A.2}$$

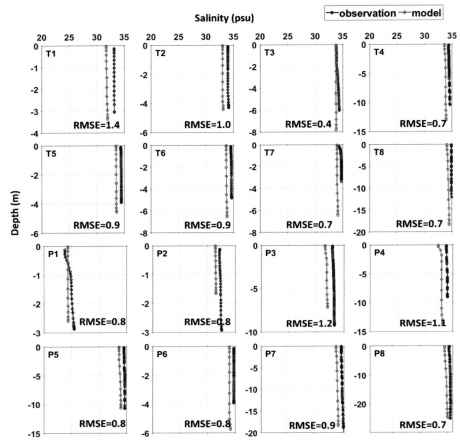

Figure 5.16 In-situ salinity profiles salinity profiles (psu) measured on 18−19 Dec 2013 for salinity calibration during low river flow conditions.

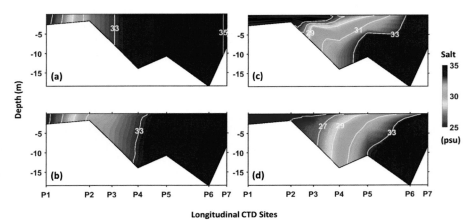

Figure 5.17 Along-estuary salinity gradient along-estuary distribution of salinity (psu) from axial CTD surveys during a period from flood to ebb over spring tide (*left*) and neap tide (*right*): (**top**) observation; (**bottom**) model.

where ρ_w is clear seawater density and ρ_s sediment density. In a sediment-laden bottom boundary layer (BBL), suspended sediment induced stratification suppresses the bottom turbulence (Wang et al., 2005). By introducing a stability function $(1+AR_f)^{-1}$ into the bottom drag coefficient C_d (Wang, 2002) (where $A = 5.5$ is an empirical constant and R_f is the flux Richardson number, an index of the vertical density stratification in the Mellor-Yamada Level 2 approximation), the model includes the impact of sediment induced stratification on BBL dynamics. The equation for C_d is written as:

$$C_d = \left[\frac{\kappa}{(1 + AR_f)\ln\,(z_b/z_{0b})} \right]^2 \tag{A.3}$$

where κ is the von Karman constant, z_b the near-bottom layer thickness and z_{0b} the bottom roughness. When $R_f > 0$, suspended bottom sediments stratify the BBL; C_d is reduced as bottom turbulence is supressed. The minimum of C_d is reached when the bottom turbulence is completed shut down at a critical value of R_f (Wang, 2002).

The parameter E_b (kg/m^2/second) in the sediment module is the vertical sediment flux at the bottom due to erosion and deposition. E_b can be expressed as (Ariathurai & Krone, 1976)

$$E_b = \begin{cases} E_0\left(\dfrac{|\tau_b|}{\tau_{ce}} - 1\right) if\, |\tau_b| \geq \tau_{ce} \, erosion \\[4mm] C_b w_s \left(1 - \dfrac{|\tau_b|}{\tau_{cd}}\right) if\, |\tau_b| < \tau_{cd} \, deposition \end{cases} \tag{A.4}$$

where E_0 (kg/m^2/second) is the empirical erosion coefficient, C_b (kg/m^3) the SSC near the bottom boundary layer, τ_b (kg/m/second^2) is the bottom shear stress, τ_{ce} and τ_{cd} (kg/m/second^2) the critical shear stress for erosion and deposition, respectively, and w_s (m/second) the particle settling velocity, positive upward, negative downward. The parameters E_0, w_s, τ_{ce} and τ_{cd} may vary widely in different regions with different sediment types and physical environments. Flocculation and deflocculation process were not considered, as SSC in the SHE (generally <0.1 kg/m^3) is much less than the threshold value of 1 kg/m^3 for making a significant contribution to the sediment settling velocity (Van Rijn, 1993).

Irvine (1980) showed that the sediment type and proportions have a spatial variability in the SHE. There is a lack of field measurements of E_0 parameters for sediment erosion and deposition process in the SHE. A range of E_0 (2×10^{-5} to 1×10^{-7} kg/m^2/second), w_s (5×10^{-4} to 1×10^{-6} m/second), τ_{ce} and τ_{cd}($0.01-1.0$ kg/m/second^2) were tested against model skill score values. Constant values of $w_s = 2 \times 10^{-5}$ m/second, $E_0 = 2 \times 10^{-5}$ kg/m^2/second and $\tau_{ce} = \tau_{cd} = 0.2$ kg/m/second^2 were used in the model; these gave the best model skill score for cohesive sediment simulation (SS $= 0.45$, Table 5.2).

Bottom turbidity was measured about 2 m above the estuary bed at 5-min sampling intervals by a CTD-mounted turbidity sensor at the BH station. Water samples were collected to obtain in-situ SSC by filtering and drying 2 L of seawater in a total of 30 samples. The turbidity data were converted to SSC by correlating the in-situ SSC (mg/ L) with the turbidity data (NTU) (Fig. 5.1d, SSC = 1.3166×Turbidity+0.1358; $R^2 = 0.89$). The bottom SSC comparison showed good agreement on tidal variations but overestimation of the peaks by the model during every second spring ebb (Fig. 5.1e). The latter is attributed to the fact that the model underestimates the bottom periodic mixing; the generating mechanism is investigated in Xiao et al.(2020). The turbidity profiles collected from monthly shipboard surveys are compared with model in Figs. 5.18 and Fig. 5.19. The RMSE for magnitude were within 10% of the observed SSC. From the along-estuary SSC gradient (Fig. 5.20), both model and observation data showed a high SSC discharged by the river in the upper estuary, moving toward estuary mouth in a well-mixed manner.

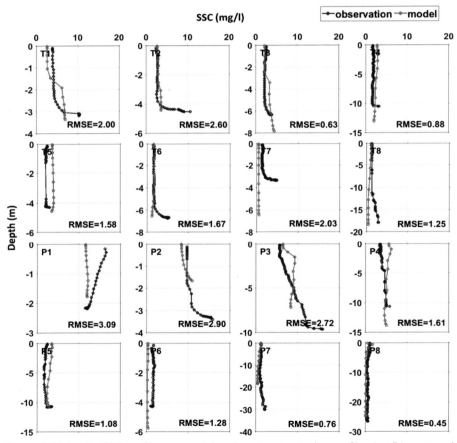

Figure 5.18 In-situ SSC profiles suspended sediment concentration profiles (mg/L) measured on 20–Nov 21, 2013 for sediment calibration during high river flow conditions.

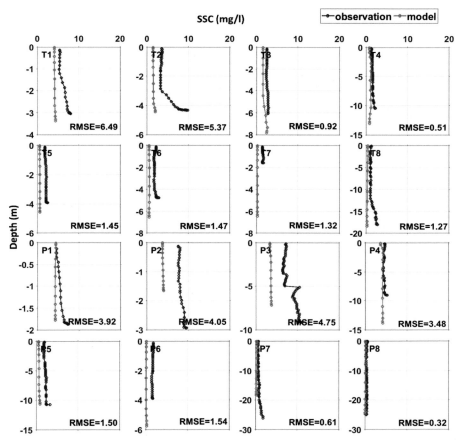

Figure 5.19 In-situ SSC profiles. Suspended sediment concentration profiles (mg/L) measured on 18−19 Dec 2013 for sediment calibration during low river flow conditions.

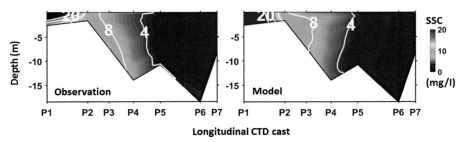

Figure 5.20 Along-estuary SSC gradient. Tidally averaged along-estuary distribution of SSC (mg/L) from axial CTD surveys (*left*) in comparison with model results (*right*). The survey was conducted on 20−21 November 2013.

Acknowledgments

Thanks to Gary Batman from Port Authority of New South Wales for providing ADCP measurement data and Dr Daniel Harrison from University of Sydney for allowing us to use the CTD survey data. This paper benefited from editorial reviews by Julie Kesby from UNSW Canberra. This work was supported by the National Computational Infrastructure National Facility at the Australian National University. This is publication no. 27 of the Sino-Australian Research Centre for Coastal Management. A. Xiao was supported by an Australian Postgraduate Award scholarship and X.H. Wang was supported by the UNSW Special Study Program.

References

Ariathurai, R., & Krone, R. B. (1976). Finite element model for cohesive sediment transport. *Journal of the Hydraulics Division, 102*(3), 323–338.

Birch, G. F., & Rochford, L. (2010). Stormwater metal loading to a well-mixed/stratified estuary (Sydney Estuary, Australia) and management implications. *Environmental Monitoring and Assessment, 169*(1–4), 531–551. https://doi.org/10.1007/s10661-009-1195-z, 15732959.

Boyd, M. J., Rigby, E. H., VanDrie, R. C., Liu, H., & Beardsley. (1996). An unstructured grid, finite-volume, three-dimensional, primitive equations Ocean Model: Application to Coastal Ocean and estuaries. *Environmental Software, 11*, 159–186. https://doi.org/10.1016/S0266-9838(96)00042-1Chen

Canuto, V. M., Howard, A., Cheng, Y., & Dubovikov, M. S. (2001). Ocean turbulence. Part I: One-point closure model—momentum and heat vertical diffusivities. *Journal of Physical Oceanography, 31*(6), 1413–1426. https://doi.org/10.1175/1520-0485(2001)031<1413:OTPIOP>2.0.CO;2

Chen, C., Liu, H., & Beardsley, R. C. (2003). An unstructured grid, finite-volume, three-dimensional, primitive equations Ocean Model: Application to Coastal Ocean and estuaries. *Journal of Atmospheric and Oceanic Technology, 20*(1), 159–186. https://doi.org/10.1175/1520-0426(2003)020<0159:AUGFVT>2.0.CO;2

Cheng, P., Valle-Levinson, A., & De Swart, H. E. (2010). Residual currents induced by asymmetric tidal mixing in weakly stratified narrow estuaries. *Journal of Physical Oceanography, 40*(9), 2135–2147. https://doi.org/10.1175/2010JPO4314.1, 00223670.

Das, P., Marchesiello, P., & Middleton, J. H. (2000). Numerical modelling of tide-induced residual circulation in Sydney Harbour. *Marine and Freshwater Research, 51*(2), 97–112. https://doi.org/10.1071/MF97177

Friedrichs, C. T., & Aubrey, D. G. (1988). Non-linear tidal distortion in shallow well-mixed estuaries: A synthesis. *Estuarine, Coastal and Shelf Science, 27*(5), 521–545. https://doi.org/10.1016/0272-7714(88)90082-0, 02727714.

Galperin, B., Kantha, L. H., Hassid, S., & Rosati, A. (1988). A quasi-equilibrium turbulent energy model for geophysical flows. *Journal of the Atmospheric Sciences, 45*(1), 55–62. https://doi.org/10.1175/1520-0469(1988)045<0055:AQETEM>2.0.CO;2

Giddings, S. N., Fong, D. A., & Monismith, S. G. (2011). Role of straining and advection in the intratidal evolution of stratification, vertical mixing, and longitudinal dispersion of a shallow, macrotidal, salt wedge estuary. *Journal of Geophysical Research: Oceans, 116*(3). https://doi.org/10.1029/2010JC006482, 21699291.

Hansen, D. V., & Rattray, M., Jr. (1965). Gravitational circulation in straits and estuaries. *Journal of Marine Research, 23*, 104–122.

Irvine, I. (1980). *Sydney harbour: Sediments and heavy-metal pollution*. Ph.D. thesis. University of Sydney.

Jay, D. A., & Smith, J. D. (1990). Residual circulation in shallow estuaries: 2. Weakly stratified and partially mixed, narrow estuaries. *Journal of Geophysical Research: Oceans, 95*(C1), 733—748. https://doi.org/10.1029/jc095ic01p00733

Kantha, L., & Clayson, C. (2000). *Small scale processes in geophysical fluid flows*.

Lee, S. B., Birch, G. F., & Lemckert, C. J. (2011). Field and modelling investigations of freshwater plume behaviour in response to infrequent high-precipitation events, Sydney Estuary, Australia. *Estuarine, Coastal and Shelf Science, 92*(3), 389—402. https://doi.org/10.1016/j.ecss.2011.01.013, 02727714.

Lee, S. B., & Birch, G. F. (2012). Utilising monitoring and modelling of estuarine environments to investigate catchment conditions responsible for stratification events in a typically well-mixed urbanised estuary. *Estuarine, Coastal and Shelf Science, 111*, 1—16. https://doi.org/10.1016/j.ecss.2012.05.034, 02727714.

Lerczak, J. A., & Geyer, W. R. (2004). Modeling the lateral circulation in straight, stratified estuaries. *Journal of Physical Oceanography, 34*(6), 1410—1428. https://doi.org/10.1175/1520-0485(2004)034<1410:MTLCIS>2.0.CO;2

Li, M., & Zhong, L. (2009). Flood-ebb and spring-neap variations of mixing, stratification and circulation in Chesapeake Bay. *Continental Shelf Research, 29*(1), 4—14. https://doi.org/10.1016/j.csr.2007.06.012, 02784343.

Mellor, G., & Blumberg, A. (2004). Wave breaking and ocean surface layer thermal response. *Journal of Physical Oceanography, 34*(3), 693—698. https://doi.org/10.1175/2517.1

Mellor, G. L. (1998). User's guide for a three-dimensional, primitive equation numerical ocean model. *Princeton University Rep.*, 41 [Available from Program in Atmospheric and Oceanic Science, Princeton University, Princeton, NJ.].

Mellor, G. L., & Yamada, T. (1982). Development of a turbulence closure model for geophysical fluid problems. *Reviews of Geophysics, 20*(4), 851—875. https://doi.org/10.1029/RG020i004p00851, 19449208.

Nepf, H. M., & Geyer, W. R. (1996). Intratidal variations in stratification and mixing in the Hudson estuary. *Journal of Geophysical Research: Oceans, 101*(C5), 12079—12086. https://doi.org/10.1029/96jc00630

Pawlowicz, R., Beardsley, B., & Lentz, S. (2002). Classical tidal harmonic analysis including error estimates in MATLAB using T_TIDE. *Computers & Geosciences, 28*(8), 929—937. https://doi.org/10.1016/s0098-3004(02)00013-4, 00983004.

Pritchard, D. W. (1956). The dynamic structure of A coastal plain estuary. *Journal of Marine Research, 15*, 33—42.

Robinson, I. S. (1981). Tidal vorticity and residual circulation. *Deep-Sea Research, Part A: Oceanographic Research Papers, 28*(3), 195—212. https://doi.org/10.1016/0198-0149(81)90062-5, 01980149.

Scully, M. E., & Geyer, W. R. (2012). The role of advection, straining, and mixing on the tidal variability of estuarine stratification. *Journal of Physical Oceanography, 42*(5), 855—868. https://doi.org/10.1175/JPO-D-10-05010.1, 15200485.

Simpson, J. H., Brown, J., Matthews, J., & Allen, G. (1990). Tidal straining, density currents, and stirring in the control of estuarine stratification. *Estuaries, 13*(2), 125—132. https://doi.org/10.2307/1351581, 01608347.

Simpson, J. H., & Souza, A. J. (1995). Semidiurnal switching of stratification in the region of freshwater influence of the Rhine. *Journal of Geophysical Research: Oceans, 100*(C4), 7037—7044. https://doi.org/10.1029/95jc00067

Smagorinsky, J. (1963). General circulation experiments with the primitive equations: I. The basic experiment. *Monthly Weather Review, 91*(3), 99−164. https://doi.org/10.1175/1520-0493(1963)091<0099:GCEWTP>2.3.CO;2

Smolarkiewicz, P. K. (1984). A fully multidimensional positive definite advection transport algorithm with small implicit diffusion. *Journal of Computational Physics, 54*, 325−362. https://doi.org/10.1016/0021-9991(84)90121-9

Song, D., & Wang, X. H. (2013). Suspended sediment transport in the Deepwater Navigation Channel, Yangtze River Estuary, China, in the dry season 2009: 2. Numerical simulations. *Journal of Geophysical Research: Oceans, 118*(10), 5568−5590. https://doi.org/10.1002/jgrc.20411

Stacey, M. T., Burau, J. R., & Monismith, S. G. (2001). Creation of residual flows in a partially stratified estuary. *Journal of Geophysical Research: Oceans, 106*(8), 17013−17037. https://doi.org/10.1029/2000jc000576, 21699291.

Van Rijn, L. C. (1993). *Principles of sediment transport in rivers.* Aqua Publication.

Wang, X. H. (2002). Tide-induced sediment resuspension and the bottom boundary layer in an idealized estuary with a muddy bed. *Journal of Physical Oceanography, 32*(11), 3113−3131. https://doi.org/10.1175/1520-0485(2002)032<3113:TISRAT>2.0.CO;2

Wang, X. H., Byun, D. S., Wang, X. L., & Cho, Y. K. (2005). Modelling tidal currents in a sediment stratified idealized estuary. *Continental Shelf Research, 25*, 655−665. https://doi.org/10.1016/j.csr.2004.10.013

Wang, X. H., & Pinardi, N. (2002). Modeling the dynamics of sediment transport and resuspension in the northern Adriatic Sea. *Journal of Geophysical Research: Oceans, 107*(C12), 3225. https://doi.org/10.1029/2001JC001303

Xiao, Z. Y., Wang, X. H., Song, D., Jalón-Rojas, I., & Harrison, D. (2020). Numerical modelling of suspended-sediment transport in a geographically complex microtidal estuary: Sydney Harbour Estuary, NSW. *Estuarine, Coastal and Shelf Science, 236*, 106605. https://doi.org/10.1016/j.ecss.2020.106605

Zimmerman, J. T. F. (1978). Topographic generation of residual circulation by oscillatory (tidal) currents. *Geophysical & Astrophysical Fluid Dynamics, 11*(1), 35−47. https://doi.org/10.1080/03091927808242650

Recent methods for sediment provenance in coastal areas: Advancements and applications

Zhixin Cheng [1,2], Zhaopeng Du [1] and Xiao Hua Wang [2]
[1]College of Environmental Science and Engineering, Dalian Maritime University, Dalian, China; [2]The Sino-Australian Research Consortium for Coastal Management, School of Science, University of New South Wales, Canberra, ACT, Australia

1. Introduction

The complex dynamics of sediment transport in estuarine environments, particularly in clay coastal zones, are of paramount importance to researchers and coastal managers. In these zones, fine-grained cohesive sediments are readily transported by marine hydrodynamics such as tidal waves. Diverse marine regions exhibit vast differences in hydrodynamic characteristics. The variations in these hydrodynamic fields, combined with changes in the sediment flux from major rivers flowing into different sea areas, result in distinct sediment transport patterns near typical ports in different marine regions. These differences can significantly influence the degree of siltation near the coastal infrastructures.

Sediment accumulation can restrict the movement of ships, and periodic dredging during port operations can lead to substantial economic losses. One of the defining features of sediment accumulation in offshore areas is the complex hydrodynamic conditions, which include factors like tides, ocean currents, and waves. These factors influence the distribution and morphology of sediment deposition. The topography of offshore areas is typically intricate, with features such as shoals, sea ridges, seamounts, and straits affecting sediment movement. The recurring tides further complicate research on sediment sources. In estuarine regions, the primary sediment is either suspended or resuspended sediment particles, which move differently from dissolved substances. Therefore, studying the hydrodynamic characteristics and sediment sources of major pile docks in different sea areas is of great significance and forms the essential foundation for research on the siltation evolution mechanism of high-pile docks.

Estuarine environments are characterized by high suspended-sediment concentration (SSC); these suspended sediments have unique features that differ from those in the open sea: (1) they have a large range of grain sizes; (2) their transport is strongly influenced by shoaling the effects of tides and waves in the estuary. Estuarine sediments are mainly from two sources: terrigenous sediments discharged by rivers and flood events and neritic sediments, including the local resuspended sediments from

Current Trends in Estuarine and Coastal Dynamics. https://doi.org/10.1016/B978-0-443-21728-9.00006-5

the seabed (Gao et al., 2019; Prajith et al., 2016; Venkatesan et al., 2013). The deposition and resuspension process of sediments in the long term can gradually change the coastline and topography in coastal regions, which in turn influences the hydrodynamics (Karunarathna et al., 2016; Wang et al., 2018).

A central question in coastal or estuarine sediment research is: Where does the sediment come from, and where does the sediment from the catchment go? Tracing the origin of coastal sediments has garnered widespread attention. Existing studies have found that changes in the transport of sediments from the Yellow River can affect the degree of siltation in Laizhou Bay (You & Chen, 2018). Sediments from the Yangtze River can settle in different locations in the delta and even enter Hangzhou Bay (Gao et al., 2019). To trace the source of sediments deposited in coastal areas, scholars both domestically and internationally often employ marine geology and geochemistry methods (Brito et al., 2018; Orani et al., 2018). Rare earth elements (REEs), in particular, have been widely used in tracing estuarine sediments due to their ability to retain their original characteristics in the environment (Lim et al., 2014; Song et al., 2017). However, these sediment tracing methods are more suitable for longer time scales (decades or even centuries) and larger spatial scales (entire basins). For short-term tracking of sediments in a specific range of estuaries, the resolution of these methods is insufficient, and the time scale is too long. Physical experiments in the field of sediment movement are generally realized through indoor long flumes, but due to indoor space constraints, it is impossible to establish a flume with a water flow environment close to the research sea area. Actual measurements of sediment movement usually involve the release of tracers to determine the trajectory of sediment movement, but this method has certain pollution risks.

In conclusion, estuaries and coastal areas emerge as dynamic interfaces where land and sea converge, with complex interactions that shape the balance of ecosystems and human activities. The study of sediment provenance in these coastal environments holds profound significance, illuminating the origins, pathways, and transformations of sediment particles that ultimately influence the vitality of these regions. From an ecological perspective, sediment source and transport patterns influence the distribution of nutrients, contaminants, and organic matter, which in turn shape the habitats and biodiversity of these ecosystems. Additionally, sediment provenance studies are crucial for coastal management and development, as they inform decisions related to dredging, land reclamation, and infrastructure planning. Furthermore, understanding sediment movement helps address concerns related to erosion and sedimentation, safeguarding both natural environments and human settlements. This chapter presents an in-depth exploration of recent methods for sediment provenance in coastal areas.

2. Geochemical method

A widely used sediment provenance analysis technique, the geochemical method employs elemental and isotopic signatures to trace the origin and history of sediment particles. This approach involves the systematic analysis of various elements, such as major and trace elements, stable isotopes, and radiogenic isotopes, to discern unique geochemical fingerprints. By comparing these signatures with those of potential source areas, researchers can reconstruct sediment provenance and transport history.

The geochemical method follows a rigorous procedure that encompasses sample collection, preparation, and subsequent analysis. Sediment samples are carefully collected from estuarine environments, with meticulous attention to minimizing contamination and preserving the chronological order of deposition. Subsequently, laboratory techniques like X-ray fluorescence (XRF), inductively coupled plasma mass spectrometry (ICP-MS), and stable isotope analysis are employed to quantify the elemental and isotopic compositions of the sediments.

2.1   Heavy minerals analysis and fingerprints techniques

Heavy minerals have been extensively used in sediment provenance studies. Heavy minerals, by definition, are denser than the common quartz and feldspar grains, and their presence can be indicative of specific source rocks or tectonic settings. For instance, the presence of certain heavy minerals can indicate specific source rock types or tectonic settings, allowing geologists to trace back the sediment's journey from its source to its place of deposition. The potential of heavy minerals studies in provenance analysis can be enhanced by using state-of-the-art protocols for sample preparation in the laboratory.

The gravimetric separation of heavy minerals in sediments and rocks is a technique that focuses on the properties of detrital minerals, primarily their grain size and density. The efficiency of this method depends on the procedure followed and the technical skills of the operator. Heavy mineral studies have traditionally been focused on the sand fraction, but recent advancements have emphasized the importance of analyzing a broader grain-size range, including suspended load in rivers, loess deposits, and marine muds. The accurate quantitative characterization of different size fractions is crucial for a comprehensive understanding of sediment dynamics (Andò, 2020).

Another innovative approach in sediment provenance studies is the use of finger-printing techniques. Using sieving and sample "unknowns" for instructional grain-size analysis in undergraduate sedimentology courses has advantages over other techniques. Students learn to calculate and use statistics, visually observe differences in the grain-size fractions, and understand how grain composition and properties are a function of size. These techniques aim to develop methods that enable the identification of the apportionment of sediment sources from sediment mixtures. The FingerPro tool, for instance, is an R package that unmixes sediment samples after selecting the optimum set of tracers. This tool provides a deeper understanding of the unmixing procedure through the use of graphical and statistical tools, offering a broader and easier application of the technique. Fingerprinting techniques have shown potential in better understanding sediment transport to water ecosystems and reservoirs and its detrimental effect on water quality and aquatic habitats (Lizaga et al., 2020).

2.2   Rare earth elements method

Of the various components in marine sediments, rare earth elements (REE) contain important information about sediment provenance due to the stability of their fractionation pattern during denudation (Brito et al., 2018; Hathorne et al., 2014; Jung et al., 2012; Orani et al., 2018). The REE consists of the lanthanides from La to Lu and can be divided into two groups: light REE (La, Ce, Pr, Nd, Sm, and Eu) and heavy REE (Gd, Tb, Dy, Ho, Er, Tm, Yb and Lu). Both light REE (LREE) and heavy REE (HREE) have stable chemical properties during the weathering and diagenetic processes from parent rock to detrital sediments, which can indicate the sources and composition of coastal sediments. Accordingly, recent progress in defining sediment provenance has been made using the REE as a sediment provenance tracer, especially in dynamic systems like the Yellow Sea (Lim et al., 2014; Li, Li, et al., 2014, Li, Wang, et al., 2014; Song & Choi, 2009; Song et al., 2017). REE fingerprints from the Northern Yellow Sea (NYS) suggested the major sediments of the NYS are of multiorigin with Korean river sediments derived from granitic rocks and Chinese river sediments originating from sedimentary rocks (Lim et al., 2014). The distribution of REE in estuaries can be influenced by salinity (Shynu et al., 2011), redox condition (Auer et al., 2017; Prajith et al., 2016), and hydrodynamics (Hannigan et al., 2010; Shynu et al., 2013). Therefore, REE has not only been used as a provenance tracer but also as a signal reflecting historical environmental changes in estuaries (Brito et al., 2018; Lim et al., 2014; Martins et al., 2012; Shi et al., 2018; Shynu et al., 2013; Um et al., 2013).

One of the primary strengths of geochemical method is its capacity to reconstruct the historical condition of sediments. By analyzing the elemental composition of sediments, researchers can trace sediment pathways back to their source regions. This offers insights into past environmental conditions, sedimentary processes, and even climatic shifts. For instance, the study by (Anaya-Gregorio et al., 2018) in the southwestern Gulf of Mexico utilized texture, mineralogy, and geochemistry to infer sediment provenance and depositional conditions. The findings highlighted the importance of weathering indices such as the Chemical Index of Alteration (CIA) in revealing the intensity of weathering and the origins of sediments.

Furthermore, the geochemical landscape of sediments often presents a mosaic of signatures, each echoing a different source. Through detailed trace element analyses, sediments can be linked to specific geological formations or rock types. (Gholami et al., 2019) emphasized the significance of Monte Carlo fingerprinting using geochemical tracers to identify terrestrial sources of coastal sediment deposits.

However, while the geochemical method offers valuable information, it has limitations. A recurring concern in geochemical provenance studies is the potential nonuniqueness of geochemical signatures. Different terrains, despite being geographically distant, can sometimes exhibit eerily similar geochemical profiles, complicating the task of pinpointing a specific origin. This is particularly true for trace elements, which can be influenced by multiple geological processes, making their interpretation in a provenance context nontrivial. Additionally, sediments experience transformation during their journey from mountains to the sea. Diagenetic alterations postdeposition can mask or modify their original geochemical signatures. For instance, the transformation of unstable minerals to more stable forms during diagenesis can alter the original trace element and isotopic compositions of sediments. Such changes necessitate caution in interpretation, ensuring that the readings are reflective of the sediment's origin and not its postdepositional modifications.

Sampling, the foundation of geochemical analyses, also affects the accuracy of this method. Sedimentary deposits are often heterogeneous, and capturing a sample that is truly representative could be difficult. Variabilities in sampling can introduce biases, potentially skewing the results and leading to erroneous interpretations.

In summary, the geochemical method has furthered our understanding of sediment provenance. However, it has several limitations in its application, especially in the complex terrains of coastal areas.

3. Numerical model method

Numerical models play a pivotal role in understanding sediment provenance in coastal areas. These models simulate the movement of sediment particles within estuarine systems based on principles of fluid dynamics, sediment mechanics, and transport mechanisms. By inputting information about hydrodynamics, sediment characteristics, and source regions, models can simulate sediment pathways and predict their potential source.

3.1 Coastal dynamics model

Accurately simulating coastal dynamic processes is crucial for understanding sediment dynamics and sediment origins and ensuring safe and efficient maritime operations. In the current numerical models for hydrodynamic calculations, spatial discretization methods are categorized into the finite difference method (FDM), finite element method (FEM), and finite volume method (FVM).

The fundamental concept of FDM is to replace derivatives with finite differences and differential equations with difference equations. By appropriately setting the initial and boundary conditions for the discretized algebraic equations, one can obtain approximate solutions on the grid, thus achieving the process of discretization and algebraization (Jeon & Choi, 2020). FDM is theoretically straightforward and computationally efficient, making it widely used in three-dimensional tidal current models. However, its primary limitation is its adaptability to irregular regions. Recent introductions of orthogonal and nonorthogonal curvilinear coordinate systems, as well as boundary-fitted coordinate methods, have addressed these shortcomings.

FEM is an integral approach to solving differential equations and has been applied to hydrodynamics since the 1970s. The principle involves approximating the solution within individual elements, minimizing the weighted residual of the spatial integral of the differential equation. By applying variational or weighted residual methods within each element, finite element discretized equations are generated. Combining boundary and initial conditions, one can then obtain the approximate solution of the overall equation system composed of the discretized equations from each element (Adibhusana et al., 2023). The most significant advantage of FEM is its flexible grid partitioning, which can accurately fit complex coastlines. Its computational accuracy surpasses FDM, but FEM requires more storage and is slower, limiting its practical engineering applications.

The core idea of FVM is to divide the computational domain into a series of nonoverlapping control volumes, centered around grid points. By integrating the differential equation over each control volume, a set of discrete equations is obtained. Combined with boundary and initial conditions, a numerical solution is derived. FVM's strengths lie in its clear physical meaning, ease of understanding, and the conservation properties of its derived discrete equations. It offers high computational accuracy and is well-suited for boundaries. Rahman et al. (2022) applied the FVM in their study on hydrodynamics in the waters of Balikpapan Bay, highlighting its effectiveness in capturing complex tidal dynamics.

Currently, there are several mature three-dimensional models, such as POM, ECOM, DELFT3D, MIKE3, EFDC, and FVCOM (Table 6.1). Some of these models, like POM, ECOM, and FVCOM, are open-source. Developed by Blumberg, Mellor, and others at Princeton University, the POM model is a three-dimensional numerical model for estuaries and oceans (Mellor, 1996). Its governing equations are based on hydrostatic pressure and the Boussinesq approximation. It uses σ-coordinates vertically and curvilinear orthogonal coordinates horizontally. The horizontal time discretization is explicit, while the vertical is implicit.

Table 6.1 Basic information for commonly used coastal models.

Model	Institution	Numerical method	References
POM	Princeton University	Sigma-coordinate, hydrostatic, primitive equation model	Mellor (1996)
ECOM	Johns Hopkins University	Sigma-coordinate, hydrostatic, primitive equation model	Blumberg and Mellor (1987)
DELFT3D	Deltares	Flexible-mesh, hydrostatic, finite difference method	Lesser et al. (2004)
MIKE3	DHI Water & Environment	Flexible-mesh, hydrostatic, finite volume method	Huang et al. (2008)
EFDC	U.S. Environmental Protection Agency	Cartesian or curvilinear grid, hydrostatic, finite difference method	Hamrick (2012)
FVCOM	University of Massachusetts	Unstructured-grid, hydrostatic, finite volume method	Chen et al. (2003)

The model employs a mode-splitting technique, with the external mode being two-dimensional, using short time steps to compute water levels and vertically averaged velocities. The internal mode computes three-dimensional velocities, temperature, salinity, and turbulence parameters.

Established in 2000 by Dr. Changsheng Chen and his team from the University of Georgia's School of Marine Sciences and the University of Massachusetts' School of Marine Science and Technology, FVCOM is an ocean circulation and ecological model. It adopts the FVM numerically and uses an unstructured triangular mesh in the horizontal direction, combining the advantages of both FEM and FDM. This allows for accurate fitting of complex coastlines and islands while ensuring high computational efficiency. FVCOM uses vertical σ-coordinates, the hydrostatic assumption, and the Boussinesq approximation. It employs the Smagorinsky parameterization method for horizontal mixing and the Mellor-Yamada 2.5-level turbulence closure scheme for vertical mixing. The model also incorporates a dry-wet grid discrimination dynamic boundary treatment technique and a mode-splitting numerical discretization method.

To date, the FVCOM model not only includes hydrodynamic field calculations but also modules for sediment transport and sea ice. The model excels in coastline fitting and boasts a comprehensive suite of application modules. Research using FVCOM for ocean circulation, tidal asymmetry, typhoons, and storm surges is increasing.

3.2 Lagrangian particle-tracking model

The LPT model is a widely used numerical method to track potential sources, trajectories, and fates of Lagrangian particles following the velocity field (Amoudry & Souza, 2011; Jalón-Rojas et al., 2019). The trajectories of Lagrangian particles are functions of both space and time, whereas the Eulerian method uses the movement of fluid within a spatially fixed framework (Geyer & Signell, 1992; Swanson et al., 2007).

In terms of the tracking of suspended sediment particles, the trajectories of sediments can be reproduced in LPT models, whereas Eulerian models treat the suspended sediments as a group and describe them using concentrations (Amoudry & Souza, 2011). Therefore, the entire history of the trajectories of the particles in suspension can be simulated by the LPT method instead of their piecewise concentrations. By using the LPT method, we can follow how a certain packet of suspended particles travels in an estuary and the probability density as it passes a given location. There is another advantage in the offline mode of LPT (to date most LPT models are offline LPT): with an existing velocity field, one can do backward tracking to find out from where the particles at a particular location have come (Van Sebille et al., 2018).

Hydrodynamic models, such as the FEM and the Reynolds-Averaged Navier–Stokes (RANS) equations, simulate water flow patterns which involves the movement of sediment particles. Thus, the core of coastal models is the hydrodynamic model, with different modules (wave, sediment, or ecosystem models) as complements. While the Eulerian method is used in these hydrodynamic models, LPT models coupled with these coastal models have been applied to track various marine particulates, including plastic debris, oil patches, dust particles and suspended sediments (Lane, 2005; Lackey & Macdonald, 2007; Li & Yao, 2015; Liubartseva et al., 2018; MacDonald et al., 2006; Tsurumura et al., 2013). Most of these models have a 2D approach, in which marine debris is assumed to be drifting in the surface layer. However, 2D LPT models cannot be applied in many situations, especially in tracking suspended sediments, as sediments sink vertically and resuspend due to the effects of gravity and turbulence.

Improvements have been made in LPT models to meet the need for 3D simulations. For example, the hydrological simulation software, MIKE includes 3D Lagrangian particle-tracking modules in MIKE 21 and MIKE 3, and suspended particles perform a random walk through the dispersion term in the vertical direction. However, MIKE is a closed system that does not allow different formats of inputs. On the other hand, a recent LPT model, the Track Marine Plastic Debris (TrackMPD) model also adopts a 3D approach, which performs better than its 2D version (Jalón-Rojas et al. 2019). As in many other LPT models, Eulerian velocity fields are the basis in TrackMPD. However, TrackMPD is compatible with a wide range of different formats of current inputs. Some complex physical processes, including windage, washing-off, degradation, biofouling, and deposition of the marine particulates, have been added to TrackMPD to extend its application (Jalón-Rojas et al. 2019). It uses a zeroth-order Markov model (random walk) to solve small-scale physical processes such as turbulence with modifiable, constant vertical, and horizontal diffusivities (Argall et al., 2004; Visser, 1997).

One of the primary strengths of the numerical method is its predictive capability. By adjusting parameters based on in-situ data, these models can forecast potential future trajectories of sediment movement. For instance, a study by Rasheed. Rasheed et al. (2020) highlighted the impact of coastline modification on sediment grain size distribution (GSD) across the atoll basin in Malé Atoll, Maldives, using the coastal model. Furthermore, numerical models can study sediment dynamics at various spatial scales, from localized coastal stretches to entire continental shelves.

However, like all methodologies, the numerical method has pros and cons. One of the primary concerns is the accuracy of input parameters. The reliability of model predictions relies on the quality of input data (Carniel et al., 2011). Additionally, while numerical models are adept at simulating sediment transport under various scenarios, they often require significant computational resources. The complexity of coastal processes, coupled with the need for high-resolution data, can make numerical modeling computationally intensive. This requires the use of advanced computational infrastructures and optimization techniques to ensure efficient and timely model outputs.

Therefore, the numerical method has advantages in simulating complex sediment transport processes, and offering insights into sediment sources. However, its efficacy can be hindered by the accuracy of input data, computational techniques, and our understanding of coastal processes.

4. Other sediment provenance methods

While geochemical and numerical models are commonly employed methods for sediment provenance, there are several other techniques that have been developed and refined over the years.

4.1 Grain size distribution analysis

One of the foundational methods in sediment provenance studies is the analysis of GSD. GSD data are pivotal in understanding sediment provenance and depositional regimes. The technique of end-member analysis (EMA) has been particularly influential in unmixing GSDs into geologically meaningful components. EMA estimates end-members based on variability within a dataset, offering a nonparametric approach. Recent advancements have introduced algorithms inspired by hyperspectral image analysis, which have shown promise in improving the unmixing of grain size data. Furthermore, the development of parametric EMA allows for the unmixing of an entire dataset into unimodal parametric end-members, aiding in the identification of individual grain size subpopulations in highly mixed datasets (Paterson & Heslop, 2015).

4.2 Luminescence as a sediment tracer

Luminescence has emerged as a potent sediment tracer and provenance tool. The method is rooted in the luminescent properties of minerals, primarily quartz and feldspar, which respond to environmental changes. Luminescence offers several advantages, including its applicability to geologically ubiquitous minerals and its relatively low cost. The method has been employed in a range of applications, from identifying sediment source locations based on unique luminescence characteristics to estimating transport rates and mixing rates in sediments. Recent research has also explored the potential of luminescence in tracing coastal longshore drift, understanding wind-blown deposit provenance, and estimating fluvial transport rates (Gray et al., 2019).

4.3 Independent component analysis (ICA)

Independent component analysis is a powerful statistical method used in sediment provenance analysis to separate mixed signals into their underlying independent source components. Originally developed in the field of signal processing and blind source separation, ICA has found applications in a wide range of scientific disciplines, including geology (Yasukawa et al., 2019).

The core idea behind ICA is to transform a set of observed signals into a new set of statistically independent components by finding a mixing matrix that represents the linear relationships between the observed signals and the source components (Wu et al., 2020). ICA can help identify the source contributions to sediment samples, allowing researchers to distinguish between various sediment sources in a mixed sedimentary record. By analyzing the composition of source components, ICA can provide insights into past environmental conditions and climate changes. For example, changes in the relative contributions of source components can be indicative of shifts in erosion patterns. ICA can be applied to petrographic data to identify and quantify the mineralogical and textural characteristics of sediment sources, aiding in the interpretation of sedimentary records.

The advantages of ICA in method can be summarized as follows: ICA is particularly effective at separating mixed signals when the source components are statistically independent. This is advantageous in scenarios where other methods may struggle to differentiate sources with correlated compositions. ICA is a blind source separation technique, meaning it does not require prior knowledge of the source components or their mixing ratios. This makes it suitable for cases where the sediment sources are not well-characterized. ICA can handle nonlinear mixing processes, which are common in geological systems, making it versatile for a wide range of sediment provenance applications.

There are of course limitations of this method. ICA requires a sufficient amount of high-quality sediment composition data from multiple sampling sites to reliably estimate the mixing matrix and source components. ICA assumes that the source components are statistically independent, which may not hold true in all geological contexts. Deviations from this assumption can lead to inaccurate results. ICA can be computationally intensive, especially when dealing with large datasets. The choice of algorithms and parameters can significantly impact the computational efficiency and accuracy of the analysis.

4.4   Q-mode factor analysis

Q-mode Factor Analysis, often referred to as QFA, is a multivariate statistical technique used in sediment provenance analysis to extract underlying factors or end-members from sediment composition data. QFA is based on the idea that sediment samples can be represented as linear combinations of a few dominant end-members, each of which corresponds to a distinct source or process (Papatheodorou et al., 2006; Voudouris et al., 1997).

QFA can identify and quantify the end-members in sediment samples, providing valuable insights into the composition and contributions of different sediment sources. By analyzing the factor loadings (F), QFA can help characterize the mineralogical, geochemical, or textural properties of the identified end-members, aiding in source discrimination. QFA can be used to track changes in the relative contributions of end-members over time, offering insights into past environmental conditions and sedimentary processes.

QFA explicitly identifies end-members, making it easier to interpret the sediment provenance results in terms of distinct source contributions. QFA allows for the incorporation of prior knowledge about potential end-members, which can enhance the accuracy of the analysis. QFA can handle a variety of data types, including sediment composition, mineralogy, and geochemistry, making it adaptable to different sedimentary environments.

As a traditional method, QFA has its limitations. QFA is sensitive to data quality, and noisy or incomplete data can lead to inaccurate results. Preprocessing and data quality assessment are critical. The choice of the number of end-members to extract can be subjective and may impact the results. Objective criteria and expert judgment are often used to address this challenge. QFA assumes linear combinations of end-members, which may not accurately represent all sedimentary processes. Nonlinear processes may require alternative methods.

4.5 Quantitative provenance analysis

Quantitative Provenance Analysis primarily focuses on the physical and mineralogical characteristics of sediments. One of its key strengths is its ability to provide information on sediment transport processes based on GSD. This is crucial as the GSD can reveal details about the energy levels of the transporting medium, depositional environment, and postdepositional modifications. Furthermore, this method can identify specific source rock types or tectonic settings based on mineralogy. Such insights are invaluable in reconstructing the geological history of a region. For instance, the presence of certain heavy minerals can indicate specific source rock types or tectonic settings, allowing geologists to trace back the sediment's journey from its source to its place of deposition.

However, Quantitative Provenance Analysis is not without its limitations. One of its major drawbacks is its potential ineffectiveness for fine-grained sediments. Fine-grained sediments, such as clays, often pose challenges in terms of sample preparation and analysis. Moreover, the method requires extensive sample preparation and analysis, which can be time-consuming and may not always yield definitive results. For instance, the presence of certain minerals might be indicative of multiple potential source regions, making it challenging to pinpoint a single source.

4.6 Statistical provenance analysis

On the other hand, Statistical Provenance Analysis (SPA) has emerged as a powerful tool, especially in the era of big data. One of its primary advantages is its ability to handle large datasets, allowing for a comprehensive analysis. With the advent of modern analytical techniques, large datasets are often generated in provenance studies. Statistical tools for provenance analysis have been developed to handle and interpret these large datasets. These tools can help in deciphering complex provenance signals and identifying multiple source regions, providing a more holistic understanding of sediment origins.

However, like all methods, SPA has its set of challenges. A significant limitation is the requirement of familiarity with statistical tools and software. Not all geologists or researchers might be well-versed with statistical software, which can pose a barrier. Moreover, there is a potential for misinterpretation if the tools are not used correctly. Statistical results can sometimes be misleading, and without proper understanding or context, one might draw incorrect conclusions.

In light of the above discussion, it is evident that both quantitative and statistical provenance analysis methods offer unique insights and challenges. While the former provides a more hands-on, physical approach to understanding sediment origins, the latter offers a data-driven, comprehensive analysis. The choice of method often depends on the research question at hand, available resources, and the specific challenges posed by the sediment samples.

In conclusion, while the geochemical and numerical modeling methods have their place in sediment provenance studies, alternative techniques like GSD analysis, luminescence offer fresh perspectives and tools. As research in this field continues to evolve, it is likely that a combination of these methods will be the way forward.

5. Comparison of methods for sediment provenance in coastal areas

The quest to understand sediment provenance, especially in coastal regions, has led to the development and refinement of various methodologies. Among these, the geochemical method and numerical models have emerged as prominent tools for sediment provenance in coastal areas, each with its unique set of advantages and challenges.

The geochemical method revolves around analyzing the mineralogical and elemental composition of sediments. By tracing specific geochemical signatures, researchers can infer the origins and pathways of sediments. One of the primary strengths of the geochemical method is its capacity to reconstruct the historical backdrop against which sediments were deposited. Through detailed trace element analyses, sediments can be linked to specific geological formations or rock types. For instance, REEs and their patterns have been extensively utilized to determine sediment provenance, as they are relatively immobile during chemical weathering and diagenesis (Navas et al., 2022). Furthermore, isotopic systems, such as strontium (Sr) and neodymium (Nd), have been employed to decipher sediment origins, linking them to the age and type of source rocks (Ngueutchoua et al., 2017). However, the geochemical method is not without its challenges. The method assumes a steady-state system, which may not accurately reflect the intricate variations in sediment transport and deposition within estuarine environments. Different terrains can sometimes exhibit similar geochemical profiles, complicating the task of pinpointing a specific origin. Additionally, diagenetic alterations postdeposition can mask or modify original geochemical signatures, necessitating caution in interpretation (O'Sullivan et al., 2018).

Numerical models offer predictive power by simulating dynamic processes and enabling "what-if" scenarios. Grounded in mathematical and physical principles, these models allow researchers to predict how sediments might move under varying environmental conditions. For instance, in the coastal area between Mount Lavinia and Negombo, Sri Lanka, the Delft3D-FLOW model has been employed to estimate longshore sediment transport rates, revealing variable sediment sources of different parts of the study zone (Jayathilaka & Fernando, 2019). However, numerical models require accurate input data, including hydrodynamic conditions and sediment characteristics, which might be challenging to obtain. The models' complexity also demands expertise in computational fluid dynamics and sediment transport dynamics.

A multifaceted approach to sediment provenance research may offer the most robust insights. The combined use of geochemical methods and numerical modeling allowed researchers to capture both the historical context and the current dynamics of sediment transport. A notable application of multiple sediment provenance methods can be observed in the Chudao Island, China (Zhou et al., 2023). They combined high-precision bathymetric surveys, high-density sediment sampling, grain-size trend analysis, and hydrodynamic numerical modeling to investigate sediment transport trends and influencing factors (Zhou et al., 2023).

In conclusion, both approaches offer unique strengths and challenges. While the geochemical method provides detailed insights into historical sediment origins, the numerical method offers predictive insights into sediment dynamics. Their judicious application can provide invaluable insights into sediment dynamics in coastal regions.

6. A case study in the Yalu River Estuary, China

6.1 Introduction

To exemplify the integration of these methods, consider a case study of an estuarine system facing sedimentation issues due to urbanization and industrial activities. Researchers combine traditional geochemical analyses of sediment cores with remote sensing imagery depicting sediment plumes and patterns. These data are used to calibrate a numerical model that simulates sediment transport dynamics and predicts future deposition areas.

By combining the strengths of both geochemical and numerical methods, the case study provides a comprehensive understanding of sediment provenance in the estuarine system. The traditional geochemical analysis elucidates historical sediment sources, while remote sensing reveals current patterns. The numerical model validates these insights and forecasts potential deposition scenarios. This approach informs sustainable sediment management strategies.

This study takes the Yalu River Estuary (YRE) as an example. The Yalu River is at the northernmost point of the coastline of China, on the border with North Korea, and is the largest river flowing into the NYS, carrying with it a large quantity of terrigenous material. Material deposited in the YRE contributes to the bottom sediments in the Yellow Sea (Chen, Beardsley, et al., 2013, Chen, Li, et al., 2013; Shi et al., 2018). Therefore, a robust understanding of the sedimentation processes in the YRE is of significance in studying material exchange and transport in the Yellow Sea (Fig. 6.1).

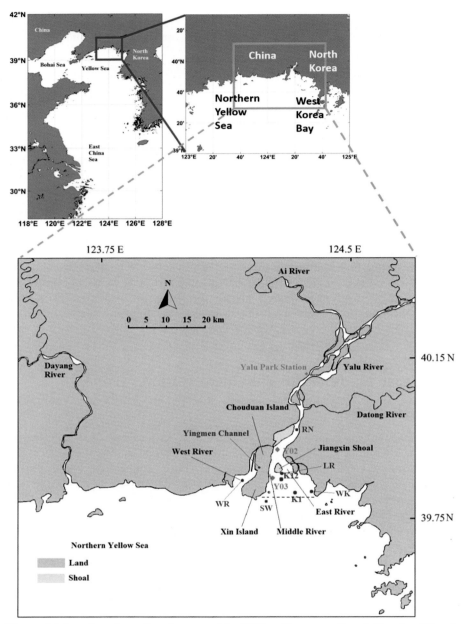

Figure 6.1 Location of sampling sites around the YRE and its surrounding shelf region. K1 and K12 indicate the core-sample collecting sites. Y02 and Y03 are the current measurement sites. Yalu Park Station is the upstream hydrology station. The black dashed line shows the estuarine entrance line. Points in *dark blue* show the source areas for the particle-tracking model: RN is a small area where the river flows into the estuary; WR a small area downstream of the West River; SW is an area in shallow water near the entrance; WK stands for Western Korean Bay; LR is in the middle region of the estuary, with the sediment here from local resuspension in the main estuary branch. From Cheng, Z., Jalon-Rójas, I., Wang, X. H., & Liu, Y. (2020). Impacts of land reclamation on sediment transport and sedimentary environment in a macro-tidal estuary. *Estuarine, Coastal and Shelf Science*, *242*. https://doi.org/10.1016/j.ecss.2020.106861

6.2 Study area and field surveys

6.2.1 Study area

The Yalu River originates in the Changbai Mountain with a stream-gradient ratio of 3%. The average annual freshwater and sediment discharge of the river are 2.67×10^{10} m3 and 1.59×10^6 T, respectively (Cheng et al., 2016). The YRE can be divided into two branches with three waterways: the West Branch on the west side of Chouduan Island contains the West River and the East Branch consists of the Middle and East rivers with Jiangxin Shoal in the middle. The underlying tidal ridge system in the YRE has an area of 4000 km^2, extending from the river entrance to the 40 m isobaths (Gao et al., 2012), and is associated with parallel tidal ridge areas in the western Korean Bay. Governed by that massive tidal ridge, the direction of the flood current in the YRE is north-east, that of the ebb current is west-south. Previous study revealed that the sediment source in the YRE was a mixture of materials from different regions (fluvial and neritic materials), and the parent rock type of surface sediments in the YRE was mainly sedimentary rock and granite (Liu et al., 2013, 2015). The YRE is a macro-tidal estuary, with a mean tidal range of 4.5 m at the entrance (Cheng et al., 2016). The estuary is dominated by regular semi-diurnal tide outside the entrance and irregular semi-diurnal tide with a tidal-river reach of 54 km inside the entrance (Yu et al., 2014). Due to its strong estuarine hydrodynamics, YRE is well-mixed for most of the time except during wet seasons, when there is an extremely large river flow (Cheng et al., 2016; Yu et al., 2014). According to field measurements during 1994, 1996, and 2009, the salinity in the YRE is well-mixed with no obvious vertical stratification (Cheng et al., 2016; Gao et al., 2004; Yu et al., 2014). Maximum salinity in the upper estuary was around 15 psu during high slacks. Winds and wave action in this area are relatively weak, as the average significant wave height outside the estuary is only 0.5 m (Yu et al., 2014), and the annual mean wind speed is 3.2 m/second (Cheng et al., 2016). The complex dynamic environment in this area may induce frequent regime shifts, which will affect the estuarine sedimentation process.

The sediment distribution in the YRE is mainly controlled by tides and there is a TMZ driven by tidal pumping is present in the lower estuary (Yu et al., 2014). There was a large-scale land reclamation finished in the early 1970s, connecting the Xin Island and Chouduan Island together.

6.2.2 Sediment-samples collection

Two sediment cores (K1 and K12) were collected in August 2014 in the YRE. K1 (124°20′25″E, 39°48,15″N) is located in the shallow sea area of the YRE while K12 (124°19′4″E, 39°50′20″N) is located in the lower estuary, at the southern edge of the TMZ observed by previous studies (Gao et al., 2004; Yu et al., 2014). The gravity corers were 4 m-long PVC tubes with an inner diameter of 85 mm and an outer diameter of 90 mm. The core samples were collected by pushing the corers into the bottom sediments with the help of laboratory technicians. Sampling was carried out during low slack water, when the water depth was less than 1 m. The major lithologic characteristic of the two samples is clayey silt, with a clay content of 20%−40%.

The core samples were then divided into sub-samples at 4—6 cm interval for REE detection, radioactive chronology testing and grain size analysis.

Surface sediment samples were collected from four sections around the YRE using clamshell-type samplers. Forty surface sediment samples from the Middle and West waterways and adjacent area (Sections A, B, C and E in) were collected from June to August 2006; 11 surface sediment samples were collected from the tidal flats outside the entrance in July 2010 (Section D).

6.2.3 In-situ hydrodynamics data collection

A field survey for hydrodynamic observations was conducted in the YRE during the wet season (August) of 2009. Current velocity and direction were measured at a ship-based anchor station Y03 (124°16′43″E, 39°50′22″N) over a continuous 25 hours period during spring tide (8—9 August) and neap tide (14—15 August) and at station Y02 (124°18′10″E, 39°55′12″N) during spring tide (7—8 August) and neap tide (13—14 August). The current data were measured by Acoustic Doppler Current Profilers (ADCP) with a frequency of 1200 kHz and a current-speed resolution of 0.001 m/second. The receivers of the ADCPs were placed 0.1 m below the water surface, facing downward, with a bin-layer depth of 0.25 m.

6.3   Data and methods

6.3.1   REE geochemistry

The collected samples were first air dried at room temperature and then ground using an agate mortar. The ground materials were filtered through a nylon sieve with a diameter of 0.107 mm to eliminate the effect of size differences on the REE measurements. In order to remove the authigenic carbonate, iron and manganese hydroxides, and residual organic matter, the filtered samples were ignited at a temperature of 450°C, acid pickled using HCl, and then dissolved in a mixed acid of HF and HNO_3. Finally, the dissolved sample solutions were transferred into polyethylene test tubes, and the REE concentrations were then determined using ICP-MS. To ensure the accuracy of the testing, repeat samples were tested concurrently with the standard samples. For all 14 elements, the relative errors and the differences between the repeat samples and standard samples were less than 6% and 7.7%, respectively, indicating the satisfactory data reliability.

6.3.2   Elemental analysis of REE

The REE enrichment values in the YRE samples were normalized by corresponding values from the Upper Continental Crust (UCC) and North American Shale Composite (NASC); this reveals the fractionation patterns of the REE during sedimentation (Hannigan et al., 2010; Um et al., 2013). Sectional-averaged REE concentrations from surficial sediment samples in the YRE and that of two reference samples are shown in Table 6.2.

Table 6.2 Sectional-averaged REE concentrations (μg/g) in the YE and REE concentrations of reference samples (UCC and NASC) obtained from Taylor and McLennan (1985).

REE (μg/g)Location	La	Ce	Pr	Nd	Sm	Eu	Gd	Tb	Dy	Ho	Er	Tm	Yb	Lu
Section A	37.81	74.04	8.27	32.40	5.43	1.00	5.29	0.67	3.51	0.74	1.98	0.29	1.83	0.27
Section B	46.72	94.19	10.28	39.79	6.66	1.12	6.31	0.80	4.45	0.92	2.49	0.36	2.30	0.34
Section C	49.43	102.05	11.44	41.84	7.23	1.28	6.08	0.72	4.26	0.86	2.39	0.34	2.16	0.34
Section D	36.48	78.70	8.21	28.48	5.01	1.13	4.55	0.64	3.14	0.53	1.52	0.24	1.46	0.22
Section E	27.95	52.67	5.94	23.08	3.89	0.86	3.65	0.44	2.41	0.51	1.37	0.22	1.27	0.18
NASC	32	73	7.9	33	5.7	1.24	5.2	0.85	5.8	1.04	3.4	0.5	3.1	0.48
UCC	30.00	64.00	7.10	26.00	4.50	0.88	3.80	0.64	3.50	0.80	2.30	0.33	2.20	0.32

From Cheng, Z., Jalon-Rójas, I., Wang, X. H., & Liu, Y. (2020). Impacts of land reclamation on sediment transport and sedimentary environment in a macro-tidal estuary. *Estuarine, Coastal and Shelf Science, 242*. https://doi.org/10.1016/j.ecss.2020.106861

Some characteristic coefficients of REE, known as fractionation factors, are widely used to demonstrate the fractionation properties among elements. NASC-based normalized REE concentrations were used to calculate the REE fractionation factors δEu_{NASC}, δCe_{NASC}, $(La/Sm)_{NASC}$, $(Gd/Yb)_{NASC}$ and $(La/Yb)_{NASC}$ in this study.

This study focuses on the lanthanides of REEs from La to Lu and can be divided into two groups: light REE (La, Ce, Pr, Nd, Sm, and Eu) and heavy REE (Gd, Tb, Dy, Ho, Er, Tm, Yb, and Lu). $(La/Yb)_N$ indicates the fractionation property between the light REE (LREE) and heavy REE (HREE): the larger the ratio, the higher the relative enrichment level of the LREE. $(La/Sm)_N$ indicates the internal fractionation property in the LREEs: the larger the ratio, the more obvious the fractionation is. $(Gd/Yb)_N$ indicates the internal fractionation property in the HREEs: the larger the ratio, the more obvious the fractionation is.

6.3.3 Hydrodynamic simulation

A 3D hydrodynamic model, the Finite Volume Coastal Ocean Model (FVCOM) (Chen et al., 2003) was used to explore the sediment provenance in the YRE. FVCOM adopts an unstructured, triangular grid to depict land boundaries with wet/dry treatments (Ge et al., 2012), which provides an accurate simulation of the irregular coastline in the YRE region. The flux Richardson number was introduced into the bottom friction coefficient C_d to consider the effects of the sediment-induced BBL in the model (Wang, 2002).

6.3.4 Three-dimensional particle resuspension and tracking model

A LPT model, TrackMPD, was adopted in this study to investigate the sediment provenance in the YRE. TrackMPD is a newly developed transport model that can consider the particular behavior of marine debris but that can be also used for the transport of sediments. It includes processes such as advection, dispersion, windage, sinking, settling, beaching, and refloating of independent particles (Jalón-Rojas et al., 2019). The horizontal position (X) of a particle at time $t + \Delta t$ in TrackMPD can be expressed as:

$$X(t + \Delta t) = X(t) + U\Delta t + R\sqrt{2K_h\Delta t}, \tag{6.1}$$

where $U = (u, v)$ and K_h are horizontal current vector and diffusion coefficient, respectively. i and j denote unit vectors in the zonal (x) and meridional (y) directions, respectively. R is a random number (from -1 to 1) generated at each time step with an average and standard deviation from 0 to 1, generated at each time step. $X(t)$ is the original horizontal location of the particle at time t. $U\Delta t$ is the advective displacement and $R\sqrt{2K_h\Delta t}$ is the random displacement due to horizontal turbulent diffusion.

The vertical position (Z) at time $t + \Delta t$ is computed as follows:

$$Z(t + \Delta t) = Z(t) + w(t)\Delta t, \tag{6.2}$$

$$w(t) = W(t) - w_s(t) + \frac{R\sqrt{2K_z\Delta t}}{\Delta t}, \tag{6.3}$$

where w, w_s, and K_z are the vertical velocity, settling velocity, and vertical diffusion coefficient, respectively. $X(t)$ is the original vertical location of the particle at time t. Two diffusion coefficients (K_h and K_z) were used in this model: the final term on the right-hand side in Eqs. (6.1) and (6.3) is from stochastic motion during turbulent diffusion.

It has always been challenging to reproduce the resuspension of particles in an LPT model. In this study, a resuspension module is added to TrackMPD. A tracking particle is resuspended from the seabed when the bottom shear stress at its sink location is larger than the critical erosion stress. Bedload transport for a particle is not considered in this model. The movement of this particle on resuspension is controlled by both sediment settling under gravity and vertical turbulence in the BBL (Ji, 2006). Accordingly, when a deposited particle meets the criteria for resuspension (Eq. 6.4), the logical of resuspension variable is true (equals 1) and it is placed at the bottom sigma level; its vertical location after resuspension is determined by w_s and K_z as in Eqs. (6.2) and (6.3). Correspondingly, the particle will deposit and stay at the bottom when the relating bottom shear stress is smaller than the critical value.

$$Resuspension = \begin{cases} 1, \text{if} |\tau_b| > \tau_c \\ 0, \text{if} |\tau_b| < \tau_c \end{cases} \tag{6.4}$$

6.4 REE geochemistry of sediments in the YRE

6.4.1 Profiles of REE fractionation factors

Values for the characteristic REE parameters, ΣREE, ΣLREE/ΣHREE, $(La/Sm)_{NASC}$, $(Gd/Yb)_{NASC}$, $(La/Yb)_{NASC}$, δCe_{NASC}, and δEu_{NASC} were calculated in this study as additional indicators of historical changes in sedimentation process in the YRE. The profiles of these parameters for K1 and K12 confirm the variation patterns found in the REE concentration profiles, with the 1975 layer as the split line (Fig. 6.2).

For K1, ΣREE ranged from 180 to 286 µg/g, with a mean value of 227 µg/g ΣLREE/ΣHREE (b) ranged from 12.0 to 15.4, with a mean value of 12.5. ΣREE was significantly larger than the average before 1975, then decreased to and remained at the average in the years after. A similar trend was found in the ratio ΣLREE/ΣHREE, with higher values before 1975 and lower values after. For K12, ΣREE ranged from 147 to 238 µg/g, with a mean value of 202 µg/g (slightly lower than that of K1). ΣLREE/ΣHREE ranged from 10.4 to 13.5, with a mean value of 12.0. In contrast to K1, ΣREE in K12 was smaller than the average before 1975, then increased to the average, close to that of K1, after 1975. ΣLREE/ΣHREE in K12 showed an increasing trend before 1975 and came close to the K1 value after 1975.

In order to eliminate the potential effects of other factors on REE fractionation, we compared the geometric shapes of the normalized REE patterns instead of the abundance of the REE. The REE concentrations of representative sedimentation layers from cores K1 and K12, together with those of surficial sediments of five sections, were normalized by values from the UCC and NASC, respectively.

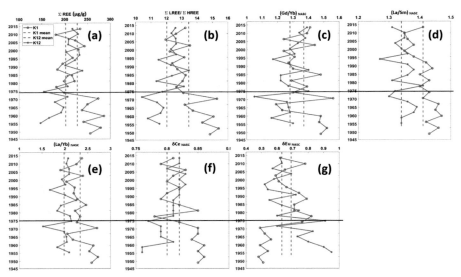

Figure 6.2 Profiles of (a) ΣREE, (b) ΣLREE/ΣHREE, (c) (Gd/Yb)$_{NASC}$, (d) (La/Sm)$_{NASC}$, (e) (La/Yb)$_{NASC}$, (f) δCe $_{NASC}$, and (g) δEu $_{NASC}$ in K1 and K12; *dashed straight lines* show the average values.
From Cheng, Z., Jalon-Rójas, I., Wang, X. H., & Liu, Y. (2020). Impacts of land reclamation on sediment transport and sedimentary environment in a macro-tidal estuary. *Estuarine, Coastal and Shelf Science, 242.* https://doi.org/10.1016/j.ecss.2020.106861

All UCC-normalized REE curves (a, b) were right-dipping lines with significant enrichment in the LREE relative to HREE, especially for REE in K1 and K12. Normalized ratios of northeastern sections (B and C) were larger than other sections (especially Section E). The UCC-normalized REE patterns of K1 and K12 showed a weak depletion of Ce and a significant negative Eu anomaly. Compared to the fractionation patterns from the five sections with relatively flat curves, the final weathered deposits at K1 and K12 showed greater discrepancies from those of the UCC. The NASC-normalized REE curves (c, d) in the YRE had similar characteristics, with the UCC-based curves having slightly stronger negative Ce and Eu anomalies. Section E, again, showed a distinct curve with significant depletion in Dy, implying REE upstream area may originated from a different source compared to other samples around the estuary. It appears evident that both UCC- and NASC-normalized REE patterns of representative layers in K1 and K12 were remarkably different before and after the 1975, these patterns tended to become relatively closer to those of Sections B and C after 1975.

6.4.2 Spatial distribution pattern of REE sedimentation in the YRE

The chart of ΣREE versus ΣLREE/HREE was used here to identify the differences in REE sedimentation between K1 and K12 and to compare the sediment property at these two sites with sediments from surrounding areas.

The patch of sedimentation units (scatter points in) at K1 before 1975 (in the solid-line circle) was distinct from other samples around the estuary and characterized by prominently higher ΣREE and ΣLREE/HREE, suggesting a unique sediment supply contributed to sediments at K1 before 1975, and this particular sediment source no longer affected K1 after 1975. After 1975 (dash-line circle), this patch of sediment units in K1 moved to an area with a center point corresponding to a total REE of 220 μg/g and a ΣLREE/HREE around 13 (the patch of sediment units in K12 moved to a similar area as well after 1975). Compared to the surface sediment samples from five sections around the YRE, deposits in K1 after 1975 were more similar to the samples from the LP muddy coast (Section C) (Fig. 6.3).

According to the relationship between ΣREE and ΣLREE/HREE at K12 and the surface sediment samples (b), the sedimentation units also became more concentrated after 1975, implying a purer supply from possibly one or two regions. The change of REE at K12 was not as prominent as K1, but a distinct change can still be identified: sediment in K12 before 1975 had more complex feature and was relatively similar to the samples from Section A, whereas units after 1975 were more similar to Section C. In summary, although the sediment sources of K1 and K12 were different before 1975, they became similar with a closer property to the surface sample from the LP coast after 1975.

Fig. 6.4 shows the UCC-normalized REE patterns (a, b) and NASC-normalized REE patterns (c, d) of K1, K12, and surface samples from surrounding shallow waters. It is noteworthy that the similarity of REE property after 1975 between sediments at two cores and surface sample from Section C does not necessarily indicate sediments were from Section C after 1975, it can only implied that the post-1975 deposits at K1, K12 and Section C may from a common sediment source. Here we concluded that the sediment property at K1 and K12 changed from diverse to homogenous, while the specific sediment source (terrigenous or neritic inputs) of the estuary before and after 1975 requires further investigation.

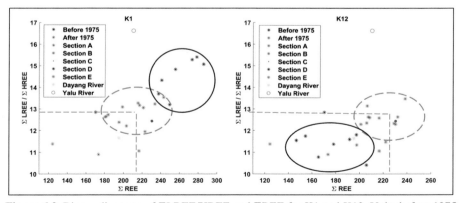

Figure 6.3 Binary diagrams of ΣLREE/HREE and ΣREE for K1 and K12. Units before 1975 were in solid-line circle, those after 1975 were in dash-line circle with a center.
From Cheng, Z., Jalon-Rójas, I., Wang, X. H., & Liu, Y. (2020). Impacts of land reclamation on sediment transport and sedimentary environment in a macro-tidal estuary. *Estuarine, Coastal and Shelf Science, 242*. https://doi.org/10.1016/j.ecss.2020.106861

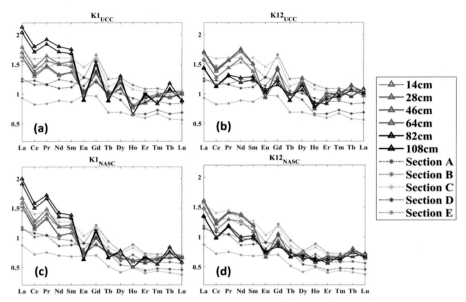

Figure 6.4 UCC-normalized REE patterns (a, b) and NASC-normalized REE patterns (c, d) of K1, K12, and surface samples from surrounding shallow waters. Black and *red* lines represent layers pre- and post-1975, respectively.
From Cheng, Z., Jalon-Rójas, I., Wang, X. H., & Liu, Y. (2020). Impacts of land reclamation on sediment transport and sedimentary environment in a macro-tidal estuary. *Estuarine, Coastal and Shelf Science, 242.* https://doi.org/10.1016/j.ecss.2020.106861

6.5   Numerical simulations of sediment dynamics in the YRE

6.5.1   Hydrodynamic model

The simulation domain covered the entire YRE area, together with part of the NYS. Two experiments with different coastlines, in 1956 (a) and 2011 (b), were designed to simulate the hydrodynamic and sediment-transport conditions before and after land reclamation, respectively. Simulation meshes were built using the surface-water modeling system (SMS), based on coastline and bathymetry data extracted from sea charts from the Navigation Guarantee Department of the Chinese Navy Head-quarters. Maximum and minimum grid sizes of the meshes were 3000 m at the open boundary and 150 m near shore. A uniformed sigma-stretched coordinate system with 20 vertical layers was applied in this model to study the barotropic hydrodynamics in the YRE (Fig. 6.5).

To eliminate errors generated in the numerical calculations and to isolate the effects of the land reclamation, the same bathymetry and initial boundary conditions in 2011 were used for Experiments 1 and 2 (prereclamation and postreclamation). This should be considered in the interpretation of prereclamation simulations. Nevertheless, previous studies in similar systems have concluded that the changes in the bathymetry after

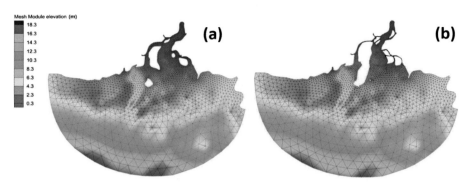

Figure 6.5 Grids of the model domain for (a) 1956 (before land reclamation) and (b) 2011 (after reclamation).
From Cheng, Z., Jalon-Rójas, I., Wang, X. H., & Liu, Y. (2020). Impacts of land reclamation on sediment transport and sedimentary environment in a macro-tidal estuary. *Estuarine, Coastal and Shelf Science, 242*. https://doi.org/10.1016/j.ecss.2020.106861

land reclamation had a limited effect on the hydrodynamics compared to the changes in coastline (Gao et al., 2014; Guo et al., 2017). Experiment 3 was designed for model validation using the same mesh as in Experiment 2 but with the boundary conditions in 2009 according to period of the field trip, validation results can be found in related research (Cheng et al., 2020).

As mentioned in Section 1, the hydrodynamics in the YRE region are dominated by tides; therefore, tidal elevation is adopted as the open boundary condition to drive the model. The tidal elevations at the open boundary were calculated based on selected tidal components including four diurnal (K_1, Q_1, P_1, Q_1), four semidiurnal (M_2, S_2, N_2, K_2), three shallow-water (M_4, MS_4, MN_4), and two long-period components (Mf, Mm). The Harmonic constants for all tidal components were retrieved from the global tidal model TPXO 7.2.

Given that river discharge also impacts the estuarine dynamics, a constant river discharge of 700 m^3/second with a suspended-sediment flux of 0.1 kg/m^3/second from the river (typical value for the Yalu River during the wet season) at 1 hour interval was used as another boundary forcing in this model. According to field observations, the YRE is well-mixed for most of the time as a consequence of the limited water depth and strong tidal action (Cheng et al., 2016; Yu et al., 2014). Therefore, only the barotropic effects of the freshwater input are considered in this model. The initial conditions for sea temperature and salinity in this model used the typical values in the YRE during wet season of 25°C and 33 psu, respectively. The influence of wind and wave action in this area is weak and transient, as discussed in Section 5(0.3; therefore, wind and wave forcing were not considered in the model.)

Experiments 1 and 2 ran for 92 days from May 01, 2011 to July 31, 2011, Experiment three for 62 days from July 1, 2009 to August 31, 2009. All experiments have run for 15 days to warm up. Model parameters are shown in Table 6.3.

Table 6.3 Key parameters of the model configuration.

Model parameters	Value/method
Model external time step	1.0 second
Bottom friction coefficient	0.0025
Horizontal diffusion	Smagorinsky scheme
Vertical eddy viscosity	M-Y 2.5 turbulent closure
Settling velocity	1.25×10^{-4} m/second
Critical bottom stress for erosion	0.1 N/m^2
Critical bottom stress for deposition	0.08 N/m^2
Erosion rate	$5 \times 10^{-6} \text{ kg/m}^2\text{/second}$
Number of mesh nodes and elements	3363, 6138 (Experiment 1)2988, 5348 (experiments 2 & 3)

From Cheng, Z., Jalon-Rójas, I., Wang, X. H., & Liu, Y. (2020). Impacts of land reclamation on sediment transport and sedimentary environment in a macro-tidal estuary. *Estuarine, Coastal and Shelf Science, 242*. https://doi.org/10.1016/j.ecss.2020.106861

Fig. 6.6 shows a comparison between the water-level measurements at 1 hour intervals from the Yalu Park Hydrology Station and the tidal elevation from the model at the north boundary point of the study domain during June 2011. The hydrology station is located upstream in the river (64 km from the estuary entrance), 15 km north of the model northern boundary. According to the comparison, modeled surface elevation agreed reasonably well with the measurements. The slight difference between the two time-series for the surface elevation around 500 hours is acceptable, considering the effects of river flow and the spatial separation between the two sites (Fig. 6.6).

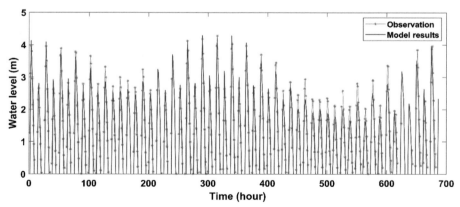

Figure 6.6 Comparison of the measured and modeled water elevation during June 2011. From Cheng, Z., Jalon-Rójas, I., Wang, X. H., & Liu, Y. (2020). Impacts of land reclamation on sediment transport and sedimentary environment in a macro-tidal estuary. *Estuarine, Coastal and Shelf Science, 242*. https://doi.org/10.1016/j.ecss.2020.106861

The model results from Experiment 3 were further validated against the observations from the field trip in August 2009. (a) shows good agreement between the modeled and observed water elevations during spring and neap tides at Stations Y02 and Y03. (b, c) show good agreement in the streamwise current speed and current direction, with mean correlation coefficients between model results and observations of 0.87 and 0.85, respectively. Validation of SSC is shown in (d), indicating the model has captured the magnitude and fluctuations over time of the observed SSC. As the field trip was conducted in August during the wet season, the relatively larger deviations in the SSC comparison at Y02 were probably caused by short-term river discharges. Simulation of the SSC has always been challenging due to the uncertainty in setting the local parameters (Song & Wang, 2013) (e.g., settling velocity and critical bottom stress; Xing et al., 2012). The simulated SSC in this study is already more accurate than in the previous study in the YE (Yu et al., 2014). Although the model did not capture the precise values of SSC, it reproduced its first-order variation and is therefore suitable for studying the changes in sediment transport after land reclamation in at least a qualitative manner (Fig. 6.7).

6.5.2 Particle resuspension and tracking model

Two scenarios (pre- and postreclamation) were designed for TrackMPD to investigate the change in sediment source after the land reclamation in the YRE. Using forward tracking to determine their subsequent movement, particles were released every hour at the bottom layer from five potential source areas around the estuary (dark-blue points in Fig. 6.1). Particles were released from the near-bottom layer over an area of 0.048 km^2 for each release area. A total number of 1512 particles was released under different tidal conditions (7 days from UTC 0000 May 22, 2011 to UTC 0000 May 28, 2011), and all particles tracked for 14 days after release. The hydrodynamic inputs (u, v, w, water elevation, bottom shear stress, and vertical eddy viscosity) for the two scenarios were from FVCOM simulations. Settling velocity and critical shear stress for resuspension were set to be the same as for the FVCOM sediment model. A particle touches the lateral boundary will beach at its final location; a particle sinks at the bottom will be resuspended into the water column when the bottom stress at its sinking location exceeds the critical shear stress for erosion.

6.5.3 Sediment provenance in the YRE under human impacts

As discussed in the previous section, the sediment sources at two sites (K1 and K12) changed significantly and composition of their REE properties became similar after 1975. To investigate this change in sediment source in the area of interest in this study (the East Branch area) after the completion of land reclamation, particle resuspension and tracking simulations were conducted using TrackMPD; 7560 particles in total were released from the same source areas before and after reclamation. Five different release areas, RN, LR, WK, WR, and SW in Fig. 6.1, indicate five potential sediment source which represent fluvial sediments from the river load (RN), local resuspended sediments from the East Branch (LR), sediments transported from the Western Korean

Figure 6.7 Comparison of depth-averaged: (a) Water elevation; (b) streamwise velocity; (c) current direction; and (d) suspended-sediment concentration between the FVCOM model results and field observations in August 2009 at Stations Y02 and Y03 during spring and neap tides.
From Cheng, Z., Jalon-Rójas, I., Wang, X. H., & Liu, Y. (2020). Impacts of land reclamation on sediment transport and sedimentary environment in a macro-tidal estuary. *Estuarine, Coastal and Shelf Science, 242.* https://doi.org/10.1016/j.ecss.2020.106861

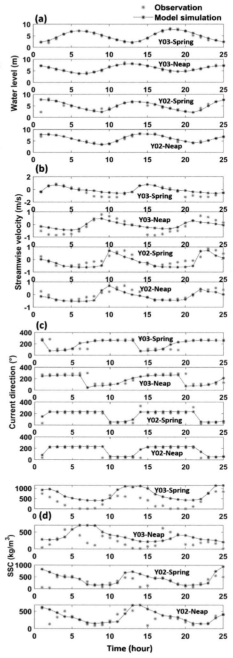

Bay (WK), sediments transported from the West Branch (WR) and sediments from the shallow waters in the NYS (SW). Fig. 6.8 shows the trajectories of these simulated particles (Fig. 6.8).

Fluvial particles released upstream (RN in) could be transported through the Yingmen Channel and West River before reclamation but distributed mainly in the East Branch afterward. The fluvial particles therefore contributed significantly more to the estuarine deposits after reclamation, corresponding to the increased seaward sediment flux (Cheng et al., 2020).

A similar pattern was also found in particles from the middle region of the East Branch (representing local resuspended particles; LR in). Due to the stronger mixing postreclamation, the trajectories of the LR particles were more spread out.

The particles released in shallow water in the NYS (SW) did not show any obvious changes in their trajectories after land reclamation. These particles did reach further north to the inner estuary after reclamation but still had a relatively low probability of being transported to K1 and K12. There was a tendency for these particles to accumulate around Xin Island before reclamation; they were more widely distributed in the shallow water away from the estuary after reclamation. A previous study using indicative minerals for provenance tracking treated this particular shallow-water area as a sediment source (termed "neritic sediments" in their study), together with the fluvial source (Liu et al., 2013).

Prereclamation particles released in WR can be washed away into the NYS, whereas it generally remained around WR postreclamation. Additionally, particles from the fluvial area (RN) reached WR before reclamation but no longer transported to this area afterward. These results suggested the sediment source of WR changed from fluvial materials to local deposits after land reclamation. The reason for this phenomenon is that the West Channel (WR in Fig. 6.8) was transformed over a relatively short time from a main water outlet into an abandoned tidal waterway without a gradual process. After reclamation, the WR area received only a very small amount of river discharge and turned into a gulf-like environment in which deposits were mainly from local resuspended sediments controlled by tides. Before the expansion of the islands in the early 1970s, the bulk of these local sediments were fluvial deposits from upstream in the river. The previous study also found that the deposits in the West Channel were old fluvial materials using indicative minerals as sediment provenance tracer (Liu et al., 2013). This would provide an explanation to why the sediments in the East Branch at K1 and K12 after reclamation had similar REE properties to the surface sediments from the Liaodong muddy coast (WR).

In terms of particles released from Western Korean Bay (WK), the trajectories of these particles covered a larger part of the estuary before reclamation but were restricted to a smaller area afterward.

Several conclusions can be drawn from these particle-tracking results. (1) Before reclamation, site K1 mainly received materials from the Western Korean Bay region; particles from other areas did not reach K1. Sediments at K12 originated from multiple sources, which contained mainly fluvial materials (from RN), partly local resuspended particles from the East Branch (LR) and a small number of neritic sediments (SW). (2) After reclamation, the bulk of the sediments at both K1 and K12 were derived

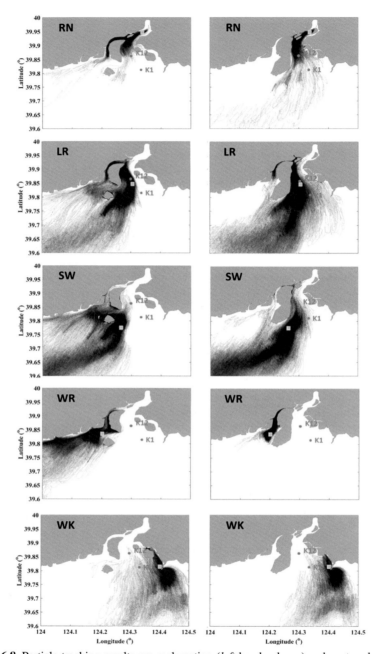

Figure 6.8 Particle-tracking results pre-reclamation (*left*-hand column) and post-reclamation (*right*-hand column). Transparent gray lines mark the trajectories of the released particles; *yellow squares* the source areas (release positions of the particles).

From Cheng, Z., Jalon-Rójas, I., Wang, X. H., & Liu, Y. (2020). Impacts of land reclamation on sediment transport and sedimentary environment in a macro-tidal estuary. *Estuarine, Coastal and Shelf Science, 242.* https://doi.org/10.1016/j.ecss.2020.106861

predominantly from two sources: fluvial materials (from RN); and local resuspended sediments from main-branch deposits (LR).

These conclusions help explain the REE chemistry from core samples: bulk of the deposits at K1 had a distinct source before reclamation, suggesting a single distinct sediment source from the Western Korean Bay. The REE property of sediments at K1 dramatically changed after reclamation, becoming similar to those from Section C (the WR area). At K12 before reclamation, sediments had diverse REE properties, corresponding to the multiple sediment sources implied by the particle-tracking results. Then, similar to K1, the REE properties became closer to those from Section C as well after reclamation.

Essentially, after land reclamation, the main branch (East Branch) of the estuary received more materials from fluvial inputs due to the increased seaward flux caused by reduced flood dominance; the sediments became more homogeneous, with a common composition of deposits contained fluvial and local resuspended sediments were found in this area as a result of stronger mixing; the materials from the Western Korean Bay no longer transported to the inner estuary.

Results from this study demonstrate the possibility of a regime shift in a medium-scale estuary after a large-scale land reclamation. Conclusions from this study also contribute to a better understanding of the link between changes in sediment source and their relationship with suspended-sediment transport. Furthermore, this study gives an approach to future study of a medium-to-long term estuarine sedimentation process by combining numerical simulations and geochemical results of sedimentology.

7. Conclusion and future perspective

The study of sediment provenance, particularly in coastal regions, stands out as a pivotal area of focus. Sediment provenance investigates the origins and movement patterns of sediments, is not just of academic interest. It has profound implications for coastal management, ecological conservation, and our broader comprehension of the dynamic processes that shape our coastlines.

The geochemical method has been a foundational approach in sediment provenance studies. One of its primary advantages is its ability to offer insights into the historical context of sediment deposition. By analyzing the mineralogical and elemental composition of sediments, researchers can trace back to potential source regions, shedding light on past environmental conditions and sedimentary processes. Additionally, geochemical analyses are adept at identifying unique signatures from different sediment sources, making it a valuable tool for distinguishing between multiple potential origins. Numerical modeling offers another robust avenue for exploring sediment provenance in coastal settings. Rooted in mathematical and physical tenets, these models empower researchers to simulate diverse sediment transport scenarios. By fine-tuning model parameters, they can forecast potential sediment movement pathways, enhancing the precision of sediment provenance studies.

The case study in the YRE shows a valuable example of the application of the new method for sediment provenance. Simulations from a newly developed Lagrangian 3-D particle resuspension and tracking model demonstrated that the sediment source in the estuary altered after reclamation as well, sediments near site K1 at the mouth of the estuary were more likely to have come from the Western Korean Bay before the reclamation but consisted of fluvial and local resuspended sediments after that. The source of deposits near site K12, further up the estuary, also changed after the reclamation. The sediments at the two sites and in the East Branch area became similar after land reclamation, indicating the estuarine environment had become more homogenous and well-mixed.

The future of sediment provenance research will undoubtedly benefit from interdisciplinary collaboration. By bridging the expertise of geologists, oceanographers, ecologists, and data scientists, a more holistic understanding of sediment dynamics can be achieved. Furthermore, the integration of machine learning and artificial intelligence (AI) into sediment provenance research holds immense promise. These computational tools can handle vast datasets, discerning patterns and trends that might be challenging to detect manually. As sensor technology continues to advance, we can anticipate more granular, real-time data collection on sediment dynamics. When these data are processed using AI-driven algorithms, the potential for real-time analysis and prediction becomes a tangible reality.

Copyright declaration

References

Adibhusana, M. N., Hendrawan, I., & Ryu, Y. (2023). Numerical study on tide and tidal current along bali strait, Indonesia using finite volume coastal ocean model (FVCOM). *Korean Society of Coastal and Ocean Engineers, 10*.

Amoudry, L. O., & Souza, A. J. (2011). Deterministic coastal morphological and sediment transport modeling: A review and discussion. *Reviews of Geophysics, 49*(2).

Anaya-Gregorio, A., Armstrong-Altrin, J. S., Machain-Castillo, M. L., Montiel-García, P. C., & Ramos-Vázquez, M. A. (2018). Textural and geochemical characteristics of late pleistocene to holocene fine-grained deep-sea sediment cores (GM6 and GM7), recovered from southwestern gulf of Mexico. *Journal of Palaeogeography, 7*(1), 1–19. https://doi.org/10.1186/s42501-018-0005-3

Andò, S. (2020). Gravimetric separation of heavy minerals in sediments and rocks. *Minerals, 10*(3), 273.

Argall, R., Sanders, B.F., & Poon, Y.K. (2004). Random-walk suspended sediment transport and settling model. In Estuarine and Coastal Modeling (2003) (pp. 713-730).

Auer, G., Reuter, M., Hauzenberger, C. A., & Piller, W. E. (2017). The impact of transport processes on rare earth element patterns in marine authigenic and biogenic phosphates. *Geochimica et Cosmochimica Acta, 203*, 140−156. https://doi.org/10.1016/j.gca.2017.01.001

Blumberg, A. F., & Mellor, G. L. (1987). *A description of a three-dimensional coastal ocean circulation model* (pp. 1−16). Wiley. https://doi.org/10.1029/co004p0001

Brito, P., Prego, R., Mil-Homens, M., Caçador, I., & Caetano, M. (2018). Sources and distribution of yttrium and rare earth elements in surface sediments from Tagus estuary, Portugal. *Science of the Total Environment, 621*, 317−325. https://doi.org/10.1016/j.scitotenv.2017.11.245

Carniel, S., Sclavo, M., & Archetti, R. (2011). Towards validating a last generation, integrated wave-current-sediment numerical model in coastal regions using video measurements. *Oceanologia, 40*.

Chen, C., Beardsley, R., Cowles, G., Qi, J., Lai, Z., Gao, G., Stuebe, D., Xu, Q., Xue, P., Ge, J., Hu, S., & Ji, R. (2013). An unstructured grid, finite-volume Coastal Ocean Model FVCOM – user manual. In *Technical Reports, SMAST/UMASSD-13-0701*. New Bedford. C: Sch. Mar. Sci. Technol., Univ. Mass. Dartmouth. https://doi.org/10.1017/CBO9781107415324.004, 416 pp.

Chen, C., Liu, H., & Beardsley, R. C. (2003). An unstructured grid, finite-volume, three-dimensional, primitive equations ocean model: Application to coastal ocean and estuaries. *Journal of Atmospheric and Oceanic Technology, 20*, 159−186.

Chen, X., Li, T., Zhang, X., & Li, R. (2013). A Holocene Yalu River-derived fine-grained deposit in the southeast coastal area of the Liaodong Peninsula. *Chinese Journal of Oceanology and Limnology, 31*, 636−647. https://doi.org/10.1007/s00343-013-2087-1

Cheng, Z., Jalon-Rójas, I., Wang, X. H., & Liu, Y. (2020). Impacts of land reclamation on sediment transport and sedimentary environment in a macro-tidal estuary. *Estuarine, Coastal and Shelf Science, 242*. https://doi.org/10.1016/j.ecss.2020.106861

Cheng, Z., Wang, X., Paull, D., & Gao, J. (2016). Application of the geostationary ocean color imager to mapping the diurnal and seasonal variability of surface suspended matter in a macro-tidal estuary. *Remote Sensing, 8*(3). https://doi.org/10.3390/rs8030244

Gao, G. D., Wang, X. H., & Bao, X. W. (2014). Land reclamation and its impact on tidal dynamics in Jiaozhou Bay, Qingdao, China. *Estuarine, Coastal and Shelf Science, 151*, 285−294. https://doi.org/10.1016/j.ecss.2014.07.017

Gao, J., Gao, S., Cheng, Y., Dong, L., & Zhang, J. (2004). Formation of turbidity maxima in the Yalu River Estuary, China. *Journal of Coastal Research, 43*, 134−146.

Gao, J.h., Li, J., Wang, H., Bai, F.l., Cheng, Y., & Wang, Y.p. (2012). Rapid changes of sediment dynamic processes in Yalu River Estuary under anthropogenic impacts. *International Journal of Sediment Research, 27*(1), 37−49. https://doi.org/10.1016/S1001-6279(12)60014-6

Gao, J. H., Shi, Y., Sheng, H., Kettner, A. J., Yang, Y., Jia, J. J., Wang, Y. P., Li, J., Chen, Y., Zou, X., & Gao, S. (2019). Rapid response of the Changjiang (Yangtze)River and East China Sea source-to-sink conveying system to human induced catchment perturbations. *Marine Geology, 414*, 1−17. https://doi.org/10.1016/j.margeo.2019.05.003

Ge, J., Chen, C., Qi, J., Ding, P., & Beardsley, R. C. (2012). A dike-groyne algorithm in a terrain-following coordinate ocean model (FVCOM): Development, validation and application. *Ocean Modelling, 47*, 26−40. https://doi.org/10.1016/j.ocemod.2012.01.006

Geyer, W. R., & Signell, R. P. (1992). A reassessment of the role of tidal dispersion in estuaries and bays. *Estuaries, 15*, 97−108.

Gholami, H., Jafari TakhtiNajad, E., Collins, A. L., & Fathabadi, A. (2019). Monte Carlo fingerprinting of the terrestrial sources of different particle size fractions of coastal sediment deposits using geochemical tracers: Some lessons for the user community. *Environmental Science and Pollution Research, 26*(13), 13560−13579. https://doi.org/10.1007/s11356-019-04857-0

Gray, H. J., Jain, M., Sawakuchi, A. O., Mahan, S. A., & Tucker, G. E. (2019). Luminescence as a sediment tracer and provenance tool. *Reviews of Geophysics, 57*(3), 987−1017. https://doi.org/10.1029/2019RG000646

Guo, W., Wang, X. H., Ding, P., Ge, J., & Song, D. (2017). A system shift in tidal choking due to the construction of Yangshan Harbour, Shanghai, China. *Estuarine, Coastal and Shelf Science, 206*, 49−60. https://doi.org/10.1016/j.ecss.2017.03.017

Hamrick, J. H. (2012). *Modeling the physical and biochemical influence of ocean thermal energy conversion plant discharges into their adjacent waters.*

Hannigan, R., Dorval, E., & Jones, C. (2010). The rare earth element chemistry of estuarine surface sediments in the Chesapeake Bay. *Chemical Geology, 272*, 20−30. https://doi.org/10.1016/j.chemgeo.2010.01.009

Hathorne, E. C., Stichel, T., Brück, B., & Frank, M. (2014). Rare earth element distribution in the Atlantic sector of the Southern Ocean: The balance between particle scavenging and vertical supply. *Marine Chemistry, 177*, 157−171. https://doi.org/10.1016/j.marchem.2015.03.011

Huang, H., Chen, C., Cowles, G. W., Winant, C. D., Beardsley, R. C., Hedstrom, K. S., & Haidvogel, D. B. (2008). FVCOM validation experiments: Comparisons with ROMS for three idealized barotropic test problems. *Journal of Geophysical Research: Oceans, 113*(7). https://doi.org/10.1029/2007JC004557

Jalón-Rojas, I., Wang, X. H., & Fredj, E. (2019). A 3D numerical model to track marine plastic debris (TrackMPD): Sensitivity of microplastic trajectories and fates to particle dynamical properties and physical processes. *Marine Pollution Bulletin, 141*, 256−272. https://doi.org/10.1016/j.marpolbul.2019.02.052

Jayathilaka, R. M. R. M., & Fernando, M. C. S. (2019). Numerical modelling of the spatial variation of sediment transport using wave climate schematization method - a case study of west coast of Sri Lanka. *Journal of the National Science Foundation of Sri Lanka, 47*(4). https://doi.org/10.4038/jnsfsr.v47i4.9679

Jeon, J., & Choi, W. (2020). Prediction accuracy of reservoir break flood simulation model using finite volume method and uav. *International Journal of Agricultural and Biological Engineering, 13*(6), 7−15. https://doi.org/10.25165/j.ijabe.20201306.4909

Ji, Z.-G. (2006). *Hydrodynamics and water quality - modelling rivers, lakes and estuaries.*

Jung, H. S., Lim, D., Choi, J. Y., Yoo, H. S., Rho, K. C., & Lee, H. B. (2012). Rare earth element compositions of core sediments from the shelf of the South Sea, Korea: Their controls and origins. *Continental Shelf Research, 48*, 75−86. https://doi.org/10.1016/j.csr.2012.08.008

Karunarathna, H., Horrillo-Caraballo, J., Burningham, H., Pan, S., & Reeve, D. E. (2016). Two-dimensional reduced-physics model to describe historic morphodynamic behaviour of an estuary inlet. *Marine Geology, 382*, 200−209.

Lackey, T., & Macdonald, N. (2007). The particle tracking model: Description and processes. *Proceedings of XVIII World Dredging Congress*, 551−566.

Lane, A. (2005). Development of a Lagrangian sediment model to reproduce the bathymetric evolution of the Mersey Estuary. *Ocean Dynamics, 55*, 541−548. https://doi.org/10.1007/s10236-005-0011-8

Lesser, G. R., Roelvink, J. A., van Kester, J. A. T. M., & Stelling, G. S. (2004). Development and validation of a three-dimensional morphological model. *Coastal Engineering,* *51*(8–9), 883–915. https://doi.org/10.1016/j.coastaleng.2004.07.014

Li, L., Wang, X. H., Andutta, F., & Williams, D. (2014). Effects of mangroves and tidal flats on suspended-sediment dynamics: Observational and numerical study of Darwin Harbour, Australia. *Journal of Geophysics Research Ocean, 119,* 5854–5873. https://doi.org/ 10.1002/2014JC009987

Li, Y., Li, A. C., Huang, P., Xu, F. J., & Zheng, X. F. (2014). Clay minerals in surface sediment of the north Yellow Sea and their implication to provenance and transportation. *Continental Shelf Research, 90,* 33–40. https://doi.org/10.1016/j.csr.2014.01.020

Li, Y., & Yao, J. (2015). Estimation of transport trajectory and residence time in large river–lake systems: Application to Poyang Lake (China) using a combined model approach. *Water,* *7*(10), 5203–5223.

Lim, D., Jung, H. S., & Choi, J. Y. (2014). REE partitioning in riverine sediments around the Yellow Sea and its importance in shelf sediment provenance. *Marine Geology, 357,* 12–24.

Liu, Y., Cheng, Y., Li, H., Liu, J., Zhang, C., Zhang, L., Zheng, C., & Gao, J. (2013). Provenance tracing of indicative minerals in sediments of the Yalu River Estuary and its adjacent shallow seas. *Journal of Coastal Research, 290*(5), 1227–1235. https://doi.org/10.2112/ JCOASTRES-D-12-00269.1

Liu, Y., Cheng, Y., Liu, J., Zhang, L., Zhang, C., & Zheng, C. (2015). Provenance discrimination of surface sediments using rare earth elements in the Yalu River Estuary, China. *Environmental Earth Sciences, 74*(4), 3507–3517. https://doi.org/10.1007/s12665-015-4391-x

Liubartseva, S., Coppini, G., Lecci, R., & Clementi, E. (2018). Tracking plastics in the mediterranean: 2D lagrangian model. *Marine Pollution Bulletin, 129*(1), 151–162. https:// doi.org/10.1016/j.marpolbul.2018.02.019

Lizaga, I., Latorre, B., Gaspar, L., & Navas, A. (2020). FingerPro: An R package for tracking the provenance of sediment. *Water Resources Management, 34*(12), 3879–3894. https:// doi.org/10.1007/s11269-020-02650-0

MacDonald, N. J., Davies, M. H., Zundel, A. K., Howlett, J. D., Demirbilek, Z., Gailani, J. Z., & Smith, J. (2006). *PTM: Particle Tracking Model Report 1: Model Theory, Implementation, and Example Applications.* ERDC/CHL TR-06-20. Vicksburg, MS: US Army Engineer Research and Development Center. http://acwc.sdp.sirsi.net/client/search/asset/1000777.

Martins, V., Figueira, R. C. L., França, E. J., Ferreira, P. A. D. L., Martins, P., Santos, J. F., Dias, J. A., Laut, L. L. M., Monge Soares, A. M., Silva, E. F. D., & Rocha, F. (2012). Sedimentary processes on the NW iberian continental shelf since the little ice age. *Estuarine, Coastal and Shelf Science, 102–103,* 48–59. https://doi.org/10.1016/j.ecss. 2012.03.004

Mellor, G. L. (1996). *Users guide for a three-dimensional, primitive equation, numerical ocean model.*

Navas, A., Lizaga, I., Santillán, N., Gaspar, L., Latorre, B., & Dercon, G. (2022). Targeting the source of fine sediment and associated geochemical elements by using novel fingerprinting methods in proglacial tropical highlands (Cordillera Blanca, Perú). *Hydrological Processes, 36*(8). https://doi.org/10.1002/hyp.14662

Ngueutchoua, G., Ngantchu, L. D., Youbi, M., Ngos III, S., Beyala, V. K. K., Yifomju, K. P., & Tchamgoué, J. C. (2017). Geochemistry of cretaceous mudrocks and sandstones from douala sub-basin, kumba area, south west Cameroon: Constraints on provenance, source rock weathering, paleo-oxidation conditions and tectonic environment. *International Journal of Geosciences, 08*(04), 393–424. https://doi.org/10.4236/ijg.2017.84021

O'Sullivan, G. J., Chew, D. M., Morton, A. C., Mark, C., & Henrichs, I. A. (2018). An integrated apatite geochronology and geochemistry tool for sedimentary provenance analysis. *Geochemistry, Geophysics, Geosystems, 19*(4), 1309–1326. https://doi.org/10.1002/2017 gc007343

Orani, A. M., Vassileva, E., Wysocka, I., Angelidis, M., Rozmaric, M., & Louw, D. (2018). Baseline study on trace and rare earth elements in marine sediments collected along the Namibian coast. *Marine Pollution Bulletin, 131*, 386–395. https://doi.org/10.1016/j.marpolbul.2018.04.021

Papatheodorou, G., Demopoulou, G., & Lambrakis, N. (2006). A long-term study of temporal hydrochemical data in a shallow lake using multivariate statistical techniques. *Ecological Modelling, 193*(3–4), 759–776.

Paterson, G. A., & Heslop, D. (2015). New methods for unmixing sediment grain size data. *Geochemistry, Geophysics, Geosystems, 16*(12), 4494–4506. https://doi.org/10.1002/2015GC006070

Prajith, A., Rao, V. P., & Chakraborty, P. (2016). Distribution, provenance and early diagenesis of major and trace metals in sediment cores from the Mandovi estuary, western India. *Estuarine, Coastal and Shelf Science, 170*, 173–185. https://doi.org/10.1016/j.ecss.2016.01.014

Rahman, S. E., Yusuf, M., & Nasution, Y. N. (2022). Numerical study of hydrodynamics in the waters of Balikpapan Bay using finite volume method. *Geosains Kutai Basin, 18*.

Rasheed, S., Warder, S., Plancherel, Y., & Piggott, M. (2020). Response of tidal flow regime and sediment transport in North Male' Atoll, Maldives to coastal modification and sea level rise. *Ocean Science, 17*.

Shi, Y., Gao, J. H., Sheng, H., Du, J., Jia, J. J., Wang, Y. P., Li, J., Bai, F. L., & Chen, Y. N. (2018). Cross-front sediment transport induced by quick oscillation of the Yellow Sea warm current: Evidence from the sedimentary record. *Geophysical Research Letters*, 1–9. https://doi.org/10.1029/2018GL080751

Shynu, R., Rao, V. P., Kessarkar, P. M., & Rao, T. G. (2011). Rare earth elements in suspended and bottom sediments of the Mandovi Estuary, central west coast of India: Influence of mining. *Estuarine, Coastal and Shelf Science, 94*(4), 355–368. https://doi.org/10.1016/j.ecss.2011.07.013

Shynu, R., Rao, V. P., Parthiban, G., Balakrishnan, S., Narvekar, T., & Kessarkar, P. M. (2013). REE in suspended particulate matter and sediment of the Zuari estuary and adjacent shelf, western India: Influence of mining and estuarine turbidity. *Marine Geology, 346*, 326–342. https://doi.org/10.1016/j.margeo.2013.10.004

Song, D., & Wang, X. H. (2013). Suspended sediment transport in the deepwater navigation channel, Yangtze River Estuary, China, in the dry season 2009: 2. Numerical simulations. *Journal of Geophysical Research: Oceans, 118*(10), 5568–5590. https://doi.org/10.1002/jgrc.20411

Song, H., Shin, W. J., Ryu, J. S., Shin, H. S., Chung, H., & Lee, K. S. (2017). Anthropogenic rare earth elements and their spatial distributions in the Han River, South Korea. *Chemosphere, 172*, 155–165. https://doi.org/10.1016/j.chemosphere.2016.12.135

Song, Y. H., & Choi, M. S. (2009). REE geochemistry of fine-grained sediments from major rivers around the Yellow Sea. *Chemical Geology, 266*, 337–351. https://doi.org/10.1016/j.chemgeo.2009.06.019

Figure 7.1 RBR pressure-type tidal level meter (https://rbr.cn/products/standard-loggers#tide-and-wave).

accuracy is high with low power consumption. It is currently the most popular tidal level meter in the world. However, the buoy-type tidal level meter is expensive to apply and must rely on special logging wells to work. Therefore, it is more flexible to use pressure-type tidal level meter (Fig. 7.1), such as pressure-type tidal level meter and acoustic tidal level meter. The following will introduce the tide station.

The pattern of tidal changes is closely related to the apparent motion of the earth and the moon. However, the patterns of tidal changes also vary due to different factors such as topography and geomorphology. Therefore, it is essential to select the location of the tidal station before measuring tidal levels. The site conditions for the tide station are as follows. (1) Data should be representative. One should choose a representative area with significant tidal fluctuations in a coastal region to establish a tide gauge station. Tide gauge stations generally face an open sea free from dense islands and shallow obstructions in order to avoid the impact of runoff on tide measurements. Even if the tide station is located in a bay, it is still necessary to ensure that the currents inside and outside the bay is unobstructed. (2) The geological characteristics near the tidal station are relatively stable, with no obvious seismic activity, scouring, or siltation. (3) Ensure accurate observation, facilitate convenient living conditions, and prioritize personal safety. The existing coastal structures were generally taken advantage of tidal observation points.

2.2.2 Observation interval

Tidal level is typically observed once every half an hour. Tidal gauge stations often record the tidal data every 10 min for half an hour before and after high and low tides. In some extreme weather events, it often continues to observe at 10-minute intervals until the water level returns to normal in case of abnormal water level fluctuations (National Standards of the People's Republic of China, 2007a).

2.2.3 Establishing a benchmark level

Once the tidal station is set up, the sea surface height can be easily determined by the water gauge. This height is calculated from the zero point of the water gauge. After establishing a benchmark level, it is necessary to connect it to the national level network in order to accurately determine the elevation of the zero point of the water gauge in the national network and maintain it over a long period of time in China. The benchmark should be located close to the station, in a sturdy and stable position that is not susceptible to being submerged by tides.

2.2.4   Observation of tide gauge wells

To minimize the impact of sea surface fluctuations on tidal level observations, the common practice is to install tide gauges in tide gauge wells. In the late 1980s, the advancement of electronic technology led to notable progress in the development of automatic tide gauge stations. The shore-based tide gauge wells are constructed on land, and the wells and instrument rooms are connected to the sea surface through a water pipeline (Wang et al., 2023).

There are many types of self-recording water level gauges, which can be classified into float-type water level gauges, pressure type water level gauges, and acoustic type water level gauges. Currently, most countries abroad primarily use float-type water level gauges and pressure type water level gauges, whereas in China, the float-type water level gauges are also commonly utilized (Fig. 7.2).

2.2.5   Remote sensing observations

People have established numerous tidal stations along the coast in order to master the tidal characteristics for centuries. However, little data were obtained in deep sea due to very few tidal stations in deep sea. Satellite altimetry can provide comprehensive spatial data of deep sea, which also provide the global tidal data to validate global tidal models (Hu, 2009). The Centre National D'Etudes Spatiales (CNES) and the National Aeronautics and Space Administration (NASA) jointly launched two satellites called TOPEX/POSEIDO (Fig. 7.2) and JASON into space to measure sea surface height with two state-of-the-art radar altimeter systems. These data are applied to study the global ocean circulation and its associated patterns. The ocean sea level refers to the stationary and uniform surface of the ocean that is only affected by gravity and the Earth's rotational force. Due to the uneven distribution of land and water on earth obviously, the geoid is not an ideal ellipsoid. If the ocean sea level is established, the deviation between the measured sea surface and the ocean sea level can be used to calculate the effects of

Figure 7.2   A tide gauge station in China (http://scs.mnr.gov.cn/scsbpad/fjdt/202201/e3725c98d7a841cab028e4666c6d5fd2.shtml).

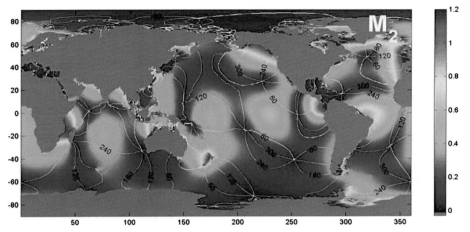

Figure 7.3 TOPEX global ocean tidal distribution (http://g.hyyb.org/archive/Tide/TPXO/ TPXO_WEB/global.html).

ocean currents, storm surges, and other factors. Therefore, precise observation of the ocean sea level has great significance. By analyzing data from altimeters and integrating it with observations from traditional tide gauges, accurate ocean sea level that encompass the entire ocean can be derived (±20 cm). The height of the ocean sea level is measured relative to the ideal reference ellipsoid of the earth (Fig. 7.3).

3.   Current and wave observation method

3.1   Introduction

The tidal current is the cyclical movement of sea water accompanied by the rise and fall of tides, which is caused by the tidal force of celestial body from the moon and the sun. Its change period mainly includes semidiurnal tidal and diurnal tidal. The movement of the sea comprises various components, including turbulence, fluctuations, periodic currents, and more stable currents. These components differ in scale, velocity, and period and are influenced by factors such as wind, season, and long-term changes. Generally, the intensity of these movements tends to decrease from the surface to deeper sea. Tides and currents are both influenced by the gravitational forces of the moon and the sun. While tides refer to the vertical rise and fall of the sea level, currents, on the other hand, refer to the horizontal movement of water. The currents consist of three types: diurnal currents, semi-diurnal currents, and mixed currents.

3.2   Current observation methods

When conducting ocean currents observations, it must last continuously for one or more days at certain time intervals. The observation results obtained are a combination of residual current and tidal current. These two parts can be separated through calculation.

There are generally two methods about ocean current observation, fixed-point observation, and navigational observation (Zhang & Zhao, 2013; Qiao et al., 2022). Fixed-point observation relies on anchored ships, buoys, offshore platforms, or special fixed frames as carrying tools to carry ocean current meters. Navigation observation is to carry out ocean current observation, while the ship is sailing.

3.3   The current meter

A current meter is an instrument used to measure the velocity and direction of the currents. There are several types of commonly used current meters as follows (National Standards of the People's Republic of China, 2007a, 2007b, 2007c; Shi, 1983).

3.3.1   The mechanical current meter

These meters use rotating cups or propellers to measure water velocity (Fig. 7.2). The basic principle of the mechanical propeller current meter is to determine the rotation speed based on the number of revolutions of the propeller blades driven by the water flow and use a magnetic compass to determine the flow direction. Mechanical current meters are widely used and vary in design depending on the desired depth range and accuracy. According to the characteristics of this type of current meters, it can be roughly divided into Ekman current meter, printing current meter, photographic current meter, etc (Fig. 7.4).

3.3.2   The electromagnetic current meter

In mechanical current meters, the propellers are indispensable. However, it can be affected by floating debris and attached organisms, leading to inaccurate observation

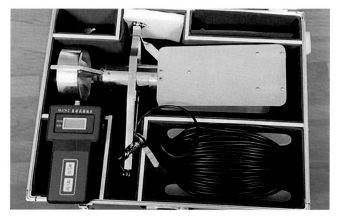

Figure 7.4 A mechanical current meter (https://www.hi1718.com/product/201651894657881. html).

results. This led to the invention of the electromagnetic current meter (Fig. 7.2). The electromagnetic current meter applies Faraday's electromagnetic induction principle to measure ocean currents by measuring the induced electromotive force generated when seawater flows through a magnetic field. According to different sources of magnetic fields, they can be divided into geomagnetic field electromagnetic current meters and artificial magnetic field electromagnetic current meters. At present, most of the world's electromagnetic current meters are produced in Japan(Fig. 7.5).

3.3.3 The Doppler current meter

Doppler current meters use the Doppler shift principle to measure the velocity. A sound source emits a continuous wave that is reflected back by particles in the water. The velocity can be determined by measuring the frequency shift between the emitted and received waves. The frequency of a train whistles gradually decreases as it moves away, while the frequency of an approaching train whistle increases and becomes sharper and more piercing in daily life. As a result, small particles and bubbles, which move with the water flow in front of the ultrasound probe and can emit ultrasound waves, will also reflect the ultrasound waves, causing a change. This change is directly proportional to the velocity at which suspended particles move in the water. Consequently, the velocity can be determined by measuring the Doppler frequency shift. The Doppler current meters do not have any moving parts. Therefore, there is not friction and hysteresis for it. It does not introduce any disturbances to the current field and does not exhibit mechanical inertia or wear. As a result, it can accurately reflect the true state of the current. Underwater instruments themselves are ultrasonic transmitters, so energy emission, battery life, and sound attenuation are limitation factors.

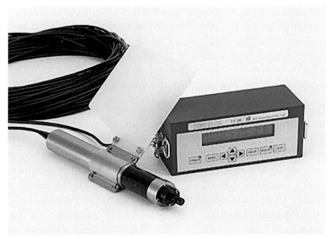

Figure 7.5 Handheld 2-D EM current meter (https://www.jfe-advantech.co.jp/eng/products/ocean-ryusoku.html).

3.3.4   The acoustic Doppler current profiler

The ADCP has the same principle as Doppler current meters (Fig. 7.2). The ADCP can measure currents at multiple depths simultaneously and are often used for profiling currents over a wide range of depths (Jiang, 2012). The difference is that there is usually four transducers for emitting ultrasound waves (the Doppler current meter usually has two or three transducers), which can measure the horizontal and vertical velocities. After combining these vectors, it determines the true velocity vector at a specific point. It can measure the three-dimensional components and absolute current direction in several layers of a cross-section at once (Li, 2022, pp. 31–45).

ADCP has the characteristics of time saving, high resolution, high precision, massive and complete information, and low energy consumption, especially suitable for the complex current condition. It can carry out three-dimensional measurement of current velocity and direction. It can automatically eliminate the influence of environmental factors and automatically eliminate poor quality data. The profile concentration of suspended sediment can be obtained by the ADCP, which provides a reliable data source for calculating sediment transport. The ADCP measurement is a dynamic current measurement method. Compared with traditional current meters, the ADCP has its superiority in the data acquisition on complex natural environment, wide section, and large flow rate. However, the disadvantage of ADCP is that there is a blind area around the transducers, which usually is depended on its resolution and emitting frequency.

There are mainly three fixed types for ADCP, river or coastal bank fixed type, ship bottom type and seabed type, which refer to lateral, downward, and upward installation for transducers. The fixed ADCP installed on the coastal bank, river bank, wall, and pier is used to measure the cross-sectional flux between. The bottom-mounted ADCP installed at the bottom of the ship or the other platforms is often used to measure the current velocity and direction along the routes or of fixed points. Modern marine scientific research ships usually equipped with ADCP at the bottom of ships. The seabed-mounted ADCP moored on the sea bed is often used to measure the current velocity and direction for a long time, even for several years within a reliable external power supply (Fig. 7.2). The data quality obtained by the seabed bottom-mounted ADCP is the best because there is rarely disturbance (Figs. 7.6 and 7.7).

3.3.5   Remote sensing current measurement

Ocean currents are primarily propelled by various factors such as wind, tidal forces, and the uneven distribution of ocean density. The motion of fluid on earth, due to its rotation, causes the surface to tilt in relation to the horizontal plane. The magnitude of this tilt is directly proportional to the velocity of the fluid flow. However, the oceanic upwelling and downwelling currents, caused by topographic effects and wind stress, are usually very slow, and they cannot be sensed by space remote sensing. However, space remote sensing can identify these phenomena and determine their locations.

The primary instrument used for measuring ocean currents is a radar altimeter, which is an advanced active microwave radar system capable of accurately

Figure 7.6 Sentinel V ADCP (https://www.teledynemarine.com/brands/rdi/sentinel-v-adcp).

Figure 7.7 A seabed-base with an ADCP.

determining the undulations of the sea surface. The measured distance of the object is approximately 800 km, with a measurement error of 2 cm and a precision of 2.5×10^{-8}. The geostrophic current velocity can be determined by applying the geostrophic balance equation. The precision required for observing ocean currents is significantly higher than ± 10 cm/second, while the positional error of these currents is typically on the order of several kilometers.

4.   Wave observation method

4.1   Introduction

Storms and swells contain tremendous amounts of energy, which can cause ships to lose stability, reduce speed, and deviate from their intended course. They can even result in serious accidents, taking a great risk to navigation, fishing, and offshore operations. The impact of storms and surges can cause significant damage to coastal protection, port facilities, and breakwaters. Waves and surf have the ability to transport sediment, which can lead to the silting of harbors, shallowing of channels, and impact the movement of ships in and out of ports (Liu, 2005).

The main objects of wave observation are wind-generated waves and swells. There are various types of waves. Their time scales range from fractions of seconds to several days, months, or even years in long-period fluctuations. Wavelengths can range from a few centimeters to several thousand kilometers. The waves which are generated by local winds and continue to be affected by the wind at the time of observation are called wind waves. The development of wind waves depends on the wind speed, wind fetch, and duration. Wind waves are characterized by a turbulent, intricate appearance, whitecaps, and foam. In addition, the wave direction is usually in line with the wind. The waves that continue to propagate in a still or nearly calm water after the wind has left their influencing area, maintained by inertia, are referred to as swells. Their shapes are relatively regular, with smooth undulations. The period of these waves is greater than that of the original wind waves. It gradually increases with increasing propagation distance. In addition, surges can also be formed in wind-driven waters where the original wind waves are in a state of decay due to a significant reduction in wind power. Mixed waves often occur when waves from different sources combine.

4.2   Selection of observation sites

Wave observation can be carried out either at shore-based stations or on the open sea or aboard ships. Wave observations are conducted at shore-based stations in order to collect representative wave data in the coastal area. Therefore, observation stations should face the open sea to avoid the influence of obstacles such as islands and sandbars, and a relatively flat sea floor is better. The water depth of the buoy should be no less than half of the typical wavelength of waves in that area. Wave buoys may be dragged by strong currents and firmly pressed against the sea surface, which may affect the accuracy. It is recommended to use self-recording instruments instead of relying on visual estimations.

4.3   Observation methods and contents

4.3.1   Methods

Wave observations involve both visual and instrumental methods. Visual observations refer to directly observing and recording wave characteristics by manual identification, such as wave height and shape. At present, the instrument can measure wave height, wave direction, and wave period. Other parameters such as wavelength and shape are still estimated through manual visual observation. The unit of wave height is meters (m), and the unit of period is seconds (s). The observed data are taken to one decimal place (National Standards of the People's Republic of China, 2007a, 2007b, 2007c; Zuo, 2008).

The wave observation instrument can be divided into three methods according to where the instrument is placed, such as underwater, water surface, on-water, acoustic, and space measurement (Long, 2005). Underwater measurement includes pressure measurement, acoustic measurement, and optical measurement. Water surface measurement includes buoy surveys and shipboard wave measurement systems. On-water measurement includes the gas displacement method, aerial photography technology, laser technology, and radar technology. Acoustic measurement in the air— water interface is a method that involves transmitting ultrasonic waves from above the water surface and receiving the waves after they have reflected off the water surface. By calculating the time taken for the wave to travel, and if the speed of sound is known, the distance can be determined. Space measurement technology unlike traditional measurement methods, space measurement techniques primarily refer to using satellite microwave remote sensing technology to measure waves. Currently, there are three satellite microwave remote sensing instruments capable of observing sea surface wind and wave information. Among these instruments, the satellite altimeter is capable of measuring the significant wave height. A divergence meter can measure the sea surface wind field. The information of waves can also be obtained by certain inversion algorithms. Synthetic aperture radar (SAR) can measure the significant wave height and direction spectrum to determine the propagation direction of ocean waves (Sun, 2005, pp. 45–51; Zhou et al., 2016).

4.3.2 Contents

The main content of wave observation includes the spatial and temporal distribution of wind waves and swell waves, as well as their appearance. The observation project aims to monitor different elements, including wave types, wave directions, wave periods, and wave heights. By analyzing these data, we can determine the wavelength, wave velocity, as well as wave heights, and other parameters.

During low wind speed, the waveforms exhibit a high degree of irregularity, with a steeper profile on the leeward side and a gentle slope on the windward side. When affected by wind of 4–5 m/second, the wave crests flip over and break, creating whitecaps. The wave direction generally aligns with the average wind direction, sometimes deviating by around 20°.

Usually wave shape as wind waves is denoted as F, and wave shape as swell is denoted as U. When wind waves and swell waves coexist and retain their original appearance, waveforms can be classified into three types. When the wind waves and swell waves have similar heights, it is denoted as FU. When the height of wind waves exceeds the height of swell waves, it is denoted as F/U. When the height of the wind waves is smaller than the height of the swells, it is denoted as U/F.

The wave direction is divided into 16 directions. The observer stands at a higher position on the boat and uses a compass to align the sighting line parallel to the wave crests that are farther away to observe the direction. After rotating the compass 90 degrees, the observer aligns it toward the direction of the waves and reads the measurement on the dial, which represents the wave direction. When using a magnetic compass to measure wave direction, it is necessary to correct for magnetic deviation and then convert the readings to azimuth.

4.3.3   Instruments

There are many various wave-measuring instruments. The pressure wave meter, gravity wave meter, and ultrasonic wave meter are widely used for on-site wave observation. Pressure wave meters are deployed directly on the seafloor or in the form of submerged beacons, which tend to observe short wave periods. The pressure wave meter indirectly calculates the fluctuation of the sea surface by recording the pressure change of seawater. Gravity wave meters are widely used on buoys, which are recognized for their capability to measure acceleration. The existing accelerometer-based wave measurement methods for buoys can be classified into two categories. One category is based on the accelerometer measurements along three axes (x, y, and z), which allows for the calculation of wave parameters such as wave height, period, and direction. Another type of sensor is based on gravity. However, these instruments are less efficient in strong currents, and their safety level is compromised. The ultrasonic wave meters are commonly situated on the seafloor and emit ultrasonic waves to transmit signals to the surface. It measures the time for waves to reflect to calculate the wave height. However, this type of instrument has a high demand for battery and can result in significant errors when there is more foam on the surface.

The acoustic wave and current (AWAC) measures the profile current by acoustic Doppler method. It uses acoustic surface tracking (AST) technology to measure the wave height and wave period (Fig. 7.2). The AWAC is mainly used for long-term wave and current observation in shallow water. It is usually moored on the sea bed to avoid damage from ships and extreme weather (Fig. 7.2). There are three working frequencies to adapt water depth ranging from 5 to 100 m measuring the period of waves from 1 to 100 s (Figs. 7.8 and 7.9).

Buoy wave measurement uses buoys to measure surface waves (Tang & Kang, 2014). The wave buoy usually contains wave sensors, a communication system, a power supply unit, a GPS positioning system, and a data acquisition system. The principle of the wave buoy is mainly to collect surface wave signals when the buoy floats on the sea surface and then convert the signal obtained by the buoy into the response value after processing, to determine the parameters of the sea surface wave, including wave height, wave speed, wave direction, effective wave energy, and so on. There are main two types wave buoys, fixed-point wave buoys (Fig. 7.10) and drifting wave buoys (Fig. 7.11).

The fixed-point wave buoy is fixed at the bottom of the water through an anchor chain. The anchor system is a composite anchor system. The anchor chain of the fixed iron anchor is not directly connected to the wave buoy, but it is connected to the anchor chain through the buoy. The buoy and the wave buoy are connected by a torque-free

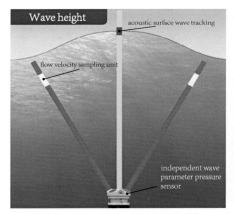

The AWAC employs a unique acoustic surface wave tracking (AST) technology to measure wave height and period. Its vertical transmitting transducer emits a brief acoustic pulse towards the water surface. The time elapsed from the emission of the pulse until its reflection back from the surface generates a time series of water surface elevations.

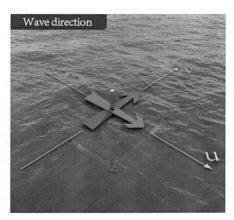

Wave direction calculation with the AWAC combines AST data with the velocity motion trajectory array close to the water surface, employing the MLMST method to process the four-point array data, thereby generating an accurate wave directional spectrum. An AWAC installed on a deep-water moored buoy can utilize the patented SUV processing method to achieve similar results.

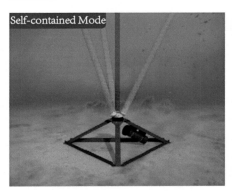

The self-contained mode is powered by an external battery pack. Data is stored in the memory of the device, which can be downloaded to a computer after the instrument is retrieved. The maximum deployment time is determined by the capacity of the memory and the number of batteries used.

The AWAC can transmit raw or processed data in real-time to shore via a cable or a data modem. The data can be visualized using the SeaState software and can be published live through customized web services.

Figure 7.8 AWAC operation schematic diagram (http://www.haiyanec.com/hlgcsb1/345).

wire rope, which can eliminate the effect of currents on the wave buoy and make the measurement of wave direction more accurate.

The floating wave buoy is connected to the oceanic survey ship through a neutral cable. It works during the voyage.

4.3.4   Satellite remote sensing

Synthetic aperture radar is an imaging radar that utilizes coherent processing to create high-resolution images. It is one of the remote sensing technologies. It is able to obtain physical characteristics of ground targets by calculating the echo properties of the targets.

Figure 7.9 A seabed-base with AWAC.

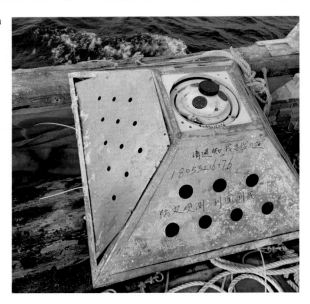

Figure 7.10 The fixed-point wave buoy (http://www.ruihaichina. com/datapage/10.html).

Figure 7.11 The drifting wave buoy (http://www.ruihaichina.com/datapage/9.html).

In order to eliminate random errors, the radar altimeter emits 1000 pulses per second downward, and the average value of the echo signals is used as the basis for calculating and inverting oceanic physical quantities. The time difference of receiving signal can be used to calculate the distance between the satellite and the sea level. When the surface of the sea is calm, the echo signal of the pulse continuously strengthens within the duration of the pulse and reaches its maximum value. The duration of the echo pulse signal's maximum amplitude is longer compared to calm sea conditions in rough sea conditions. Therefore, the wave height of the sea surface can be inferred by analyzing the slope of the pulse echo reflected from the sea surface.

5. Temperature and salinity observation

5.1 Seawater temperature introduction

Seawater temperature is a physical quantity that represents the thermal condition of seawater. The unit of measurement is generally Celsius (°C) in oceanography. The magnitude of seawater temperature primarily depends on the heat balance from various factors, such as the absorption of solar radiation, the long wave radiation, the heat loss through evaporation, the turbulent heat exchange between the sea surface and atmosphere, and the internal circulation within the water. The temperature of seawater generally decreases with increasing depth.

5.2   Seawater temperature observation methods

There are two types of seawater temperature observations: surface seawater temperature observation and seawater temperature observation below the surface. The seawater temperature should be measured at standardized levels.

Coastal seawater temperature stations can only observe surface seawater temperature at 02:00, 08:00, 14:00, and 20:00 daily. The seawater temperature observation begins when the ship arrives at a station or a section during a marine survey. Continuous seawater temperature observation is conducted every 2 hours. It is best to have continuous records, and shorter intervals are recommended.

Remote sensing of sea surface temperature (SST) is one of the most widely used applications in the ocean. At present, the algorithms for retrieving SST are highly developed. The extraction of SST information is generally based on statistical models. In cloud-free oceanic regions, the absolute accuracy of SST remote sensing can reach $1°C$, while the relative accuracy can reach $\pm0.5°C$. The currently remote sensing instruments primarily include airborne infrared sensors, SST scanners, ocean color scanners, and high-resolution radiometers.

1. Seawater temperature observation instruments

Liquid thermometer: The representatives of liquid thermometers are surface thermometers and inverted thermometers.

Thermoelectric thermometer: Thermoelectric temperature sensors use a thermocouple as their sensing element. One end of the thermocouple is connected to a cable, which directly detect the temperature of the seawater, while the other end is kept at a constant temperature.

Resistance thermometer: This thermometer uses metal wire resistors or thermistors as temperature-sensing elements.

Electronic thermometer: The sensing element of thermometer is the same as that of a resistance temperature detector. It works by using the sensing element as the frequency-adjusting element of a resistance-capacitance oscillation circuit.

Crystal oscillating thermometer: This thermometer uses a quartz crystal as the sensing element and the oscillation frequency of the quartz crystal changes with temperature.

Remote SST radiation detection: In the past decade or so, the technology of calculating SST based on measured radiation values in the infrared spectrum has been widely applied. Chapter 2 has a very detailed description of current T measurements including remote sensing measurements. Refer to Chapter 2 for more readings.

Conductivity, temperature, and depth (CTD) is currently the most widely used (Fig. 7.12, Fig. 7.13). It is an accurate, reliable, and fastest measuring instrument for measuring temperature, salinity, and depth.

XBT is also a commonly used instrument for measuring seawater temperature. It consists of a probe, a signal transmission cable, and a receiving system. The probe is deployed using a launcher, and the temperature sensed by the probe is transmitted to the receiving system through the cable. The depth value is then determined based on the descent speed of the instrument.

Figure 7.12 RBR CTD (https://rbr.cn/products/standard-loggers#large-multi-channel).

Figure 7.13 The investigators are using the RBR CTD.

Expendable conductivity, temperature, and depth (XCTD) is an instrument used for profiling the conductivity, temperature, and depth of water (Fig. 7.14). It uses a thermistor to measure temperature and a resistivity induction sensor to measure conductivity.

5.3 Salinity observation introduction

The definition of seawater salinity is always related closely to its measurement method. Initially, seawater salinity is the total mass of solutes contained in 1 kg of seawater. Based on the law of constancy of seawater composition, the salinity was calculated by measuring the chlorine content of seawater according to Knudsen formula method (chemical method). The ocean water with a chlorinity value of 19.374% (salinity value of 35.000%) is used as the standard seawater internationally. The salinity value is calculated by measuring the conductivity of seawater according to the conductivity method. The practical salinity standard is generally used as the international standard.

A precise concentration of potassium chloride (KCI) solution was selected as a repro-
ducible conductivity standard, and the conductivity ratio of seawater to KCI solution
was used to determine the salinity of seawater. Therefore, the measurement of salinity
has nothing to do with chlorine. The standard seawater is still used as a reference point.

The seawater salinity is generally observed simultaneously with temperature and
hydrology. The seawater salinity observation begins when the ship arrives at a station
or a section during a survey. Continuous seawater salinity observation is conducted
every 2 hours. The standard levels and other pertinent provisions for measuring
salinity are identical to those for temperature.

5.4 Salinity observation method

There are two methods, physical methods and chemical methods. Chemical methods:
The chemical method is also known as the silver nitrate titration method. The principle
of the method is as follows: Under the condition of constant ion ratio, silver nitrate solution is used for titration to determine the chloride content. Based on the linear relationship between chloride content and salinity, the salinity of the water sample can be
obtained. Physical methods: The physical methods mentioned are relative density
method, refractometry, and conductivity method. The method of relative density is
widely used in oceanography.

6. Sea ice observation

6.1 Introduction

Sea ice is the general term for all ice in the ocean, including saltwater ice formed by
freezing seawater and freshwater ice brought by rivers. It also includes ice floes and

icebergs that have slipped into the sea as a result of glacial breakups in the polar continents.

The heaviest amount of ice is found in the center of the Arctic Ocean and around the Antarctic continent. The sea is still frozen even in summer. The Barents Sea, Laptev Sea, Beaufort Sea, Canadian Arctic Archipelago, Baffin Bay, waters near the Labrador Peninsula, Hudson Bay, and St. Lawrence Bay in the Arctic Ocean will freeze in winter. There is also a large amount of sea ice in the Baltic Sea, Gulf of Bothnia, Gulf of Finland, Bering Sea, and the coast and offshore of the Okhotsk Sea in the Pacific Ocean.

The northern part of the Bohai Sea and the Yellow Sea in China, due to high geographical latitude, are characterized by various degrees of ice in winter. The ice conditions are not severe, which does not pose a great threat to maritime activities when there is no cold wave. Therefore, ports in these areas are ice-free ports. However, in particularly cold years, especially if the cold wave lasts for a long period, severe icing can also occur along the northern coast due to continuous cooling. Not only did the ice cover waterways, disrupting shipping, but it could even freeze ships. For example, the entire Bohai Sea was frozen because of the long duration of the cold wave between 1968 and 1969 in winter. Not only a large number of ships in Tanggu and Qinhuangdao harbors but also ships under sailing were frozen. The ice also destroys offshore structures, such as oil platforms. Due to the lack of preparation, the ice phenomenon often brought great losses. Therefore, the role of sea ice observation is to make forecasts for ice conditions and provide important ice information for offshore projects and maritime activities in harbors, so that effective countermeasures can be taken to prevent problems before they occur.

Elements of sea ice observations include the ice floe observation, the fixed ice observation, and the iceberg observation.

The ice floe observation includes ice volume, ice density, ice type, ice surface features, ice shape, ice block size, ice drift direction and speed, ice thickness, and ice edge lines. The fixed ice observation includes ice type and ice boundary. Specifically, accumulation volume, accumulation height, fixed ice width, and thickness are also included. The iceberg observation includes location, size, shape, and direction and speed of drift. Auxiliary observations of sea ice include sea surface visibility, air temperature, wind speed, wind direction, and weather phenomena.

Time of the sea ice observation: once every 2 h at continuous stations.

6.2   Sea water icing and salinity

As the salinity increases, the freezing point and the temperature at the maximum density of seawater begin to decrease, but it does not have the same process of decreasing. After both the upper and lower layers of seawater have cooled to the temperature at maximum density, at which point convection ceases, the surface seawater can freeze as long as it continues to cool to the freezing point. When the salinity is higher than 24.695, the surface seawater is cooled to the freezing point, but the temperature of the maximum density values is all below the freezing point. As a result, the surface water near the freezing point will be heavier than the warm water below, which causes convection between the upper and lower layers of warm and cold water, thus slowing down the cooling of the seawater. Icing can only occur when the upper and lower layers of seawater mix to the freezing point. When the salinity is 24.695, the maximum density temperature and freezing temperature are both $-1.33°C$.

When sea ice forms, a large amount of salt precipitates out of the ice. The salinity of the seawater below the ice has to increase, which makes it more difficult for the seawater to ice over. Generally speaking, seawater starts to freeze when it reaches the freezing point, but the freezing of seawater varies due to natural and meteorological conditions. For example, ice is less likely to form in oceans with high winds and big waves. However, icing occurs much more rapidly in windless, calm sea conditions, or during low tides (low current velocities) sea areas. In addition, shallow sea areas and estuary areas are prone to ice formation.

6.3   Types of sea ice

Based on various sea ice characteristics internationally, ice types are categorized as follows: by growth process are primary ice (including ice needles, ice waves, greasy ice, sticky ice, and spongy ice), nile ice (including dark nile ice, bright nile ice and ice skins), lotus leaf ice, early years ice (including gray ice, gray, and white ice), 1 year ice (including thin year ice, middle year ice, back year ice), and old age ice (including 2-year ice, multi-year ice). There are flat ice, stacked ice, overlapping ice, ice ridges, ice mounds, icebergs, bare ice, snowcap ice, and so on as it is classified by surface characteristics. There are primary ice, secondary ice, laminated ice, and agglomerated ice as it is classified by the crystal structure. Classified by solid form are fixed ice, initial littoral ice, ice footing, anchor ice, sitting bottom ice, stranded ice, and sitting bottom ice mounds. According to the movement pattern, there are large ice sheets, medium ice sheets, small ice sheets, ice floes, ice masses, ice floe belts, ice floe tongues, ice floe strips, ice bays, ice plugs, ice edges, etc., classified by density as having dense ice floes, very dense ice floes, dense ice floes, sparse ice floes, very sparse ice floes, open water, and ice-free areas. According to the melting process, there is puddle ice, pore ice, dry ice, honeycomb ice, water-covered ice, and so on (Ma et al., 2023).

In general, sea ice can be divided into two types, fixed ice and floating ice. In the floating ice, the huge ice blocks separated from the glaciers that enter the sea and rise above the sea level by more than 5m are called icebergs.

Fixed ice is an ice cap that is frozen to the coasts, islands, and seafloor. Most of the fixed ice in frozen sea areas is littoral ice that is frozen to the coast. Lifting and lowering movements are sometimes produced under the influence of the tides. As a result, fixed ice is not easily formed during the initial stages of sea ice formation.

In contrast to fixed ice, ice that floats on the surface of the sea and drifts with the wind, waves, and currents is called floating ice. For this reason, ice floes are also known as drifting ice floes.

6.4   Ice age and ice conditions

Ice age is related to the pattern of change in the formation, development, duration, and distribution of sea ice and its activity. Various changes in the elements of the sea ice itself in response to relevant hydrological, meteorological, topographical, and other factors.

In sea ice observations, many terms used to express and describe ice conditions are collectively referred to as ice conditions elements. An ice condition element expresses or describes the condition of only one side of the ice. The more ice condition elements are selected, the more detailed the expression of the ice condition will be. At the same time, the selection of elements of the ice situation is not the same in different sectors and according to different needs. For example, in the hydraulic construction sector, to study the physical properties of sea ice, the elements of ice conditions that are often selected are sea ice salinity, temperature, density, pressure resistance, loading force. In maritime transport sector, for research and forecasting ice conditions, all elements of ice conditions are selected.

6.5 Site selection for sea ice observatories

The site selection of sea ice observation mainly includes two aspects: the shore and the sea area. Shore-based stations should be selected at locations where a wide range of sea-ice conditions can be observed, with the requirement that the sea-ice features in the range of vision around the observation point should be representative. Generally, sites with an open sea surface and an altitude of 10m or more are selected as observation sites. High-rise structures such as lighthouses and lookouts should be used as much as possible to observe seawater features in waterways, harbor anchorages, and offshore structures. Additionally, convenience and safety conditions should be considered. Once the observation site has been selected, the elevation and baseline direction should be determined.

In addition, it is also necessary to take into account the cooperation of conventional observation sites on the shore to form an observation network, while achieving a focused comprehensive, and systematic understanding of ice conditions in the sea area.

6.6 Observations of ice volume and ice floe density

Ice volume is the area covered by sea ice in the visible sea area as a proportionate of the area of that sea area. Ice volume includes three types, such as total ice volume, floating ice volume, and fixed ice volume. The total ice volume is the number of percent of the entire visible sea surface covered by all ice. Ice floe volume is the percentage of the visible sea area covered by ice floes. Fixed ice volume is the percentage of the total visible sea surface covered by fixed ice.

Any ice that floats on the sea, drifting with the wind and currents, is known as an ice floe. Ice floe density is a physical quantity. It describes how tightly packed ice is in a group of ice floes. It is defined as the percentage of the total area of all the ice in the ice floe group over the entire ice floe area. The usual classification is dense stay ice, very dense drift ice, dense drift ice, sparse drift ice, very sparse drift ice, open water, and ice-free areas.

The total ice volume (floating ice volume, fixed ice volume) is observed by dividing the entire visible sea surface into deciles and estimating the number of percent covered by ice (floating ice, fixed ice) in each decile. It is expressed in a total of 10 numbers called levels and symbols from 0 to 10. For example, an ice level of six means that ice occupies 60% of the visible sea surface.

6.7   Observations of ice type and external characteristics of ice and ice shape

Ice type is to indicate the different stages of sea ice generation and development process. Ice type is divided into floating ice type and fixed ice type.

Common fixed ice types are classified as glacier tongues, ice shelves, littoral ice, foot ice, and stranded ice.

Sea ice formation begins when the water temperature decreases to the freezing point. As a result, the external characteristics of ice floes vary greatly depending on the weather, water temperature, topography, and other factors of the sea area in which they are located.

6.8   Observations of ice floe motion parameters and fixed ice accumulation conditions

The drift of ice floes and icebergs at sea mainly depends on the combined effects of winds and currents. The drifting direction of floating ice is biased by 30 degrees–40 degrees to the right of the wind direction in the northern hemisphere while to the left of the wind direction in the southern hemisphere. However, the drift direction of the ice is more complicated to observe in the strong tidal sea area due to the combined effects of wind and tide. The process of movement of ice floes includes discrete, agglomerating, and shearing.

Ice floe observations can be categorized into observations of ice floe block size, ice floe direction, and speed (Table 7.1).

The fixed ice accumulation condition includes the amount and height of accumulation. Fixed ice accretion is the accumulation of clumps of ice along the shoreline. There are two types of fixed ice stacking heights: general stacking height and maximum stacking height. General stacking height is the vertical distance from the ice surface to the top of the stack for most stacked ice. Maximum accumulation height is the vertical distance from the ice surface to the apex of the accumulation for most accumulations of ice.

Table 7.1 Reference table for visual drift speed of ice floes.

Ice dynamics	Slow	Apparent	Swift	Quick
Ice velocity v (m/second)	$0 < v \leq 0.3$	$0.3 < v \leq 0.5$	$0.5 < v \leq 1.0$	$1.0 < v$
Velocity class	1	2	3	4

7. Ocean color and transparency

7.1 Ocean color

The color of ocean is mainly caused by the optical properties of seawater, the absorption, reflection, and scattering of the sun's rays by seawater.

Factors affecting the color of seawater include suspended particles and plankton in seawater. There are fewer suspended particles in the oceans, and the particles are so small that they reflect blue light strongly. So that ocean color tends to appear blue. Due to the increase in suspended particles and larger particles in some coastal waters, the scattering of yellow-green light of long waves increases, and the ocean color tends to be light blue. In coastal or estuarine sea areas, due to increased sediment and particulate matter, the scattering of yellow light is significantly enhanced, resulting in the sea appearing yellowish.

7.2 Transparency

Transparency indicates the degree of transparency of seawater (the degree to which light is attenuated in seawater). Transparency is the ability of seawater to transmit light. However, it is not the absolute depth that light can reach. It depends on the light intensity and the number of suspended solids and plankton in the seawater. The light is strong, and the transparency is high. The highest transparency of seawater is about 79 m in the coast of Antarctica. The content and composition of optically active substances such as chlorophyll, suspended sediment, and yellow matter are different, which result in differences in the optical properties of seawater and makes the transparency change so obvious. The transparency of seawater in tropical waters is high in general, up to 50 m.

7.3 Ocean color and transparency observation

After observing the transparency, the transparency disk is brought up to the position where the transparency value is average, and based on the color scale of the transparency disk, the most similar color scale is identified and the corresponding color is recorded as ocean color observed (Fig. 7.15). Ocean color observations are conducted only during the day. It is generally conducted every 2 h at continuous stations.

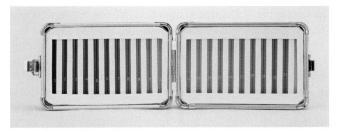

Figure 7.15 The seawater colorimeter (https://www.xusensmart.com/index.php? s=product&c=show&id=7).

Observations should be conducted in a location back to the sun, and observations should avoid the effects of the sewage from the ship. Standard ocean color grade contains 21 colors, such as light blue—yellowish green—brown.

Transparency and ocean color are two different ways of describing seawater optical characteristics. The former indicates the degree of visibility of seawater, while the latter is the color of seawater determined by the optical properties of seawater and suspended substances in the ocean. They are indicators of the degree of turbidity of seawater. Their relationship is extremely close, high ocean color (ocean color number is small), transparency is large, and small transparency, ocean color is low.

8. Marine meteorological observation

8.1 Contents of marine meteorological observation

Marine meteorological observation is continuous and systematic. The elements of observation are clouds, near-surface atmospheric conditions, and their variations visually or with the aid of instruments. The visual items include clouds, visibility, and weather phenomena (sound, light, electricity, etc.). The instrumental observation items include barometric pressure, air temperature, air humidity, wind direction, wind speed, precipitation, evaporation, sunshine, etc (National Standards of the People's Republic of China, 2007a, 2007b, 2007c).

Survey vessels acting as meteorological observers (whether traveling or spotting) are required to make four mapping weather observations per day. Observations should be conducted at 2:00, 8:00, 14:00, and 20:00 (Beijing time). The meteorological observation is generally conducted when arrival at the station in field observation.

8.2 Visibility observation

The definition of visibility is that distinguishing the outline of the target from the background of the sky. Although some details or even part of the outline of the target may not be recognizable, it is still regarded as to be visible in some cases. A target can only be regarded as to be invisible if it blends in so well with the background of the sky, and it cannot even be seen in general. Visibility is usually defined as the maximum horizontal distance which a person with normal vision can see something under the prevailing weather conditions. Effective visibility is defined as the maximum horizontal distance at which more than half of the surrounding field of view is visible.

When the ship is near the coast, the visibility in the shoreward direction should first be estimated with the aid of individual targets in the field of view (e.g., mountain ranges, promontories, lighthouses) whose distances can be measured from charts or by radar. Then, an estimate of the visibility in the seaward direction should be made in terms of the degree of clarity of the horizontal line (Table 7.2).

When the ship is in the open sea area, refer to the table below for visibility class estimation as long as it is based on the clarity of the horizontal line. When the horizontal line is not visible at all, it is estimated empirically.

Table 7.2 Sea surface visibility reference table.

Clarity of the sea-sky level	Eyes ≤7m above sea level	Eyes >7m above sea level
Crystal clear	> 50.0	—
Clear	20.0–50.0	> 50.0
Relatively clear	10.0–20.0	20.0–50.0
Barely discernible	4.0–10.0	10.0–20.0
Completely unclear.	< 4.0	< 10.0

8.3   Cloud observation

8.3.1   Cloud classification

Clouds are classified into 29 species in 10 genera in three families of low-level (Fig. 7.16), middle clouds (Fig. 7.17) and high clouds (Fig. 7.18) according to their shape characteristics, structure, features, and base height. For the average clouds base height, refer to the cloud species height table below (Table 7.3).

8.3.2   Cloud shape observation

Clouds shape is judged primarily according to clouds shape, structure, transparency, and associated weather phenomena. For example, the letters Ns are recorded as the rain cover cloud for continuous precipitation, and the letters Cb are recorded as cumulonimbus cloud for lightning and thunder. For more accurate recognized, continuous observation is necessary, especially for the development of the cloud. Individual clouds can be easily recognized when certain characteristics of the sky can be recognized whether they are stable or not.

Figure 7.16 Low clouds (https://zhuanlan.zhihu.com/p/22385960).

Figure 7.17 Middle clouds (https://zhuanlan.zhihu.com/p/22385960).

Figure 7.18 High clouds (https://zhuanlan.zhihu.com/p/22385960).

Table 7.3 Types of clouds height.

Types of clouds	Boreal zone (km)	Temperate zone (km)	Tropical zone (km)
Low clouds	From sea level to 2		
Middle clouds	2—4	2—7	2—8
High clouds	3—8	5—13	6—18

8.4 Air temperature and humidity

Air temperature indicates how hot or cold it is.

Humidity indicates the amount of water vapor in the air.

Air temperature and humidity observation on board are usually conducted by wet and dry bulb temperatures in a louver box or a ventilated wet and dry meter. Many ships are equipped with simple meteorological observation equipment, and air temperature and humidity are basic observation elements.

8.5   Wind observation

The movement of air is called wind. This chapter refers to wind as the component of wind in the horizontal direction. Average wind direction and speed are basic wind observation elements. Wind observation location is very important. It must be conducted in an open area where the wind speed and direction are not affected by buildings or other structures. The wind observation instrument should be installed at a height of 10m above the sea surface when observation.

The wind direction is the direction from which the wind is blowing. Wind speed is the distance traveled by the wind per unit of time in "m/second." When there is no wind (wind speed is about 0.0–0.2 m/second), the wind speed is recorded as "0" and the wind direction is recorded as "C." The instruments for temporary wind speed and direction observation on board are usually hand-held wind speed and direction meter. Modern ships are generally equipped with a wind speed and wind meter fixed on the upper deck.

In order to make the measured wind speed and direction data representative, the average wind speed and direction of a certain period of time are generally regarded as wind speed and direction. According to the experiments data, it shows that the average wind speed and direction within 10 min are well representative. What is more, if the wind speed and direction do not change much, the average wind speed and direction within 2–3 min are also representative.

Therefore, 2-min average wind direction and maximum wind speed are generally taken for manual observations, and 10-min average wind direction and maximum wind speed are taken for self-recording instruments.

8.6   Barometric pressure observation

Barometric pressure is the atmospheric pressure acting per unit area. The unit of air pressure is "hPa." Its resolution is 0.1 hPa, and the accuracy is ± 1.0 hPa. Mercury barometers are usually equipped on board for air pressure observation, and box barometers are also equipped on board.

9.   Conclusions

This chapter serves as an in-depth exploration of observation methods and instruments employed in the examination of key elements within physical oceanography and marine meteorology. Physical oceanography encompasses a spectrum of observation elements, comprising tidal patterns, currents, waves, water temperature, salinity, sea ice, watercolor, and transparency. On the other hand, marine meteorological observations focus on elements such as visibility, temperature, humidity, air pressure, wind speed, and wind direction.

Through marine surveys, researchers can systematically assess the spatial distribution, temporal characteristics, and changes in these vital elements. The observations derived from marine hydrological and meteorological observations serve as the bedrock for comprehensive oceanic research and the responsible utilization of marine resources. Moreover, these observations play a crucial role in marine environmental protection and the early warning systems for potential marine disasters.

Diverse survey methods, including large-area surveys, cross-sectional surveys, and continuous observations, are employed, leveraging various platforms such as satellites, ships, underwater devices, and buoys. The instrumentation used enables the acquisition of a comprehensive range of marine elements. However, the frequency of observation varies significantly due to the unique distribution patterns and changes inherent in these elements. This chapter provides an essential overview of the methodologies and instruments that form the cornerstone of scientific research in physical oceanography and marine meteorology.

References

Encyclopedia of China. (1987). *Atmospheric sciences, oceanic sciences, hydrological sciences*. Encyclopedia of China Publishing House.

Hu, J. (2009). Extraction of ocean tidal information from satellite altimeter data and climate change. *Advances in Marine Science, 21*(10), 1094−1104.

Jiang, J. (2012). Ocean observation technology. *Acoustic Technology, 31*(1), 61−63.

Li, D. L. (2022). *Research and design of acoustic Doppler current profiler*. Qilu University of Technology.

Liu, Y. (2005). Offshore Seawater depth and wave Height measurement. *Journal of Bohai University, 26*(2), 165−168.

Long, X. M. (2005). SZS3-1-pressure type wave and tide meter. *Journal of Tropical Oceanography, 24*(3), 81−85.

Ma, Y. X., Yu, F. X., & Tian, Y. (2023). Basic properties of physical and mechanical properties of sea ice on the east coast of Liaodong Bay. *Marine Environmental Science, 42*(5), 780−787.

National Standards of the People's Republic of China. (2007a). *GB/T 12763.2-2007-Marine survey specification − Part 2, Marine hydrological observations* (pp. 7−8).

National Standards of the People's Republic of China. (2007b). *GB/T 12763.3-2007-Specification for marine surveys − Part 3, Marine meteorological observations* (pp. 4−6).

National Standards of the People's Republic of China. (2007c). *GB/T 12763.5-2007-Marine Survey Specification Part 5, Marine acoustic and light element investigation* (pp. 9−10).

Qiao, Z. M., Sun, X., & Liu, X. (2022). Development and application of current observation methods. *Ocean Development and Management, 5*, 106−110.

Shi, M. C. (1983). Comparison of several methods for surface currents observation. *Oceanographic Survey, 2*, 48−58.

Sun, J. (2005). *Wave information inversion of SAR images*. Ocean University of China.

Tang, Y. G., & Kang, J. (2014). Comparison of wave buoy measurement methods. *Modern Electronic Technology, 37*(15), 121−122.

Wang, Q. Y., Zhou, T., & Hu, Y. Y. (2023). Construction practice of tide gauge station under the concept of one observation station with multiple functions. *Marine Information Technology and Application, 4*, 218−227.

Zhang, B. H., & Zhao, M. (2013). Methods for measuring seawater velocity and its application. *Acoustic Technology, 32*(1), 24−28.

Zhou, Q. W., Zhang, S., & Wu, H. (2016). Review of wave observation technology. *Marine Surveying and Mapping, 36*(2), 39−44.

Zuo, Q. H. (2008). Development and application of wave observation technology. *Ocean Engineering, 26*(2), 124−139.

Historical changes of hydro and sediment dynamics due to coastline changes in Hangzhou Bay, China

Li Li, Chenhui Fan, Yueying Zha, Yi Wan and Kai Gao
Ocean College, Zhejiang University, Zhoushan, China

1. Impacts of coastline changes on hydro and sediment dynamics

1.1 Introduction to coastline and coastline changes

1.1.1 Coastline

The coastline is the dividing line between sea and land formed by the averaged high tide of the spring tide for many years (Boak & Turner, 2005; Lui et al., 2011). It is the basis for coastal zone management, as well as the baseline for ocean navigation and sea area determination. The coastline contains rich marine resources and environmental information, possessing important ecological functions and resource value.

An idealized definition of coastline is that it coincides with the physical interface of land and water (Dolan et al., 1980). However, the coastline is constantly changing and reshaping because of cross-shore and alongshore sediment movement in the littoral zone and especially because of the dynamic nature of water levels at the coastal boundary (e.g., waves, tides, groundwater, storm surge, setup, runup) (Boak & Turner, 2005). Breakers erode cliffs, shift sand to and fro, breach barriers, build walls, and sculpt bays. Even the gentlest of ripples constantly reshape coastlines in teeny, tiny ways—a few grains of sand at a time. Therefore, the actual coastline should be a collection of countless sea-land boundaries between high and low tides, which is a strip in space, rather than a fixed geographical line (The Revision Committee of the Water Conservancy Dictionary of Hohai University, 2015).

The coastlines could be incorporated into two classes: natural and artificial coastlines. Natural coastline refers to the maintenance of the natural coastal properties, where the shape and properties of the coastline are not altered by human activities. It generally refers to the long-term interaction between sea and land and the formation of a spatial natural boundary between sea and land. In its spatial form, it has a zigzag morphology, natural direction, relatively fixed location, ecological system structure integrity, functional stability, and other characteristics. Artificial coastlines are built at the junction of land and sea for production and living needs. They are mostly

Current Trends in Estuarine and Coastal Dynamics. https://doi.org/10.1016/B978-0-443-21728-9.00008-9

constructed with cement, stone, sand, concrete, and other construction materials, including breeding berms, salt field berms, docks (ports, fishing ports), jetties (tidal dikes, antislope dikes, shore protection barriers), transportation berms, energy extraction facilities, recreational facilities, artificial island construction, and other types. Most artificial coastlines are located in the intertidal zone or the subtidal zone.

1.1.2 Coastline changes

Coastline changes refer to the retreat or advancement of the coastline caused by sea level fluctuations, crustal movements, and human activities. The characteristics of coastline change include length fluctuation, morphological evolution, positional changes, transfer of utilization types, and spatial replacement of land and sea enclosed by the coastline. When analyzing changes in the coastline, qualitative analysis or quantitative analysis can be carried out using some simple basic statistics, such as using length values, land and sea area, fractal dimension, and change rate to analyze the spatiotemporal changes in the length, shape, and position of the coastline.

1.1.2.1 When and where coastline changes

The coastline is constantly in motion and development. The historical changes of each specific shore section have their own unique characteristics.

Coastal reclamation is an important way for additional living and production space from the sea, resulting in changes to the coastline. Many countries, such as Japan (e.g., Chen et al., 2014), the Netherlands (e.g., Zhang, 2013), Singapore (e.g., Zhou et al., 2016), the United Arab Emirates (e.g., Li & Yao, 2018), and China (e.g., Huang et al., 2013), have reclaimed land from the sea. The Netherlands can be regarded as the country with the longest history of coastal reclamation. Since the 13th century, the Dutch have explored nearly 7000 square kilometers of land to the sea, accounting for approximately 20% of their current land area. From 1945 to 1975, the Japanese government reclaimed 118,000 ha from the sea. Singapore has so far reclaimed over 100 square kilometers of land from the sea, including Changi International Airport and Jurong Town.

Artificial islands are also a type of sea reclamation. Modern artificial islands have a wide range of uses, mainly concentrated in the areas of coastal engineering, deep-sea platforms, and marine energy. Modern artificial islands are often used to build airports, nuclear power plants, oil and gas processing plants, aquatic processing plants, paper mills, waste treatment plants, dangerous goods warehouses. Additionally, sea parks and even entirely new sea cities can be constructed. For example, the United Arab Emirates has built Palm Islands in Dubai to develop the tourism industry. The buildings on the island mainly consist of villas, hotels, and sea view apartments, making it a famous tourist resort (Li, Guan, et al., 2018; Li, Ye, He, & Xia, 2018).

The river sediments can affect changes in the coastline. For example, the Yellow River in China is currently the river with the highest sediment concentration in the world, with an average sediment concentration of about 37 kg per cubic meter. It pours up to 1.6 billion tons of sediment into the sea every year. Due to the large amount of sediment deposition at the entrance to the sea, the Yellow River estuary extends an average of 2—3 km toward the sea each year. At the same time, the coastline is constantly advancing toward the ocean (e.g., Cui & Li, 2011).

1.1.2.2   How coastline changes?

Firstly, the main reason for the changes in the coastline is the crustal movement. Due to the influence of crustal uplift and subsidence activities, the invasion of seawater or the retreat of seawater has caused significant changes in the coastline. This change is still happening nowadays.

Secondly, coastline changes are greatly influenced by glaciers. In the Arctic and Antarctic, land and mountains are covered with a huge number of glaciers. If temperatures rise and glaciers melt, sea levels will rise and coastlines will advance toward the land; on the contrary, if the temperature decreases relatively, the sea level will gradually decrease and the coastline will push toward the ocean.

Thirdly, coastline changes are also affected by the sediment in rivers that enter the sea. When a river brings a large amount of sediment into the ocean, it accumulates near the coast and becomes land over the years, pushing the coastline toward the ocean.

Finally, human activities can also alter the coastline. Examples of such human activities include the construction of seawalls, land reclamations, the expansion of port terminals, and the construction of artificial islands.

1.2   Calculation of coastline changes

The coastline can be obtained through field surveys. Although the data are very accurate, gathering the data is expensive, inefficient, and time-consuming. Remote sensing technology has the advantage of covering an extensive observation range, allowing for regular observations, and providing data promptly with good economic benefits. Extracting coastline from remote sensing images is currently a commonly used method.

1.2.1   Methods to extract waterlines

Due to the characteristics of instantaneous imaging in remote sensing images, most of the information directly extracted from remote sensing images is the instantaneous waterline rather than the actual coastline. Although the waterline is not a coastline, in many methods of extracting coastline, the coastline is obtained by a series of modifications to the waterline after extracting the waterline. Therefore, accurately and quickly extracting waterlines is the foundation for extracting coastline. The main problem faced when extracting water boundary lines is how to accurately and quickly obtain continuous water land boundary lines from images. The main methods for extracting waterlines include threshold segmentation, edge detection, object-oriented extraction, and region growth method (Liang et al., 2018).

The threshold segmentation method is a method of segmenting the target object and background object by setting corresponding thresholds based on their different pixel grayscale values. The commonly used threshold methods include: manual interaction threshold segmentation, density segmentation, Otsu algorithm, and maximum expectation algorithm (EM algorithm). Lin et al. (2016) used EM threshold segmentation method to segment the image. Lu et al. (2011) used Otsu threshold segmentation method to separate water and land, achieving good results. Overall, the threshold segmentation method is a relatively efficient algorithm with simple methods and fast operation speed. However, it still has certain requirements for sea land contrast and average extraction accuracy.

Edge detection is a classic method for image segmentation and extracting boundary lines. Traditional edge detection operators include Sobel operator, Canny operator, Prewitt operator, Laplace operator, and Roberts operator. Ma et al. (2007) used IDL language to implement the Canny operator and completed the automatic extraction of coastline from satellite images. Zhuang (2009) used Roberts operator to extract bedrock and artificial coast, and used sobel operator to extract sandy coast.

The object-oriented extraction method is a classification extraction method that first segments the coastline using spectral, spatial, and texture information from remote sensing images. Then, it extracts the coastline based on the segmentation results. Based on classifying coastal types, Wang et al. (2017) extracted waterlines based on the characteristics of each coastal type. Ge et al. (2014) extracted the coastline of the Pearl River based on the object-oriented idea. Ge et al. (2014) first divided the coast into four types of artificial coast, bedrock coast, sandy coast and muddy coast, then established different coastline extraction standards for these four types of coast, and then extracted the coastline. Although object-oriented methods are relatively cumbersome and time-consuming, their accuracy can be well guaranteed.

As a classic segmentation algorithm, regional growth method has many applications in extracting waterlines from remote sensing images. The basic idea of regional growth is to assemble similar pixels to form a region. Chen et al. (1995) used region growth method to extract waterlines in optical remote sensing images. Xie et al. (2007) proposed an improved method for automatically selecting growth points when extracting coastline from SAR images. The regional growth method is a relatively mature image segmentation method, which has the advantages of simple extraction algorithm, fast speed, and stable and continuous results in extracting waterlines. However, in the case of complex terrain background around the coastline, the regional growth method is easily affected and can cause deformation of the waterlines.

1.2.2   Methods to extract coastlines

Most of the directly extracted coastlines from remote sensing images are instantaneous waterlines rather than actual coastlines. To extract coastlines, other methods need to be combined. The commonly used methods for extracting coastline include waterline correction, visual interpretation, and light detection and ranging (LiDAR) technology for extracting coastline.

After extracting the waterline from remote sensing images, a series of corrections can be made to the waterline to obtain the coastline. There are generally three ways to use waterline correction to obtain coastline: (1) combine beach slope information and tidal station information, and use geometric relationships to calculate the position relationship between waterline and high tide lines; (2) based on tidal data, the elevation value of the waterline is obtained, and then the digital elevation model (DEM) is interpolated to obtain the coastline; and (3) simply determine the location of the coastline based on the distance between the waterfront and the features on the beach (Chang et al., 2004). Ma et al. (2007) used edge detection to extract the waterline, obtained the instantaneous tidal height during image imaging based on the tidal level information obtained from the tidal station, and then calculated the offset distance of the waterline using information such as the average high tidal level and coastal slope to obtain the position of the coastline. When obtaining coastline through this approach, there are certain requirements for the terrain of the coast, which can only be applied in sea areas with small terrain fluctuations, and detailed tidal level data are required as support.

Visual interpretation is the most direct method for extracting coastline. During the continuous pounding of the coast by high tides, corresponding marks will be left on the coast, such as sand ridges formed on sandy coasts, growth boundaries of some salt alkali tolerant plants on biomass coastlines, and seawater marks on artificial coastlines. These are important criteria for determining the location of coastlines. Through on-site investigation and the different characteristics reflected by different land features in remote sensing images, corresponding interpretation markers are established, and then these markers are used to interpret the coastline, obtaining a strict sense of coastline. The visual interpretation method obtains coastline with high accuracy and persuasiveness, but it requires a combination of remote sensing images and on-site inspections and requires manual interpretation to interpret the coastline, which is a huge workload.

LiDAR is generally an active remote sensing method based on aerial platforms for observing ground objects. Its fast speed, strong anti-interference ability, and high accuracy have quickly become a research hotspot for coastline extraction. Since 2002, many scholars in the United States have conducted research on coastal zones using LiDAR technology (Stockdonf et al., 2002). The research idea is to use high-precision LiDAR data to generate DEM and then combine it with the coastal elevation value of a defined tidal level surface to determine the specific position of the coastline with the use of Contouring Method or Cross Shore Profile Method.

1.3 Impacts of coastline changes on hydrodynamics and sediment dynamics

Researchers typically use the following methods to study the hydrodynamic impacts of shoreline change:

(1) Numerical simulation: Mathematical models and computer simulations are used to model the hydrodynamic effects of shoreline changes. This approach can be used to predict changes in hydrodynamic parameters such as currents, waves and tides by simulating different shoreline patterns and tidal conditions.

(2) Field observations: Hydrodynamic data are collected and the effects of shoreline changes are analyzed through in situ observations in actual marine or riverine environments. This may involve the use of instruments such as buoys, tide gauges, and current meters to measure hydrodynamic parameters in the ocean or river.

(3) Laboratory experiments: Physical models are used in the laboratory to simulate the hydrodynamic effects of shoreline changes. By simulating different shoreline morphologies and flow conditions in scaled-down flume or concrete models, researchers can observe and measure changes in parameters such as currents, waves, and tides.

(4) Satellite remote sensing: Satellite sensors are used to acquire remote sensing images of oceans or rivers and changes in hydrodynamics are inferred by analyzing shoreline changes in the images. This method can provide data on shoreline changes over large areas and long-time scales.

1.3.1   Impacts of coastline changes on tides

Hydrodynamics serves as a crucial indicator of the water environment of the sea, directly influencing the capacity for water exchange, pollutant dispersion, and self-purification, and has an important position in the basic research of the marine environment. Coastline change is an important factor that causes hydrodynamic changes in the sea. In recent years, the acceleration of coastline development and utilization has significantly changed the characteristics of the original coastline, which also affected the hydrodynamic environment of the sea to different degrees. Research on the impact of coastline changes on the hydrodynamic environment of the sea is of great scientific and practical significance for the rational development and utilization of coastline, the protection of marine ecosystems, and the safeguarding of the sustainable development of marine resources (Lee & Ryu, 2008; Ya et al., 2017). Anthropogenic coastline alterations directly lead to a reduction in sea area and changes in seabed geomorphology and further affect nearby waters and the hydrodynamic environment (Zhang et al., 2017).

The hydrodynamic changes cause the process of water exchange, as well as changes in the movement of sediment, which in turn lead to changes in the seabed. The difference between the highest tidal level and the lowest tidal level in a tidal cycle is the maximum tidal range in this tidal cycle, and the coastline changes cause the local topography to change, thus affecting the tidal range. For the Hangzhou Bay study between 1974 and 2016, coastline changes narrowed the channel of the bay, which increased tidal levels and increased differential tides (Cao, 2018). For the Pushen waterway between 1997 and 2017, the effect of coastline changes on the tidal range at different locations in the waterway was slightly different: the water south of Lujiazhi Island has a slight increase in tidal range, and the tidal range of the water on the east side decreases; the tidal range in the middle of the Pushen waterway does not change much, and the tidal range in the north decreases slightly (Huang et al., 2022). Zhu et al. (2018) used the MIKE3 model to establish a three-dimensional tidal model of the Bohai Sea and the Yellow Sea and simulated the tidal wave system of the Bohai Sea and the Yellow Sea, and the results showed that, from 1987 to 2016, the effect of the changes in coastline and topography on the amplitude of the semidiurnal tides was greater than that on the diurnal tides, and the semidiurnal tidal amplitude caused by the changes in coastline accounted for 27.76%–99.07% of the total amplitude changes, and the semidiurnal tidal amplitude caused by changes in water depth accounted for 0.93%–72.24% of the total amplitude changes (Zhu et al., 2018).

Researchers have conducted studies on the problems caused by reclamation projects, which have shown that large-scale reclamation not only increases the useable land area but also puts pressure on the hydrodynamic and marine ecosystems both locally and far field (Yao, 2021). Choi et al. (2010) carried out numerical simulations before and after reclamation in the sea area near Saemangeum, Korea, and found that the M2 tidal amplitude increased after reclamation. Kang (1999) showed that the tidal amplitude difference in the coastal waters of Mokpo, Yeongsan Estuary, Korea, increased due to reclamation activities. Han and Park (1999) simulated the effects of the reclamation project in Pusan Harbor on the tidal currents and tidal levels by using the Resources Management Association-2 (RMA-2) model and found that the tidal levels had no change but the tidal currents accelerated. Li et al. (2012) investigated the effects of the presence or absence of tidal flats in Darwin Harbor in Australia by using the finite volume coastal ocean model (FVCOM) and found that the removal of the tidal flats increased the tidal asymmetry by about 20%. Sieyes et al. (2011) found that siltation and uplift of tidal flats in the Dee Estuary of the East Irish Sea led to an increase in ebb dominance in the tidal asymmetry of the Dee Estuary. Some scholars focus on analyzing the impact of the reclamation project on the tidal system (Gao et al., 2018; Liang et al., 2019). Chen et al. (2009) showed that the overall trend of tidal asymmetry increased with the increase of reclamation intensity by comparing the tidal levels in the Tianjin Harbor in China. By analyzing the tide level data in Yangshan Harbor, Guo (2017) showed that with the project, the shallow water tide as well as the tidal asymmetry is increasing.

1.3.2 Impacts of coastline changes on flow

In recent years, due to the construction of ports and Central Business Districts, the artificial coastline has increased rapidly, while the natural coastline has decreased abruptly, which has led to changes in the velocity of the flood and ebb currents. Chen (2022) analyzed the impact of coastline changes on hydrodynamics in Liaodong Bay, China, over the past decade, and found that the changes in the coastline weakened the flow velocity in Liaodong Bay, with a decrease of more than 10% at the time of flooding and more than 25% at the time of ebbing. The coastline changes have a greater impact on the flow velocity in the near-shore water body, but the tidal field characteristics have not changed significantly. In Bohai Bay and Laizhou Bay, during the period 2003–13, the construction of the ports increased the length of the artificial shoreline, which led to a decrease in the tidal prism and current velocity (Zhang et al., 2021).

Tidal currents play an important role in water exchange and transport of substances in seawater. Qin et al. (2012) studied the impact of coastline changes on the nearshore water quality in Bohai Bay in China from 2003 to 2011 and showed that the coastline changes impacted the hydrodynamics, and then resulted in the high values of the main pollutants being shifted to the northeast of Bohai Bay (Qin et al., 2012). The coastline changes in Liaodong Bay caused the chlorophyll-a content to rise by 18.05%, the DIN content to decrease by 10.91%, and the DIP content to increase by 7.92%. Liaodong Bay is still in a phosphorus-limited state, but the N/P value has decreased (Chen,

2022). The Qiantang River Estuary in China mudflat project has led to the narrowing of the bay, which has significantly reduced the tidal volume. Then, the tidal flow is obviously reduced, and the exchange capacity between the bay and the outer sea is weakened, resulting in an increase in inorganic nitrogen, phosphates and chlorophyll-a concentrations in the bay (Zhou, 2022). Velocity variations of up to 20%–100% have been calculated in west coast of Korea by Lee et al. (2006) through simulations of flow velocities. Changes in tidal currents and other factors can also cause changes in seabed siltation and sediment transport.

The vast majority of experiments have proved that the trend of hydrodynamic changes due to the influence of human activities is unfavorable to water exchange and prone to the accumulation of pollutants discharged by runoff and outfalls in the mouth of the bay. Healy and Hickey (2002) studied the reclamation history of the Shannon Estuary in western Ireland and found that the reclamation led to the weakening of tidal dynamics in the surrounding waters, which led to a decrease in the diversity of phytoplankton and flora in the waters of the area, affecting the structure of the community. Sato and Azuma (2002) conducted a study on the Isahaya Bay seawall project in Japan, and found that the project had a large impact on the fauna in the area, with a large die-off of benthic bivalves. Wu et al. (2005) studied the biological communities in the reclaimed area affected by estuarine seaward reclamation, concluded that the reclamation had a serious negative impact on the community structure of the fauna. Park et al. (2009) found that the growth and reproduction of macroalgae in the estuary of the Nakdong River in Korea were greatly affected after the reclamation project. Groot et al. (2002) assessed the nearshore ecological impacts caused by reclamation and noted that the regulatory and information functions of nearshore ecosystems were negatively affected. Changes in water quality conditions reflect more vigorous phytoplankton growth and reproduction after coastline changes, which are more likely to lead to red tide disasters from May to October. Lu et al. (2002) analyzed the macrobenthic community in the reclaimed area of Sungei Punggol Estuary in Singapore and pointed out that the reclamation activities had damaged the benthic community structure in the nearby waters, resulting in a significant decrease in the species and abundance of benthic organisms near the project area (Lu et al., 2002).

The impact of coastline changes on tidal wave characteristics is often analyzed by means of measured data analysis and numerical simulation to analyze the impact on the surrounding hydrodynamic characteristics before and after the coastline changes (Wang & Zhang, 2023). Therefore, many scholars have conducted research on the characteristics and mechanisms of tidal dynamics changes before and after the coastline changes. The coastline changes in the south Yellow Sea have a great influence on the tidal wave, and the maximum change value of the tidal amplitude occurs in the radiation sandbar areas (Chen et al., 2009). The large-scale reclamation in Tianjin Port in China led to an increase of 3–5 cm/s in the residual currents in the offshore waters from Tianjin Port to the northern part of Huanghua Port, while the construction of breakwaters in Huanghua Port formed an anticlockwise circulation in the southern part of the vicinity, and the residual current decreased by 2–5 cm/s. The residual current of Huanghua Port increased by 3–5 cm/s from Tianjin Port to the northern part of Huanghua Port (Dong et al., 2020).

1.3.3 Impacts of coastline changes on sediment transport

Along any stretch of coastline, erosion or accretion is common, associated with local sediment budget and coastal characteristics such as the distribution of headlands, cliffs, and beaches (Montreuil & Bullard, 2012). They are slowly changing the coastline. In recent years, large-scale reclamation projects have also greatly changed the coastline. Coastline changes have significant impacts on coastal hydrodynamics, with direct consequences for sediment transport. Jun et al. (2004) analyzed the spatio-temporal features of coastline evolution during 1976–2000 based on remote sensing and established the relation between coastline changes and the runoff and sand transport. Aquino da Silva et al. (2019) used satellite images from Landsat missions 2 to 8 between 1972 and 2016 to address the decadal evolution of the delta in terms of suspended sediment transport changes in Parnaíba River Delta in Brazil, which is still developing in almost natural conditions.

Researchers studied the impacts of coastline changes on sediment transport using field investigation, data analysis, and numerical simulation. For example, Dong et al. (2015) studied the coastline change and sediment erosion caused by the reclamation. The tidal flat in Qinzhou Bay in China has transitioned from a micro-siltation to obvious-erosion. Large-scale projects have changed the hydrodynamic and erosion of Qingzhou Bay greatly. Based on ECOMSED, Zang et al. (2009) predicted the impact of coastal engineering on the suspended sediment. Gao et al. (2018) used the FVCOM to study the suspended-sediment dynamics and its change in Jiaozhou Bay due to land reclamation over the period 1935 to 2008. Xu et al. (2023) simulated the direct on local sediment erosion and deposition in Xiangshan Bay in China on a small scale. Li et al. (2019) explored the impact of reclamation in different locations on the water and sediment environment in Hangzhou Bay, with reclamation at the head of the bay having the greatest impact. Yu et al. (2021) established a three-dimensional numerical model of fine sediment transport in Hangzhou Bay. Compared with the calculation results of the coastline topographic data in 1974 and 2020, the outward sediment transport from Jintang Channel in the South decreases. The overall trend features that the sediment transport into the bay increases, with the bay mouth silting.

Coastline change is controlled by many factors, such as geology, hydrology, climate, vegetation, and environmental problems (Guariglia et al., 2009; Zhao et al., 2008). Anthropogenic impacts (i.e., construction of dams and land reclamation) have great impacts on those factors, and the interception by dams is the main cause of the decline of sediment flux into the sea (Vörösmarty et al., 2003). In recent decades, a worldwide decrease in riverine sediment supply to the ocean has been observed over because of river damming (Day et al., 2016; Luo et al., 2012). Based on datasets begun in the 1950s, Dai et al. (2008) give a quantitative evaluation of the impacts of the dams on the sediment flux of the Pearl River in China. The sediment flux has shown a drastically decreasing trend since the mid-1980s, which is attributed mainly to deposition in the upstream reservoirs. Generally, human activities as the main factor explain

>90% of the sediment load in China's large rivers (Yin et al., 2023). There is also growing evidence that the water and sediment fluxes in Asia typically displayed a significant decreasing trend on the 5- to 10-year scale, due to the construction of numerous dams (Khan et al., 2021; Lu & Chua, 2021; Walling, 2011, pp. 37−51). For instance, sediment load to the Mekong Delta in Vietnam has decreased by 74.1% from the predam period (1961−2011) to the postdam period (2012−15), 40.2% of which was caused by the six main-stem dams in the Lancang River Basin (Binh et al., 2019).

With the rapid growth of population and alleviation of land shortages, reclamation has been increasingly adopted as a means of expanding living space (Xu, 2016). Reclamation projects are concentrated along the coasts of Southeast Asia, Europe, the Americas, and the Persian Gulf. The impact of land reclamation on the dynamics of suspended sediments is one of the topics of concern due to the significant impact of suspended sediments on the maintenance of engineering projects and navigation channels, the geomorphological development, and the availability of light for marine primary productivity. The issue is worldwide and has been reported in a number of studies. For example, the San Francisco Bay in United States (Nichols et al., 1986); the Wadden Sea in Germany (Flemming & Nyandwi, 1994); the Abu Qir Bay in Egypt (Mostafa, 2012); the Darwin Harbour in Australia (Li et al., 2014); the Meizhou Bay (Guo et al., 2014), the Jiaozhou Bay (Wang, 2002), and the Hainan Island, China (Zhong, 2017). In different time intervals and dynamic environments, many hydrodynamic parameters have a significant relationship with the changes in beach morphology. Reclamation projects will directly affect the morphology of coastal areas, exacerbating the impact and damage to adjacent coasts.

2.   A case study in Hangzhou Bay, China: Coastline changes due to tidal flat reclamation

2.1   Introduction

Tidal flat in estuaries is an important area for ecosystem, environment, and socioeconomy. Aquaculture and industries are mostly developed surrounding tidal flat areas. Tidal flat usually shelters large areas in estuaries, particularly in macro-tidal estuaries. Hydrodynamics on tidal flat are highly dynamic with straining, advection, and mixing in balance with the temporal changes in stratification-induced potential energy anomaly (Pavel et al., 2013). These characteristics subject to different physical forcing, such as tides, currents, waves, winds, and drainage (Le Hir et al., 2000).

Tidal flat areas have large tidal prism and dissipate vast tidal energy. The morphology alteration of tidal flat changes the tidal energy dissipation generally dominated by friction in shallow water (Le Provost & Lyard, 1997). Changes of tidal flat areas and slopes reduce tidal prism and energy dissipation (Li et al., 2012; Shi et al., 2011; Song, Wang, Zhu et al., 2013; Suh et al., 2014), modulates tides, particularly shallow water tides (Gao et al., 2014; Li, Ye, Wang, et al., 2018; Zhu, Wang, et al. 2016; Zhang et al., 2010), and subsequently impacts sediment dynamics in estuaries (Ali et al., 2007; Gao et al., 2018; Le Hir et al., 2000; Li et al., 2014). Tidal flat with mangroves are complicated by the presence of runnel networks and plants, acting as traps for sediments when currents are reduced in these areas (Le Hir et al., 2000; Nardin et al., 2016; Stokes et al., 2010). Subsequently, tidal flat protects coastlines from erosion and marine disasters like typhoon, through trapping sediments by vegetation (Ha et al., 2018) and dissipating energy (Song, Wang, Zhu, et al., 2013; Song, Wang, Cao, et al., 2013).

Tidal flats along the Chinese coast are classified into the plain type and the embayment type, neither of which supplies sufficient fetch for wind/wave interaction due to the very gentle coastal slope. Thus, the tidal flat is dominated by tidal process and is often associated with a large sediment supply (Wang & Zhu, 1994). Such kinds of tidal flats have very important impacts on hydrodynamics and sediment dynamics in both local estuaries and remote coasts, through redistributing tidal energy. According to Song, Wang, Zhu, et al. (2013), Song, Wang, Cao, et al. (2013), different tidal patterns occurred when tidal flats around the Bohai Sea, Yellow Sea, and East China Sea (BYECS) were removed, in which the tidal range and phase are changed, and the amphidromic points are displaced. Far-field effects on tidal dynamics would be observed on the west coast of Korea following significant reclamation on the Chinese coast, and vice versa. The changes of coastlines caused by reclamation in the BYECS can result in rise of tidal amplitude and onshore sediment transport. Tidal energy extraction in the Minas Passage, Bay of Fundy, would alter the tidal elevations and tidal currents throughout the Gulf of Maine (Hasegawa et al., 2011). Hence, careful consideration should be paid to any proposed artificial changes to tidal flat, given the effects of these on both the local environment and further afield.

Hangzhou Bay is a funnel-shaped bay, located on the coast of East China Sea (Fig. 8.1a). It is semi-diurnal macro-tidal, with maximum tidal range of 8−9 m during the tidal bore, and even mean spring tidal range can reach up to 6.44 m (Fan et al.,

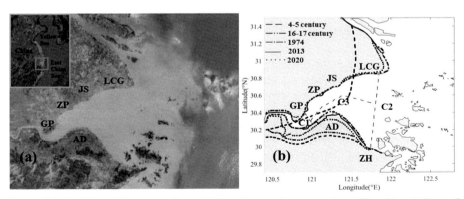

Figure 8.1 (a) Map of Hangzhou Bay. (b) Coastlines and cross sections. The abbreviations of YG, GP, ZP, JS, LCG, AD, and ZH are for Yanguan, Ganpu, Zhapu, Jinshan, Luchaogang, Andong, and Zhenhai, respectively.

Table 8.1 Time of the dynasties in China.

Dynasty	Period
Tang	618–907
Ming	1368–1644
Qing	1644–1911

2014). The width of the bay decreases from 100 km at the bay mouth to about 20 km at Ganpu section, over a distance of about 90 km. Hangzhou Bay is a typical estuary of high turbidity, with suspended sediment concentration (SSC) of about 10.6 kg/m^3 at the surface level at the head of the bay (Pan et al., 2013; Xie et al., 2017a).

Hangzhou Bay experienced severe coastline changes historically (Fig. 8.1b). In about fifth century, the bay had a narrow mouth and wide head and was located to the south of that in 2020 (Fig. 8.1b). In about 16th century, the northern bank moved much northward and developed into the shape as they are today, and the southern bank also moved northward (Fig. 8.1b). The southern tidal flat was fast silted and moved northward in about 11th~18th centuries; the main tidal channels moved northward and the northern bank was eroded (Xiong, 2011). There were three main tidal channels in the bay. In about Tang and Ming dynasties, the tides mainly propagated along the southern channel. In Ming (1643) and Qing (1645) dynasties (Table 8.1, http://www.gov.cn/test/2005-07/27/content_17445.htm), the middle channel became the main tidal channel. In 1720 (Qing dynasty), the northern channel was the main tidal channel. In 1747 (Qing dynasty, Reign of Qianlong), the middle channel was dredged and became the main tidal channel. However, after 12 years in 1759, the main tidal channel diverted from the middle channel to the northern channel. The bay was then gradually developed into its funnel shape (Wang & Lv, 2019; Xiong, 2011). About 694.4 km^2 tidal flat was reclaimed around the bay over a period of near 30 years since 1986 (Zhang et al., 2005). About 94% of the coastline around HZB was artificial coastline by 2012 (Kemp et al., 2016). The mechanism of the impacts of tidal flat reduction on the variations of sediment dynamics needs deep investigation.

As shown by the remote sensing and bathymetry data of HZB, the surface SSC and morphology varied in the near decades (Jiang et al., 2013; Shen et al., 2013; Xie et al., 2017a). Researchers paid much attention to the variation of sediment dynamics in the bay, for example, the sources of suspended sediments, the SSC values, the erosion and deposition, and the morphology (Li, Gao, et al., 2018; Lu et al., 2015; Xie et al., 2017b; Xie, Gao, et al., 2017; Yu et al., 2012). Previous studies adopted numerical models, for example, MIKE21 (Guo et al., 2012), simplified-geometry model (Xie et al., 2013), ECOMSED (Du et al., 2010), to simulate hydrodynamics or sediment dynamics in HZB. However, the numerical model accuracy in HZB still needed further improvement. Hence, little attention has been paid to the numerical simulation of sediment dynamics in the bay, particularly with a constant changing coastline and in the bottom boundary layer.

This study aims to numerically examine the characteristics and mechanics of tides and suspended sediments of the Hangzhou Bay since Tang Dynasty. Firstly, a BBL sediment model was developed and calibrated to well reproduce the sediment dynamics in the bay. Secondly, the historical changes of tides and suspended sediments due to coastline variation were examined using the numerical model. The mechanism of the tidal and sediment dynamics was then investigated. This study can provide scientific reference for the research on coastal morphology and harbor management.

2.2 Numerical model and experiments

The finite volume community ocean model (FVCOM, Version 4.0) (Chen et al., 2006) was chosen to simulate the hydrodynamics of HZB. FVCOM is a three-dimensional hydrodynamic model. FVCOM uses a 3-D unstructured mesh, which has flexible grid size and node number and is well-suited to the complex geometry of Hangzhou Bay. A σ-stretched coordinate system is applied in the vertical direction to improve the representation of the complicated bathymetry. The detailed model equations and numerical algorithm are referred to the user manual of the model. The model configuration and initial conditions for Hangzhou Bay are referred to Li et al. (2023).

The detailed information of the numerical experiments is provided in Table 8.2. Experiment 5 was the reference run. To discuss the effects of the historical coastline variations on suspended sediment dynamics, experiments 1 and 2 were designed using the coastlines of the 5th century and coastlines of the 16th century, respectively (Xiong, 2011). Experiments 3−6 were used to track the variation of the SSC from 1974 to 2020, due to the change of coastline, and the coastline change data were collected through remote sensing images. The coastline of 2020 was decided according to the land using plan of Zhejiang Province (Lu et al., 2015). The coastline data were directly extracted if the coastlines were made of concrete due to land reclamation (Sun et al., 2011). Otherwise, the location of the coastline was decided according to the salt-tolerant plants (Yan et al., 2009; Zhang et al., 2005).

As shown in Fig. 8.1b, the western section of the northern bank in HZB gradually moved northward and became concave from bulged toward the sea from the 5th century to 2020, while the eastern section gradually moved eastward toward the sea. The southern bank bulged gradually to the north during this time. Over past 1400 years, the southern bank had changed much more than the northern bank. The change of the coastlines since 1974 tended to be stable compared with that before 1974.

Table 8.2 Descriptions of the numerical experiments.

Experiments	Descriptions
Exp1	Use coastlines of 5 century A.D.
Exp2	Use coastlines of 16 century A.D.
Exp3	Use coastlines of 1974
Exp4	Reference model using coastlines of 2013
Exp5	Use coastlines of 2020

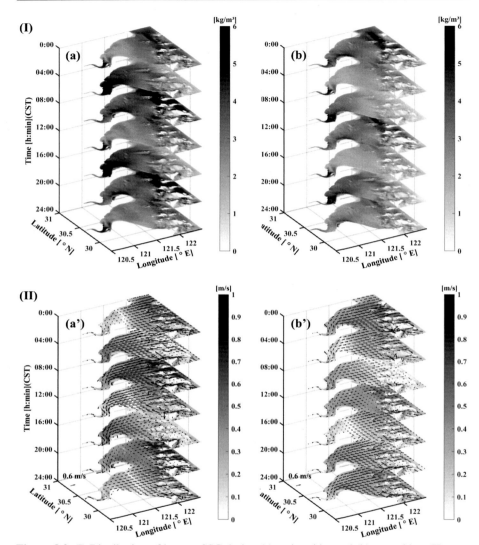

Figure 8.2 (I) Distribution of bottom SSC during (a) spring tides and (b) neap tides. (II) Distribution of the bottom currents during (a') spring tides and (b') neap tides. Gray contours and vectors indicate current magnitudes and directions, respectively.

2.3 Model results and discussions

2.3.1 Hydro and sediment dynamics of the reference model

2.3.1.1 Currents and SSC in bottom boundary layer

As the reference model showed (Exp 5, Fig. 8.2I), SSC values during spring tides were much larger (6 kg/m³, Fig. 8.2Ia) than those during neap tides (3 kg/m³, Fig. 8.2Ib). The turbid zones near the Andong tidal flat and at the bay mouth occurred at the flood and ebb currents and moved landward and seaward with flood-ebb tidal cycles

(Fig. 8.2Ia). At the high and low slack waters, the turbidity near the bay mouth reduced, while the turbid zone at the bay head still existed (Fig. 8.2Ib). As the water depth from Ganpu to Jinshan was higher near the northern bank than that near the southern bank, SSC values were smaller near the northern bank than those near the southern bank.

The spatial and temporal distribution of SSC was closely related to the current velocity, and both changed periodically during flood tides and ebb tides (Exp 5, Fig. 8.2II). Tidal velocity during neap tide was significantly lower than that during spring tide (about 50% of that during spring tide, Fig. 8.2II). During spring tides, peak SSC values occurred due to the stronger spring flood current (Fig. 8.2IIa). At the high and low water slack, bottom SSC values were much smaller than those at the peak flood and ebb currents. As currents were weak during neap tides (Fig. 8.2IIa'), the peak SSC values were half of that during spring tides. Only the bottom currents and SSC were illustrated here, as the surface SSC and currents had the similar pattern as that at the bottom level. Because of the strong tidal currents and shallow water depth in the Hangzhou Bay, the vertical water mixing was sufficient.

2.3.1.2 Residual currents and sediment fluxes in the bay

Fig. 8.3I illustrated the vertical profiles of net sediment fluxes at the three cross sections, averaged over a spring-neap tidal cycle, to examine the longitudinal and lateral flux of sediments (Fig. 8.1b, Exp 5, reference model). The net sediment fluxes averaged over a spring-neap tidal cycle were largely seaward at the shallow water area and landward at the tidal channel at C1 section (Fig. 8.3Ia). The surface sediment fluxes were landward and the bottom sediment fluxes were seaward where water depth changes dramatically near the northern bank. At C2 section (Fig. 8.3Ib), contrary to the C1 section, the net sediment fluxes near southern/northern bank were seaward/landward, which reflected the characteristics of sediment transport in the Hangzhou Bay. At the longitudinal section C3 (Fig. 8.3Ic), the net sediment fluxes were generally northward at the surface level, and southward near the bottom level. The northward sediment fluxes were obvious near Ganpu and the southward sediment fluxes were obvious near Andong. This pattern of net sediment flux was consistent with the residual circulation in the bay (Fig. 8.3II). A residual circulation in the entire bay was formed with northward residual currents near surface level and southward residual currents near the bottom level.

2.3.2 Historical changes of hydro and sediment dynamics

2.3.2.1 Tides and SSC

Time series of sea surface elevations at the sections C1 and C2 were selected to analyze the impact of coastline change on tidal range (Fig. 8.4, Table 8.3), which can illustrate the strength of the tides. The funnel shape geomorphology amplifies tidal wave energy and causes larger tidal range upstream. The tidal range varies in section C1 in different time stages (Fig. 8.4a), with smaller tidal range previously and larger tidal range later. The variations of tidal ranges were mostly due to the impact of the geomorphology. The tidal range has reached 4.9 m at C2 in the 5th century (Fig. 8.4b). After the 16th century, the maximum tidal range at section C2 trended to decrease

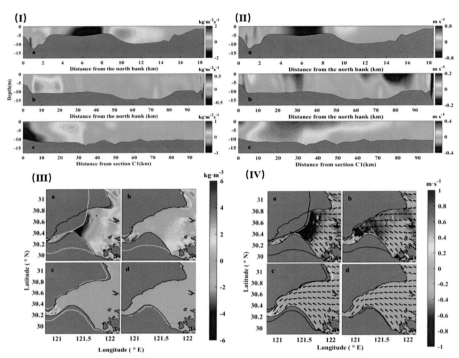

Figure 8.3 (I) Distribution of net sediment fluxes averaged over a spring-neap tidal cycle at section: (a) C1, (b) C2, and (c) C3. Positive values indicate landward fluxes at C1 and C2 sections, and northward fluxes at C3 section. (II) same as (I), but for residual currents. (III) changes of bottom SSC at peak flood during spring tides: (a) Exp1-Exp4, (b) Exp2-Exp4, (c) Exp3-Exp4, (d) Exp5-Exp4. (IV) same as (III) but for *bottom* currents. Vectors indicate current directions in Exp4.

(Fig. 8.4b). Since the 21st century, continuous construction of breakwaters, cofferdams, and other projects caused continuous increase in coastline length and reclamation area. Decreased bay width makes the tidal range tend to increase.

The moment of peak flood was selected to illustrate the currents and SSC. Fig. 8.3III showed the comparison of SSC between the historical models and the reference model at peak flood during spring tide. In the 5th century (Fig. 8.3IIIa), the bottom SSC was reduced significantly in the inner bay, where the main tidal channel in Exp4 was located and increased in the outer bay. In the 16th Century (Fig. 8.3IIIb), the southern coastlines were straighter than those in Exp4 and the curvature of southern bank was smaller. The bottom SSC was basically higher than those in Exp4 due to enhanced tidal prism. From the 16th Century to 2020, the peak values of bottom SSC decreased by about 2 kg/m^3 at the bay head but increased slightly at peak ebb. The peak values of bottom SSC increased less than 1 kg/m^3 occurred near Andong station (Fig. 8.1) in the outer bay. The peak SSC near the northern bank (Ganpu station, Fig. 8.1) decreased by 29.2%, while the SSC values near the southern bank (Zhenhai station) and in the main tidal channel (Tanhu station) increased by 21.3% and 20.8%. The variation of SSC values was consistent with the changes of tidal currents in the bay

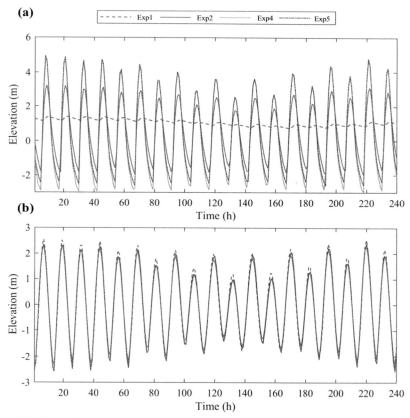

Figure 8.4 Sea surface elevation at (a) section C1 and (b) section C2.

Table 8.3 Maximum tidal ranges.

	Maximum tidal range (m)	
Experiments	**C1**	**C2**
Exp1	0.22	4.95
Exp2	5.02	4.64
Exp4	7.60	4.58
Exp5	7.40	4.90

(Li, Ye, Wang, et al., 2018). Historical changes of coastlines in Hangzhou Bay slightly decreased/increased the bottom SSC values in the inner/outer bay, respectively. The decrease/increase rate in the inner/outer bay reduced over time.

The variations of the SSC values were mainly due to the changes of tidal currents (Fig. 8.3IV) after the coastline variations (Xie et al., 2013). Tidal currents affected resuspension of SSC through modulating bottom stress. The local resuspension of the reclaimed areas also contributed a small part to the variations of the SSC values in the deep tidal channel. When the resuspension was inhibited near the southern bank in the numerical model, the locally resuspended sediment concentration was decreased subsequently.

Besides the local resuspension, the location of land reclamation also counted for the SSC changes. If coastlines changed at bay head (upstream of Ganpu), the impacts on SSC would be larger than that at the side banks. The impacts caused by land reduction at the bay head were more locally, with large impact at the bay head (1 kg/m^3) and small impact in the entire bay (0.2 kg/m^3). As tides were stronger and SSC values were large at the bay head, land increase at this area would generate large impacts.

2.3.2.2   Sediment fluxes

At the bay head (Ganpu section, C1, Fig. 8.5I), the characteristics of tidal averaged sediment fluxes distribution over the years were generally similar to the reference

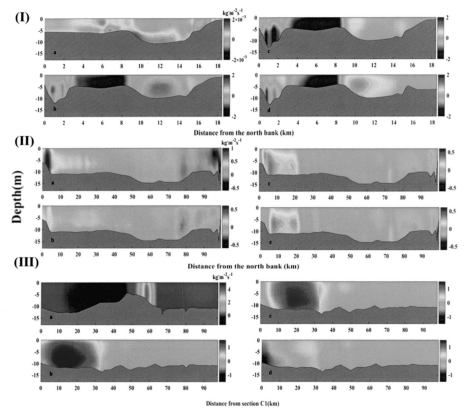

Figure 8.5 (I) Net sediment fluxes at section C1: (a) Exp1, (b) Exp2, (c) Exp3 and (d) Exp5. (II) Same as (I), but for section C2. (III) same as (I), but for section C3. Positive values indicate landward and northward fluxes.

model. In the 5th century (Fig. 8.5Ia), the surface/bottom sediment fluxes near northern bank were landward/seaward, and near the southern bank it was the opposite. However, the magnitude of the sediment fluxes was negligible. From the 16th century to 2020, the net sediment fluxes were seaward at the shallow water area and landward at the deep tidal channel. The seaward fluxes at the shallow water area increased and reached the peak in 1974 (Fig. 8.5Ic, more than 2 kg/m^{2}s), then decreased in 2020, and then increased gradually. The fluxes at the tidal channel were seaward in the 16th century, then turned landward and gradually increased until 2020.

At the bay mouth (Luchaogang section C2, Fig. 8.5II), the distribution and magnitude of sediment fluxes over 1400 years were similar to the reference model but had reversed trend compared with those at C1 section. The landward sediment fluxes near northern bank were significant in the 5th century. In the 16th century, the sediment fluxes were landward at a distance of 5 km from the northern bank and extended to 20 km in other years. The landward fluxes were stronger before the 16th century and weaker after 1974.

In the Lateral (Ganpu to Luchaogang section C3, Fig. 8.5III), the distribution of sediment fluxes over 1400 years was similar to the reference model. In the 5th century (Fig. 8.5IIIa), the sediment fluxes were generally southward, except northward near Jinshan. The northward sediment fluxes near Zhapu decreased after the 16th century, while the southward sediment fluxes near Ganpu increased. Between Zhapu and Luchaogang, the surface sediment fluxes were northward and the bottom sediment fluxes were southward. The northward sediment fluxes were mainly concentrated near Jinshan, while the southward sediment fluxes were distributed in the whole seabed.

3. Summary

Coastline is the physical interface of land and water. It is a collection of countless sea land boundaries between high and low tides, which is a strip in space, rather than a fixed geographical line. Coastline is constantly changing and reshaping, because of cross-shore and alongshore sediment movement in the littoral zone and especially because of the dynamic nature of water levels at the coastal boundary. Coastline changes refer to the retreat or advancement of the coastline, such as length fluctuation, morphological evolution, positional changes, transfer of utilization types, and spatial replacement of land and sea enclosed by the coastline.

Coastline changes interact with the hydro-sediment dynamics in estuaries and coastal zones. The case study shows that from the 5th century to 2020, the northern coastline of Hangzhou Bay (HZB) gradually concave to the north and convex to the east, while the southern coastline rapidly changed to northward. The tidal range has an increase trend since 21st century. The amplitudes of tides gradually increased from the bay mouth to the bay head, due to the decreased bay width. The suspended sediment concentration and fluxes are also impacted by the coastline changes, through changing currents. More data are needed to investigate the impacts of coastline changes on hydro-sediment dynamics, and attentions are needed when planning coastal development.

Acknowledgments

This research was supported by the National Natural Science Foundation of China (41976157, 42076177), Zhejiang Natural Science Foundation (2022C03044). Data and samples were partially collected onboard of R/V "Zheyuke2"/"Runjiang1" implementing the open research cruise NORC2021-03 supported by NSFC Shiptime Sharing Project (42049903).

References

Ali, A., Mynett, A. E., & Azam, M. H. (2007). Sediment dynamics in the meghna estuary, Bangladesh: A model study. *Journal of Waterway, Port, Coastal, and Ocean Engineering, 133*(4), 255–263.

Aquino Da Silva, A. G., Stattegger, K., Vital, H., & Schwarzer, K. (2019). Coastline change and offshore suspended sediment dynamics in a naturally developing delta (Parnaíba Delta, NE Brazil). *Marine Geology, 410*, 1–15.

Binh, D. V., Kantoush, S. A., & Sumi, T. (2019). Changes to long-term discharge and sediment loads in the Vietnamese Mekong Delta caused by upstream dams. *Geomorphology, 353*, Article 107011.

Boak, E. H., & Turner, I. L. (2005). Shoreline definition and detection: A review. *Journal of Coastal Research, 21*(4), 688–703. https://doi.org/10.2112/03-0071.1

Cao, J. (2018). *Numerical simulation on impacts of coastline changes on hydrodynamics in Hangzhou Bay*. Master, Zhejiang University. (in Chinese).

Chang, J., Liu, G. H., & Liu, Q. S. (2004). Remote sensing dynamic monitoring of the coastline in the Yellow River Delta. *Geo-Information Science*, (01), 94–98 (in Chinese).

Chen, A. J., Chen, C. F., & Chen, K. S. (1995). Investigation of shoreline change and migration along Wai-San-Ding-Zou barrier island, Central Western Taiwan. In *International geoscience and remote sensing symposium*. IEEE.

Chen, C., Cowles, G., & Beardsley, R. C. (2006). *An unstructured grid, finite-volume coastal ocean model: FVCOM user manual, SMAST/UMASSD technical report-04-0601*, 183 pp.

Chen, F., Lu, W., & Li, J. (2014). Practice and enlightenment of the construction and development of the keihin coastal industrial zone in Japan. *International Urban Planning*, (4), 109–115 (in Chinese).

Chen, K., Wang, Y., Lu, P., & Zheng, J. (2009). Effects of coastline changes on tide system of Yellow Sea off jiangsu coast, China. *China Ocean Engineering, 23*(4), 741–750.

Chen, Y. (2022). *Simulation of the effects of shoreline changes on hydrodynamics and water quality in Liaodong Bay based on FVCOM*. Master, Dalian University of Technology (in Chinese).

Choi, B. H., Kim, K. O., Lee, H. S., & Yuk, J. H. (2010). Perturbation of regional ocean tides due to coastal dikes. *Continental Shelf Research, 30*(6), 553–563.

Cui, B. L., & Li, X. Y. (2011). Coastline change of the yellow river estuary and its response to the sediment and runoff (1976–2005). *Geomorphology, 127*(1–2), 32–40.

Dai, S. B., Yang, S. L., & Cai, A. M. (2008). Impacts of dams on the sediment flux of the Pearl River, southern China. *Catena, 76*(1), 36—43.

Day, J. W., Agboola, J., Chen, Z., Elia, C. D., Forbes, D. L., Giosan, L., Kemp, P., Kuenzer, C., Lane, R. R., & Ramachandran, R. (2016). Approaches to defining deltaic sustainability in the 21st century. *Estuarine, Coastal and Shelf Science.* https://doi.org/10.1016/j.ecss.2016.06.018

Dolan, R., Hayden, B. P., May, P., & May, S. (1980). Reliability of shoreline change measurements from aerial photographs. *Shore and Beach, 48*(4), 22—29.

Dong, D. X., Li, Y. C., Chen, X. Y., Chen, B., & Ya, H. Z. (2015). Impacts of ocean engineering on shoreline, topography and deposition-erosion environment in Qinzhou Gulf. *Guangxi Sciences*, (3), 266—273 (in Chinese).

Dong, J., Sun, J., Chen, Y., Liu, C., Ma, H., & Liu, C. (2020). Numerical simulation of the hydrodynamics affected by coastline and bathymetry changes in the Bohai Sea. *Advances in Marine Science, 38*(4), 676—687. https://doi.org/10.3969/j.issn.1671-6647.2020.04.011 (in Chinese).

Du, P., Ding, P., & Hu, K. (2010). Simulation of three-dimensional cohesive sediment transport in Hangzhou Bay, China. *Acta Oceanologica Sinica, 29*(2), 98—106.

Fan, D., Tu, J., Shang, S., & Cai, G. (2014). Characteristics of tidal-bore deposits and facies associations in the Qiantang Estuary, China. *Marine Geology, 348*, 1—14.

Flemming, B. W., & Nyandwi, N. (1994). Land reclamation as a cause of fine-grained sediment depletion in backbarrier tidal flats (Southern North Sea). *Netherlands Journal of Aquatic Ecology, 28*(3—4), 299—307.

Gao, G. D., Wang, X. H., & Bao, X. W. (2014). Land reclamation and its impact on tidal dynamics in Jiaozhou Bay, Qingdao, China. *Estuarine, Coastal and Shelf Science, 151*, 285—294.

Gao, G. D., Wang, X. H., Bao, X. W., Song, D., Lin, X. P., & Qiao, L. L. (2018). The impacts of land reclamation on suspended-sediment dynamics in Jiaozhou Bay, Qingdao, China. *Estuarine, Coastal and Shelf Science, 206*, 61—75.

Ge, X. Z., Sun, X. L., & Liu, Z. Q. (2014). *Object-oriented coastline classification and extraction from remote sensing imagery.* https://doi.org/10.1117/12.2063845

Groot, R. S., Wilson, M. A., & Boumans, R. M. J. (2002). A typology for the classification, description and valuation of ecosystem functions, goods and services. *Ecological Economics, 41*(3), 393—408.

Guariglia, A., Buonamassa, A., Losurdo, A., Saladino, R., & Colangelo, A. (2009). A multisource approach for coastline mapping and identification of shoreline changes. *Annals of Geophysics, 49*(1), 295—304. https://doi.org/10.4401/ag-3155

Guo, W. (2017). *The time-varying characteristics of tidal duration asymmetry and its response to project.* Doctor, East China Normal University (in Chinese).

Guo, X., Wang, C., Chen, C., & Tang, J. (2014). Numerical simulation of the transport diffusion of suspended matter during the construction of Fengwei reclamation project at Meizhou Bay. *Journal of Applied Oceanography, 33*(1), 8 (in Chinese).

Guo, Y., Wu, X., Pan, C., & Zhang, J. (2012). Numerical simulation of the tidal flow and suspended sediment transport in the Qiantang estuary. *Journal of Waterway, Port, Coastal, and Ocean Engineering, 138*(3), 192—202.

Ha, H. K., Ha, H. J., Seo, J. Y., & Choi, S. M. (2018). Effects of vegetation and fecal pellets on the erodibility of cohesive sediments: Ganghwa tidal flat, west coast of Korea. *Environmental Pollution, 241*, 468—474.

Han, M. W., & Park, Y. C. (1999). The development of anoxia in the artificial lake shihwa, korea, as a consequence of intertidal reclamation. *Marine Pollution Bulletin, 38*(12), 1194−1199.

Hasegawa, D., Sheng, J., Greenberg, D. A., & Thompson, K. R. (2011). Far-field effects of tidal energy extraction in the Minas Passage on tidal circulation in the Bay of Fundy and Gulf of Maine using a nested-grid coastal circulation model. *Ocean Dynamics, 61*(11), Article 1845e1868. https://doi.org/10.1007/s10236-011-0481-9

Healy, M. G., & Hickey, K. R. (2002). Historic land reclamation in the intertidal wetlands of the Shannon estuary, western Ireland. *Journal of Coastal Research, 36*(sp1), 365−373.

Huang, G. Z., Zhu, T., & Cao, Y. (2013). Reflections and prospects on ecological reclamation of land from reclamation in China. *Future and Development*, (5), 12−17 (in Chinese).

Huang, S., Bo, Y., Zhu, Y., Xie, H., & Nie, H. (2022). Analysis of hydrodynamic response for shoreline changing of the pushen strait. *Journal of North China University of Water Resources and Electric Power, 43*(3), 66−74 (in Chinese).

Jiang, X., Lu, B., & He, Y. (2013). Response of the turbidity maximum zone to fluctuations in sediment discharge from river to estuary in the Changjiang Estuary (China). *Estuarine, Coastal and Shelf Science, 131*(0), 24−30.

Jun, C., Huan, L. G., & Sheng, L. Q. (2004). Analysis on spatio-temporal feature of coastline change in the Yellow River Estuary and its relation with runoff and sand-transportation. *Geographical Research, 23*(3), 339−346.

Kang, J. W. (1999). Changes in tidal characteristics as a result of the construction of sea-dike/ Sea-walls in the Mokpo coastal zone in korea. *Estuarine, Coastal and Shelf Science, 48*(4), 429−438.

Kemp, G. P., Day, J. W., Rogers, J. D., Giosan, L., & Peyronnin, N. (2016). Enhancing mud supply from the lower Missouri river to the Mississippi River Delta USA: Dam bypassing and coastal restoration. *Estuarine, Coastal and Shelf Science, 183*(Part B), 304−313.

Khan, U., Janjuhah, H. T., Kontakiotis, G., Rehman, A., & Zarkogiannis, S. D. (2021). Natural processes and anthropogenic activity in the indus river sedimentary environment in Pakistan: A critical review. *Journal of Marine Science and Engineering, 9*(10), 1109.

Lee, M. O., Park, S. J., & Kang, T. S. (2006). Influence of reclamation works on the marine environment in a semi-enclosed bay. *Journal of Ocean University of China*, (03), 219−227.

Le Hir, P., Roberts, W., Cazaillet, O., Christie, M., Bassoullet, P., & Bacher, C. (2000). Characterization of intertidal flat hydrodynamics. *Continental Shelf Research, 20*(12−13), 1433−1459.

Le Provost, C., & Lyard, F. (1997). Energetics of the M2 barotropic ocean tides: An estimate of bottom friction dissipation from a hydrodynamic model. *Progress in Oceanography, 40*(1e4), 37−52.

Lee, H. J., & Ryu, S. O. (2008). Changes in topography and surface sediments by the Saemangeum dyke in an estuarine complex, west coast of Korea. *Continental Shelf Research, 28*(9), 1177−1189.

Li, Z. L., & Yao, Y. (2018). World miracle artificial island − Dubai Palm Island. *Builders' Monthly*, (03), 59 (in Chinese).

Li, C., Gao, S., Wang, Y., & Li, C. (2018). Sediment flux from the zhoushan archipelago, eastern China. *Journal of Geographical Sciences, 28*(4), 387−399.

Li, L., Guan, W., Hu, J., Cheng, P., & Wang, X. H. (2018). Responses of water environment to tidal flat reduction in xiangshan bay: Part I hydrodynamics. *Estuarine, Coastal and Shelf Science, 206*, 14−26.

Li, L., Ye, T., Zhang, L., Yao, Y., & Xia, Y. (2019). Influences of reclamation location on the water and sediment environment in Hangzhou Bay. *Journal of Harbin Engineering University, 40*(11), 1870−1875 (in Chinese).

Li, L., Ye, T., Wang, X. H., He, Z., & Shao, M. (2018). Changes of hydrodynamics of Hangzhou Bay due to land reclamation in the past 60 years. In *Sediment dynamics of Chinese muddy coasts and estuaries: Physics, biology and their interactions, accepted, 2018.07*. Elservier.

Li, L., Ren, Y., Ye, T., Wang, X. H., Hu, J., & Xia, Y. (2023). Positive feedback between the tidal flat variations and sediment dynamics: An example study in the macro-tidal turbid Hangzhou Bay. *Journal of Geophysical Research: Oceans, 128*, e2022JC019414. https://doi.org/10.1029/2022JC019414

Li, L., Wang, X. H., Andutta, F., & Williams, D. (2014). Effects of mangroves and tidal flats on suspended-sediment dynamics: Observational and numerical study of Darwin Harbour, Australia. *Journal of Geophysical Research: Oceans, 119*, 5854−5873. https://doi.org/10.1002/2014JC009987

Li, L., Wang, X. H., Williams, D., Sidhu, H., & Song, D. (2012). Numerical study of the effects of mangrove areas and tidal flats on tides: A case study of Darwin harbour, Australia. *Journal of Geophysical Research, 117*, Article C06011.

Li, L., Ye, T., He, Z., & Xia, Y. (2018). A numerical study on the effect of tidal flat's slope on tidal dynamics in the Xiangshan Bay, China. *Acta Oceanologica Sinica, 37*(9), 29−40.

Liang, H., Kuang, C., Gu, J., Ma, Y., Chen, K., & Liu, X. (2019). Tidal asymmetry changes in a shallow mud estuary by a restoration project. *Journal of Ocean University of China, 18*(02), 339−348.

Liang, L., Liu, Q. S., Liu, G. H., Li, X. Y., & Huang, C. (2018). Overview of coastal line extraction methods based on remote sensing images. *Journal of Earth Information Science*, (12), 1745−1755 (in Chinese).

Lin, H., Xu, J., Jiang, D., Gao, Y., & Liu, J. (2016). Sand dam dynamic monitoring in coastal areas based on time-series remote sensing images. In *Igarss 2016 - 2016 IEEE international geoscience and remote sensing symposium*. IEEE.

Lu, L., Goh, B. P. L., & Chou, L. M. (2002). Effects of coastal reclamation on riverine macrobenthic infauna (Sungei Punggol) in Singapore. *Journal of Aquatic Ecosystem Stress and Recovery, 9*(2), 127−135.

Lu, S., Wu, B., Yan, N., & Wang, H. (2011). Water body mapping method with hj-1a/b satellite imagery. *International Journal of Applied Earth Observations & Geoinformation, 13*(3), 428−434.

Lu, X. X., & Chua, S. (2021). River discharge and water level changes in the mekong River: Droughts in an era of mega-dams. *Hydrological Processes*.

Lu, Y. P., Liang, S. X., Sun, Z. C., & Cong, P. F. (2015). Cumulative effects of topography change on waterway's hydrodynamic along the southern coast of Hangzhou Bay. *Marine Environmental Science, 34*(3), 384−390.

Lui, S. W., Zhang, J., Ma, Y., & Sun, W. F. (2011). Coastline extraction method based on remote sensing and DEM. *Remote Sensing Technology and Application, 26*(5), 613−618. doi: CNKI:SUN:YGJS.0.2011-05-011.

Luo, X. X., Yang, S. L., & Zhang, J. (2012). The impact of the Three Gorges Dam on the downstream distribution and texture of sediments along the middle and lower Yangtze River (Changjiang) and its estuary, and subsequent sediment dispersal in the East China Sea. *Geomorphology, 179*, 126−140.

Ma, X. F., Zhao, D. Z., Xing, X. G., Zhang, F. S., Wen, S. Y., & Yang, F. (2007). Research on the method of extracting coastline satellite remote sensing. *Marine environmental science, 26*(2), 5 (in Chinese).

Montreuil, A., & Bullard, J. E. (2012). A 150-year record of coastline dynamics within a sediment cell: Eastern England. *Geomorphology, 179*, 168−185.

Mostafa, Y. (2012). Environmental impacts of dredging and land reclamation at Abu Qir Bay, Egypt. *Ain Shams Engineering Journal, 3*(1), 1−15. https://doi.org/10.1016/j.asej.2011. 12.004

Nardin, W., Woodcock, C. E., & Fagherazzi, S. (2016). Bottom sediments affect sonneratia, mangrove forests in the prograding mekong delta, vietnam. *Estuarine, Coastal and Shelf Science, 177*, 60−70.

Nichols, F. H., Cloern, J. E., Luoma, S. N., & Peterson, D. H. (1986). The modification of an estuary. *Science, 231*(4738), 567−573.

Pan, C. H., Zeng, J., Tang, Z. W., & Shi, Y. B. (2013). A study of sediment characteristics and riverbed erosion/deposition in Qiantang Estuary. *Hydro-Science and Engineering*, (1), 1−7.

Park, S. R., Kim, J., Kang, C., An, S., Chung, I. K., Kim, J. H., & Lee, K. (2009). Current status and ecological roles of Zostera marina after recovery from large-scale reclamation in the Nakdong River estuary, Korea. *Estuarine, Coastal and Shelf Science, 81*(1), 38−48.

Pavel, V., Raubenheimer, B., & Elgar, S. (2013). Processes controlling stratification on the northern Skagit Bay tidal flats. *Continental Shelf Research, 60*(Suppl. ment), S30−S39.

Qin, Y., Zhang, L., Zheng, B., Cao, W., Liu, X., & Jia, J. (2012). Impact of shoreline changes on the costal water quality of Bohai Bay (2003−2011). *Acta Scientiae Circumstantiae, 32*(09), 2149−2159 (in Chinese).

Revision Committee of the Water Conservancy Dictionary of Hohai University. (2015). *A dictionary of water resources*. Shanghai Lexicographical Publishing House (in Chinese).

Sato, S. I., & Azuma, M. (2002). Ecological and paleoecological implications of the rapid increase and decrease of an introduced bivalve Potamocorbula sp. after the construction of a reclamation dike in Isahaya Bay, western Kyushu, Japan. *Palaeogeography, Palaeoclimatology, Palaeoecology, 185*(3), 369−378.

Shen, F., Zhou, X., Li, J., He, Q., & Wouter, V. (2013). Remotely sensed variability of the suspended sediment concentration and its response to decreased river discharge in the Yangtze estuary and adjacent coast. *Continental Shelf Research, 69*, 52−61.

Shi, J., Li, G., & Wang, P. (2011). Anthropogenic influences on the tidal prism and water exchanges in Jiaozhou bay, Qingdao, China. *Journal of Coastal Research, 27*(1), 57−72.

Sieyes, N. R., Yamahara, K. M., Paytan, A., & Boehm, A. B. (2011). Submarine groundwater discharge to a high-energy surf zone at stinson beach, California, estimated using radium isotopes. *Estuaries and Coasts, 34*(2), 256−268.

Song, D., Wang, X. H., Cao, Z., & Guan, W. (2013). Suspended sediment transport in the deepwater navigation channel, yangtze River Estuary, China, in the dry season 2009: 1. Observations over spring and neap tidal cycles. *Journal of Geophysical Research: Oceans, 118*(10), 5555−5567.

Song, D., Wang, X. H., Zhu, X., & Bao, X. (2013). Modeling studies of the far-field effects of tidal flat reclamation on tidal dynamics in the East China Seas. *Estuarine, Coastal and Shelf Science, 133*, 147−160.

Stockdonf, H. F., Sallenger, A. H., & Holman, L. R. A. (2002). Estimation of shoreline position and change using airborne topographic lidar data. *Journal of Coastal Research, 18*(3), 502−513.

Stokes, D. J., Healy, T. R., & Cooke, P. J. (2010). Expansion dynamics of monospecific, temperate mangroves and sedimentation in two embayments of a barrier-enclosed lagoon,

Tauranga Harbour, New Zealand. *Coastal Education and Research Foundation, 26*(1), 113—122.

Suh, S. W., Lee, H. Y., & Kim, H. J. (2014). Spatio-temporal variability of tidal asymmetry due to multiple coastal constructions along the west coast of korea. *Estuarine, Coastal and Shelf Science, 151*, 336—346.

Sun, W., Ma, Y., Zhang, J., Liu, W., & Ren, G. (2011). Study of remote sensing interpretation keys and extraction technique of different types of shoreline. *Bulletin of Surveying and Mapping, 3*, 41—44 (in Chinese).

Vörösmarty, C. J., Meybeck, M., Fekete, B., Sharma, K., Green, P., & Syvitski, J. P. M. (2003). Anthropogenic sediment retention: Major global impact from registered river impoundments. *Global and Planetary Change, 39*(1—2), 169—190. https://doi.org/10.1016/S0921-8181(03)00023-7

Walling, D. E. (2011). *Human impact on the sediment loads of Asian rivers.* IAHS-AISH publication.

Wang, R., & Zhang, G. (2023). Analysis of temporal and spatial variation of tidal wave characteristics in Macao under the influence of reclamation [analysis of temporal and spatial variation of tidal wave characteristics in Macao under the influence of reclamation]. *Pearl River, 44*(2), 76—80 (in Chinese).

Wang, S., & Lv, L. (2019). The hydrographic survey and tide control of the Qiantang Estuary during the reign of Qianlong of the Qing Dynasty. *Journal of Dialectics of Nature, 41*(6), 36—43 (in Chinese).

Wang, X. H. (2002). Tide-induced sediment resuspension and the bottom boundary layer in an idealized estuary. *Journal of Physical Oceanography, 32*, 3113—3131.

Wang, Y., & Zhu, D. (1994). Tidal flats in China. In D. Zhou, Y. Liang, & C. Zeng (Eds.), *Oceanology of China seas* (pp. 445—456). Dordrecht, The Netherlands: Kluwer Academic Publishers.

Wang, C. Y., Wang, Z. R., Chu, J. L., & Zhao, J. H. (2017). High resolution image coastline extraction method based on decision tree and density clustering. *Marine environmental science, 36*(4), 6 (in Chinese).

Wu, J., Fu, C., Lu, F., & Chen, J. (2005). Changes in free-living nematode community structure in relation to progressive land reclamation at an intertidal marsh. *Applied Soil Ecology, 29*(1), 47—58.

Xie, D. F., Gao, S., Wang, Z. B., & Pan, C. H. (2013). Numerical modeling of tidal currents, sediment transport and morphological evolution in Hangzhou Bay, China. *International journal of sediment research, 28*(3), 316—328.

Xie, D., Gao, S., Wang, Z. B., Pan, C., Wu, X., & Wang, Q. (2017). Morphodynamic modeling of a large inside sandbar and its dextral morphology in a convergent estuary: Qiantang Estuary, China. *Journal of Geophysical Research: Earth Surface, 122*(8), 1553—1572.

Xie, D., Pan, C., Wu, X., Gao, S., & Wang, Z. B. (2017a). Local human activities overwhelm decreased sediment supply from the changjiang river: Continued rapid accumulation in the Hangzhou bay-qiantang estuary system. *Marine Geology, 392*, 66—77.

Xie, D., Pan, C., Wu, X., Gao, S., & Wang, Z. B. (2017b). The variations of sediment transport patterns in the outer Changjiang Estuary and Hangzhou Bay over the last 30 years. *Journal of Geophysical Research: Oceans, 122*(4), 2999—3020.

Xie, M. H., Zhang, Y. F., & Fu, K. (2007). Algorithm of detection coastline from sar images based on seeds growing. *Journal of the University of the Chinese Academy of Sciences, 24*(1), 93—98 (in Chinese).

Xiong, S. (2011). *Fluvial process and regulation for tidal estuary.* Beijing: China Water&Power Press.

Xu, W. Q. (2016). *Influence of Laizhou Bay reclamation project on the surrounding hydrodynamics and sediment.* Doctoral dissertation. Dalian University of Technology.

Xu, Y., Wang, Y., Hu, S., Zhu, Y., Zuo, J., & Zeng, J. (2023). Study on the impact of the coastline changes on hydrodynamics in Xiangshan bay. *Applied sciences, 13*(14), 8071.

Ya, H., Xu, Y., Li, Y., & Dong, D. (2017). Effects of shoreline change on hydrodynamic environment in Qinzhou bay. *Guangxi Science, 24*(03), 311−315 (in Chinese).

Yan, H., Li, B., & Chen, M. (2009). Progress of researches in coastline extraction based on RS technique. *Areal Research and Development, 28*(1), 101−105 (in Chinese).

Yao, J. (2021). *Influence of reclamation project on hydrodynamic environment in Bohai Bay.* Master. Tianjin University of Science and Technology (in Chinese).

Yin, S., Gao, G., Huang, A., Li, D., Ran, L., Nawaz, M., Xu, Y. J., & Fu, B. (2023). Streamflow and sediment load changes from China's large rivers: Quantitative contributions of climate and human activity factors. *Science of the Total Environment, 876*, Article 162758.

Yu, Q., Wang, Y., Gao, S., & Flemming, B. (2012). Modeling the formation of a sand bar within a large funnel-shaped, tide-dominated estuary: Qiantangjiang Estuary, China. *Marine Geology, 299−302*(1), 63−76.

Yu, Z. Z., He, Z. G., Li, L., Ye, T. Y., & Xia, Y. Z. (2021). Exchange mechanism of the suspended sediment at the mouth of Hangzhou bay under coastline changes. *E3S Web of Conferences, 233*, 3035.

Zang, N., Sun, Y., & Li, Y. (2009). Prediction on impact of change of land boundary on distribution change of suspended sediment. *Transactions of Oceanology and Limnology*, (3), 6 (in Chinese).

Zhang, H., Guo, Y., Huang, W., & Zhou, C. (2005). A remote sensing investigation of inning and silting in Hangzhou Bay since 1986. *Remote Sensing for Land & Resources, 2*, 50−54.

Zhang, Q., Jin, Y., Li, X., Wang, L., & Ye, F. (2017). Progress in the impact of reclamation projects on offshore marine environment. *Advances in Marine Science, 35*(4), 454−461. https://doi.org/10.3969/j.issn.1671-6647.2017.04.002 (in Chinese).

Zhang, T., Sun, Z., & Niu, X. (2021). *Analysis on the influence of coastline change on hydrodynamics in bays around the Bohai Sea* (in Chinese).

Zhang, W., Ruan, X., Zheng, J., Zhu, Y., & Wu, H. (2010). Long-term change in tidal dynamics and its cause in the Pearl River delta, China. *Geomorphology, 120*(3), 209−223.

Zhang, Z. H. (2013). *The integration of national spatial planning and ecological protection: Inspiration from the Dutch land reclamation project* (Vol 05, pp. 66−71+4). China Three Gorges (in Chinese).

Zhao, B., Guo, H., Yan, Y., Wang, Q., & Li, B. (2008). A simple waterline approach for tidelands using multi-temporal satellite images: A case study in the yangtze delta. *Estuarine, Coastal and Shelf Science, 77*(1), 134−142.

Zhong, X. (2017). *Morphodynamics of the beaches around Hainan Island: The normal process, the influences of extreme events and artificial island construction.* Doctor. East China Normal University (in Chinese).

Zhou, Q. (2022). *Numerical simulation on the influence of qiantang estuary reclamation on the hydrodynamics and water quality of Hangzhou bay based on FVCOM.* Master. Dalian University of Technology (in Chinese).

Zhou, Y., Chen, T., & Zhang, H. (2016). The spatial evolution and scale change trends of reclamation areas in Singapore. *International Urban Planning, 03*, 71−77 (in Chinese).

Zhu, L., Hu, R., Zhu, H., Jiang, S., Xu, Y., & Wang, N. (2018). Modeling studies of tidal dynamics and the associated responses to coastline changes in the Bohai Sea, China. *Ocean Dynamics, 68*(12), 1625−1648.

Zhu, Q., Wang, Y. P., Ni, W., Gao, J., Li, M., Yang, L., et al. (2016). Effects of intertidal reclamation on tides and potential environmental risks: A numerical study for the southern Yellow Sea. *Environmental Earth Sciences, 75*(23), 1472.

Zhuang, C. R. (2009). Research on remote sensing dynamic monitoring of xiamen coastline. *Safety and Environmental Engineering, 16*(3), 5 (in Chinese).

Further reading

Allen, J. I., Somerfield, P. J., & Gilbert, F. J. (2007). Quantifying uncertainty in high-resolution coupled hydrodynamic-ecosystem models. *Journal of Marine Systems, 64*(1—4), 3—14.

Amiruddin. (2018). Distribution of basic sediments (bedload transport) on changes in coastal coastline Donggala, Central Sulawesi Province, Indonesia. *Journal of Physics: Conference Series, 983*(1), Article 12032.

An, B. (2016). *Numerical studies and application of tidal dynamics based on FVCOM in Hangzhou Bay and adjacent seas* (Vol 20). Shanghai Ocean University (in Chinese).

Chant, R. J., & Wilson, R. E. (1997). Secondary circulation in a highly stratified estuary. *Journal of Geophysical Research: Oceans, 102*(C10), 23207—23215.

Feng, J. L., Li, W. S., Wang, H., et al. (2018). Evaluation of sea level rise and associated responses in Hangzhou Bay from 1978 to 2017. *Advances in Climate Change Research, 9*(4), 227—233.

Guo, W., Wang, X. H., Ding, P., et al. (2018). A system shift in tidal choking due to the construction of Yangshan Harbour, Shanghai, China. *Estuarine, Coastal and Shelf Science, 206*, 49—60. J.

He, X., Bai, Y., Pan, D., Huang, N., Dong, X., Chen, J., et al. (2013). Using geostationary satellite ocean color data to map the diurnal dynamics of suspended particulate matter in coastal waters. *Remote Sensing of Environment, 133*(12), 225—239.

He, X., Bai, Y., Pan, D., Tang, J., & Wang, D. (2012). Atmospheric correction of satellite ocean color imagery using the ultraviolet wavelength for highly turbid waters. *Optics Express, 20*(18), 20754—20770.

Hu, R. J. (2009). *Sediment transport and dynamic mechanism in the Zhoushan Archipelago sea area, PhD thesis* (p. 140). Ocean University of China.

Kalkwijk, J. P. T., & Booij, R. (1986). Adaptation of secondary flow in nearly horizontal flow. *Journal of Hydraulic Research, 24*(1), 19—37.

Koutitas, C. G. (1988). *Mathematical models in coastal engineering*. London: Pentech Press, 56 pp.

Liu, C., Sui, J., He, Y., & Hirshfield, F. (2013). Changes in runoff and sediment load from major Chinese rivers to the Pacific Ocean over the period 1955—2010. *International Journal of Sediment Research, 28*(4), 486—495.

Murphy, A. H. (1992). Climatology, persistence, and their linear combination as standards of reference in skill scores. *Weather and Forecasting, 4*(7), 692—698.

Nidzieko, N. J., Hench, J. L., & Monismith, S. G. (2009). Lateral circulation in well-mixed and stratified estuarine flows with curvature. *Journal of Physical Oceanography, 39*(4), 831—851.

Sanford, L. P., & Halka, J. P. (1993). Assessing the paradigm of mutually exclusive erosion and deposition of mud, with examples from upper Chesapeake bay. *Marine Geology, 114*(93), 37—57.

Sheng, H., Guo, M., Gan, V., Xu, M., Liu, S., Yasir, M., Cui, J., & Wan, J. (2022). Coastline extraction based on multi-scale segmentation and multi-level inheritance classification. *Frontiers in Marine Science.* https://doi.org/10.3389/fmars.2022.1031417

Tang, J. H., He, Q., Wang, Y. Y., & Liu, H. (2008). Study on in-situ flocs size in turbidity maximum of the Changjiang Estuary. *Journal of Sediment Research, 02,* 27−33.

Van Prooijen, B. C., & Winterwerp, J. C. (2010). A stochastic formulation for erosion of cohesive sediments. *Journal of Geophysical Research, 115*(C1), 1−15.

Wang, Y., Zhu, D., & Wu, X. (2002). Tidal flats and associated muddy coast of China. In T. Healy, Y. Wang, & J.-A. Healy (Eds.), *Muddy coasts of the world: Processes, deposits and function* (pp. 319−346). Amsterdam, The Netherlands: Elsevier Science B.V.

Warner, J. C., Sherwood, C. R., Signell, R. P., Harris, C. K., & Arango, H. G. (2008). Development of a three-dimensional, regional, coupled wave, current, and sediment-transport model. *Computers & Geosciences, 34*(10), 1284−1306.

Winterwerp, J. C. (2001). Stratification effects by cohesive and non-cohesive sediment. *Journal of Geophysical Research: Oceans, 106*(C10), 22559−22574.

Wu, L., Chen, C., Guo, P., Shi, M., Qi, J., & Ge, J. (2011). An FVCOM-based unstructured grid wave, current, sediment transport model, I. Model description and validation. *Journal of Ocean University of China, 10*(1), 1−8.

Xiao, Z., Wang, X. H., Roughan, M., & Harrison, D. (2018). Numerical modelling of the sydney estuary, new south wales: Lateral circulation and asymmetric vertical mixing, estuarine. *Coastal and Shelf Science, 217,* 132−147.

Zhang, J., Chu, D., Wang, D., Cao, A., Lv, X., & Fan, D. (2018). Estimation of spatially varying parameters in three-dimensional cohesive sediment transport models by assimilating remote sensing data. *Journal of Marine Science and Technology, 23,* 319−332.

Zhu, L., He, Q., Shen, J., & Wang, Y. (2016). The influence of human activities on morphodynamics and alteration of sediment source and sink in the Changjiang Estuary. *Geomorphology, 273,* 52−62.

Modeling study on hydrodynamics and sediment dynamics in estuarine environment: Two case studies in Yalu River Estuary and Batemans Bay in China and Australia

9

Gang Yang
Nanjing University of Information Science and Technology School of Marine Sciences, Nanjing, China

1. Introduction

The coastal zone is an important geographical unit connecting the ocean and terrestrial areas, where frequent interactions and transformation of energy and materials occurred (Gao et al., 2014; Harff et al., 2005; Yang et al., 2020). The river discharge, tidal current, biogeochemical processes, wind-driven waves, and estuarine circulation influence the sediment dynamics in estuaries, which make coastal zones highly complex natural systems (Neal et al., 2003; Jing & Ridd, 1997; Cheng et al., 2016; Yang et al., 2018). In recent decades, the number of large-scale industries in coastal areas has increased dramatically, influencing the local sediment environment and leaded to the severe water pollution and damages to the coastal ecosystems. The coastal hydrodynamics and sediment dynamics are fundamental for estuarine environment, which can have large impact on the geomorphology, light attenuation, primary production, and ecosystems. For example, high suspended sediment concentration can degrade the water quality and cause the reduction of the phytoplankton biomass due to the low level of available light.

Suspended sediments in estuaries have unique characteristics compared to those found in open waters: (1) generally, there is a higher suspended sediment concentration in estuaries; (2) their transport is strongly influenced by baroclinic effects, especially when the river discharge is large; (3) their transport is greatly impacted by tidal and wave effects; and (4) human activities, including coastal engineering works such as land reclamation and channel deepening, can also have an important influence on the coastal sediment dynamics (Dijkstra et al., 2019; Ralston et al., 2019; Song et al., 2013; Xie et al., 2017; Yang et al., 2021). The deposition and resuspension of sediments in long-term can eventually alter the coastline and morphology of seabed in coastal areas and then change the local hydrodynamics (Cheng et al., 2020).

Current Trends in Estuarine and Coastal Dynamics. https://doi.org/10.1016/B978-0-443-21728-9.00009-0

The turbidity maximum zone (TMZ) is an important feature in estuarine sediment dynamics. The TMZ is mainly influenced by four factors: fluvial sediment supply, local sediment resuspension, tidal pumping, and estuarine circulation (Dyer, 1997, pp. 0−471; Yu et al., 2014). Firstly, river sediment supply can provide sediment source for the formation of TMZ in estuaries. Secondly, resuspension is controlled by bottom stress: strong local sediment resuspension caused by large bottom stress can also offer the sediment source for the TMZ. Thirdly, tidal pumping, denotes the barotropic suspended sediment transport caused by tidal velocity asymmetry and sediment lag effects (Brenon & Le Hir, 1999; Dyer, 1997, pp. 0−471; Uncles & Stephens, 1989; Yu et al., 2014). The convergence of tidal pumping flux can contribute to the formation of TMZ. Lastly, the estuarine circulation, which is a complex process and often occurs at the upper estuary with bottom landward flow and surface seaward flow, mainly induced by gravitational circulation and asymmetric tidal mixing (ATM) (Burchard & Baumert, 1998; Burchard & Hetland, 2010; Cheng et al., 2011; Song et al., 2013). For the gravitational circulation, freshwater discharge can generate seaward river plumes, along-channel density gradients, and occasionally strong vertical stratification. The along-channel density gradient can induce bottom-layer landward saltwater intrusion and seaward flow near the sea surface, which may dominate the sediment flux around the river entrance (Cho et al., 2016; Giddings & MacCready, 2017; Goodrich & Blumberg, 1991; Grasso et al., 2018; Pritchard, 1956; Hansen & Rattray, 1966). ATM is caused by strain-induced periodic stratification. Tidal currents can stratify the water column through the straining of the density field during ebb tides (low vertical mixing coefficient Kh) and de-stratify it during flood tides (large Kh), which can induce the ATM leading to a bottom-layer landward residual flow and seaward flow at upper layers, and a similar pattern with gravitational circulation. Based on the study of Burchard and Hetland (2010), gravitational circulation contributes one-third, and ATM contributes two-thirds to the estuarine circulation, without considering residual runoff and surface wind straining. In weakly stratified estuaries, ATM-induced flow can even be stronger than density-driven flow (Cheng et al., 2011). The enhanced estuarine circulation caused by ATM and gravitational circulation can control the TMZ formation: with upstream residual flow and higher SSC at the bottom, TMZ can be formed at a location where the residual current is converged (Burchard & Baumert, 1998; Festa & Hansen, 1976; Postma, 1967; Yu et al., 2014).

The cross-channel estuarine circulation (lateral circulation) also plays a critical role in TMZ location and sediment transport (Chen et al., 2020; Lerczak & Rockwell Geyer, 2004; Valle-Levinson et al., 2003). Lateral circulation is affected by numerous factors: density gradients induced by differential advection (Lerczak & Rockwell Geyer, 2004); curvature effect due to topography (Chant, 2002; Xiao et al., 2019); and the combined effect of advection and baroclinic pressure gradient (Scully et al., 2009).

In macro-tidal areas, strong tidal currents can have large impact on the coastal sediment dynamics. The strong tidal currents can generate large bottom stress and resuspend the sediment on the seabed, increasing the concentration of suspended sediment in the waters. As introduced before, tidal velocity asymmetry and sediment lag effects can also influence the sediment transport (Brenon & Le Hir, 1999; Dyer, 1997, pp. 0−471; Uncles & Stephens, 1989; Yu et al., 2014).

The hydrodynamic and sediment dynamics in coastal areas are complex, influenced by many factors. In addition to the estuarine circulation and tidal effects, the effect of waves is also regarded as an important process contributing to the sediment dynamics change and beach erosion in coastal areas. Based on the wave-current model of Wang and Pinardi (2002), wave-driven sediment resuspension is an essential mechanism for sediment resuspending in shallow estuaries. Wang et al. (2007) found that wave forcing can significantly influence the sediment distributions and horizontal fluxes in the northern Adriatic Sea. The wind effects under storm events on sediment dynamics have been discussed by Lettmann et al. (2009) and can enhance the sediment erosion in the East Frisian Wadden Sea. Bever and MacWilliams (2013) applied the 3-D Untrim San Francisco Bay–Delta Model coupled with the SediMorph morphological model and SWAN wave model to simulate the tidal current and wave effects on sediment dynamics in San Pablo Bay and found that the erosion caused by wave during large wave events can be 3–8 times that in experiment without wave resuspension. According to the Delft-3D wave coupled model of Li et al. (2019), wave-induced bottom shear stress can significantly influence the sediment transport in the Ribble Estuary. Using field measurements of hydrodynamics, SSC, and bed thickness variations, Christiansen et al. (2006) found that the net sediment flux was three times higher in November with stronger wave than that in summer and SSC was more controlled by wave effects than by the tidal currents. Castelle et al. (2007) explored the beach erosion of Gold Coast, Australia, caused by three storm events in 2006 and suggested the large waves can induce severe erosion and the average beach width was narrow and may not withstand extreme wave event similar to the event occurred in 1967 (eight million cubic meters of sand was removed). Through the modeling and field data analysis, Dufois et al. (2014) explored the impact of two strong storms on sediment dynamics in the Gulf of Lions and found that seabed during these two events were eroded about 2 cm, with an SSC increase (up to about 50–100 mg/L), and the sediment dynamics was changed under the effect of waves. Xu and You (2017) analyzed the tidal currents and waves on sediment resuspension in Oujiang River, China, using modified environmental fluid dynamics code (EFDC) coupled with SWAN, and found the contribution of wave on sediment concentration in shallow estuaries can be higher than 60%. Loureiro and Cooper (2018) investigated the winter storms in north-western Ireland in the past 67 years and found these storm events can increase seabed erosion and potentially extend recovery times along the north-western coast of Ireland. Brand et al. (2010) explored the driving forces of sediment dynamics at the shoals in South San Francisco Bay using field data and found the wave-current nonlinear interaction can contribute to increase of the sediment flux in this areas; Olabarrieta et al. (2011) describes the importance of wave-current interaction in Willapa Bay and found the wave–current interaction effects can influence the sediment dynamics and morphology inside the estuary; Gao et al. (2018) explored the effect of each wave–current interaction mechanism on sediment dynamics under large wave events in a semi-closed bay, Jiaozhou Bay in China; Grasso et al. (2018) simulated the wave effects on the estuarine turbidity maximum (ETM) zone in Seine Estuary in France and found wave effects can increase the mass of ETM dramatically by three times; Yang et al. (2018, 2020) explored the hydrodynamic and sediment dynamic in Darwin

Harbour during the monsoon season using the satellite data and three-dimensional Finite-volume Coastal Ocean Model (FVCOM); they found that the large wave can cause large bottom stress, resuspend large amount of sediment and increase the magnitude of the sediment flux outside the harbor; Li et al. (2019) retrieved the surface SSC using satellite images in Hangzhou Bay, China, and found that SSC was enhanced during the large wave events in dry seasons; based on the analysis of Landsat eight satellite data in Yellow River estuary by Qiu et al. (2017), sediment distribution shows significant seasonal variations under wind-wave effects. Liu et al. (2020) highlighted a strong erosion event occurred in the presence of high waves, during which the bed thickness decreased by 7.7 mm/hour in northern estuary of the Yellow River, China; according to the study of Lu et al. (2021), the gradient of wave radiation stress can cause the vertical gyres with offshore flows in the lower layers and onshore flows in the upper layers, contributing to the sediment erosion and driving the offshore eroded sediment transport. According to the study of Wang et al. (2021), strong wind surges and waves (can reach to 6.0 m) have caused catastrophic damage to coastal infrastructures and the economic loss of government in past 10 years. However, the storm impacts on tidal deltas and estuarine beaches in embayed or sheltered estuarine environment are less well understood.

The flood-tide delta (FTD) is an important morphological feature in embayed estuaries. Changes of the FTD morphology and associated tide-wave–river dynamics have strong impacts on the adjacent beach systems and leads to complex morphodynamic processes (Wang et al., 2021). FTDs are sediment deposits formed on the landward side of tidal inlets by the incoming sediment transport driven by onshore tidal current and wave action (Hayes, 1975; Hayes & FitzGerald, 2013), but the dynamic processes of FTDs are rarely explored quantitatively in sheltered estuaries. Austin et al. (2018) set up a conceptual model in the embayed Port Stephens in New South Wales (NSW), Australia, and the tidal currents and waves were assumed to be the dominant factor driving sediment transport over FTD and adjacent beaches. Harris et al. (2020) found the FTD can influence the estuarine beaches and shoreline change in Port Stephens using field data and suggested more detailed sediment transport nearshore under different processes (e.g., tide, wind, and waves) was needed. Such a conceptual model and the analysis of field data would benefit from modeling, which can quantify the morphodynamic process and evolution of these complex coastal systems (Harris et al., 2020).

Therefore, understanding the sediment dynamics, morphology change, and the management of the embayed coastal zone under strong wave events are of increasing importance for local governments and researchers.

2.   Study domain

2.1   The Yalu River Estuary

The Yalu River is located between China and North Korea discharging into YRE, which is also the largest regional river in the Northern Yellow Sea. The YRE is separated into two channels by Chouduan Island: west and middle channel (Fig. 9.1).

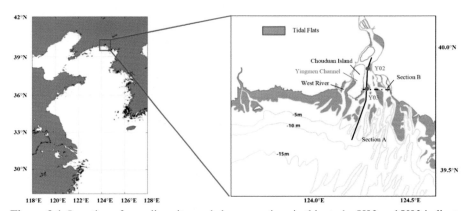

Figure 9.1 Location of sampling sites and chosen sections in this study. Y02 and Y03 indicate the sampling locations of water elevation, current, salinity and suspended sediment concentration. Section A: The lower estuary to upper estuary; Section B: Entrance section from west to east.

From Yang, G., Wang, X. H., Cheng, Z., Zhong, Y. & Oliver, T. (2021). Modeling study on estuarine circulation and its effect on the turbidity maximum zone in the Yalu River Estuary, China. Estuarine, Coastal and Shelf Science, 263, 107634. https://doi.org/10.1016/j.ecss.2021. 107634

Currently, the west channel has silted up and is no longer navigable. The middle channel has become the main passage for water and suspended sediment exchange. From 1958 to 2010, the annual averaged sediment and water discharges of the Yalu River are 1.59×10^6 T and 2.5×10^{10} m^3, respectively (Cheng et al., 2016; Yu et al., 2014). Strong seasonal variation can be found in the river discharges: 33% yearly fresh water and 80% sediment load were output into the Northern Yellow Sea during wet season (from June to September). According to the river discharge measurements at Huang-gou Hydrology Station, the peak monthly averaged Yalu River discharge is about 700 m^3/second during wet season in 2009 and can occasionally reach to $\sim$5500 m^3/ second (in August 2010, when a 50-year flooding event happened).

The YRE is a semidiurnal macro-tidal estuary. Tidal currents here can reach 2.0 m/ second, and the mean tidal range is about 4.5 m. Strong tidal asymmetry can be observed in this area: the flood current is stronger than the ebb. The sediment dynamics in the YRE are mainly dominated by tidal effects, and there is a TMZ located at the interaction area of river discharge and tidal currents. Based on previous studies, the annual mean wind speed is 3.2 m/second, and wave height is less than 0.5 m outside of the YRE, which is relatively weak (Cheng et al., 2016). Thus, local wave effects can be neglected in the hydrodynamic models in this area (Cheng et al., 2020; Yu et al., 2014).

2.2   The Batemans Bay

Batemans Bay (henceforth BB) (35.7144 °S, 150.1795 °E) is a funnel-shaped estuary facing southeast with containing numerous beaches and a FTD (Fig. 9.2), which is a complex marine system affected by waves, tides and river discharge (Wang et al.,

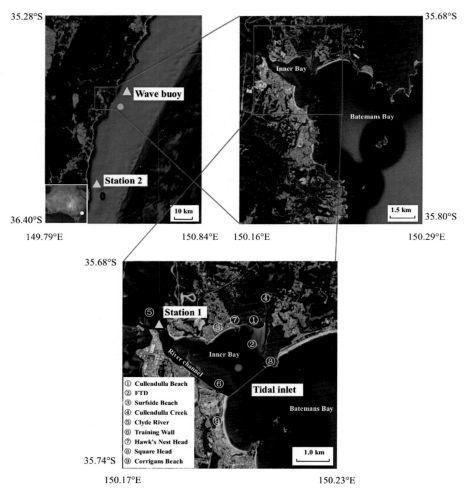

Figure 9.2 Satellite map of Batemans Bay (35.7144° S, 150.1795° E) with spot markers for site locations and notable geographical features; *yellow triangles* indicate the location of Station 1, Station 2, and wave buoy; *red* and *green dots* in the inner and outer bay indicate the stations used for drawing wave rose figures; the inset is a map of Australia with location of Batemans Bay marked with a *white dot*.
From Yang, Gang, Wang, Xiao Hua, Zhong, Yi & Oliver, Thomas S.N. (2022). Modeling study on the sediment dynamics and the formation of the flood-tide delta near Cullendulla Beach in the Batemans Bay, Australia. Marine Geology, 452. https://doi.org/10.1016/j.margeo.2022.106910. Google Earth 2020

2021; Water Research Laboratory, 2017). BB has a total area of 28 km^2 and the averaged water depth is 11.1 m. BB is a drowned river valley estuary (Roy et al., 1994) with small tidal range (1.85 m). The nearby BB wave buoy is stationed approximately 4 km offshore (35.7031°S, 150.3439°E, Fig. 9.2) and commonly records significant

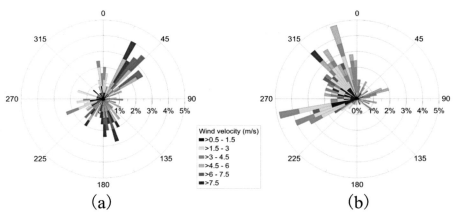

Figure 9.3 Wind rose figure in (a) January and (b) July 2018.
From Yang, Gang, Wang, Xiao Hua, Zhong, Yi & Oliver, Thomas S.N. (2022). Modeling study on the sediment dynamics and the formation of the flood-tide delta near Cullendulla Beach in the Batemans Bay, Australia. Marine Geology, 452. https://doi.org/10.1016/j.margeo.2022.106910

wave heights up between 4 and 6 m during severe storm events. The Clyde River discharges into BB and the averaged river flow rate in the last 20 years is 6.75 m³/second, although this increases dramatically during large floods.

BB experiences an oceanic climate, with 21.9°/10° annual mean maximum/minimum temperature. The annual rainfall is 912 mm, with rainfall maximums in summer. According to the field wind data at the weather station 69,148 (35.9004°S, 150.1437°E) during our study period (2018), the direction of the prevailing wind in BB is northeastly/southeastly in January and changed to northwestly/southwestly in July (Fig. 9.3). The monthly averaged wind velocity was 3.65 m/second and 2.74 m/second in January and July 2018, respectively; the gentler localized winds in winter were due to the sheltering effect of the hills and ranges west of BB (Wang et al., 2021). Based on the wave data at the wave buoy from 2014 to 2018, the monthly averaged wave height, period, and direction in summer (January)/winter (July) are 1.6/1.3 m, 9.1/10.6 seconds, and 124.6°/140.2°, respectively.

3. Methods

3.1 Field measurements for YRE

The continuous 25-hour field hydrodynamic data (water elevation, tidal current, turbidity and salinity) were measured at two anchor stations: Y03 (124.28°E, 39.84°N) during spring and neap tide (spring tide: 8−9 August, neap tide: 14−15 August), and Y02 (124.30°E, 39.92°N) during spring and neap tide (spring tide: 7−8 August, neap tide: 13−14 August) in 2009 (Fig. 9.1). Field measurements were carried out in calm days during wet season. The receivers of Acoustic Doppler

Current Profilers (ADCP) (1200 kHz frequency, 0.001 m/second current-speed resolution, 0.25 m bin-layer depth) were applied to measure the current and water elevation data. Turbidity data and salinity data were measured by a multiparameter water quality probe YSI6600. According to 1-hour in situ filtered water samples, the measured turbidity was then converted into SSC (Cheng et al., 2016). The details of field data collection were described in the study of Cheng et al. (2020).

3.2   Field measurements for BB

The 1-hourly interval field water level data in January and July 2018 were measured at two stations: (1) Station 1 at Princess Jetty inside BB (35.7038°S, 150.1778°E) and (2) Station 2 outside BB (36.1898°S, 150.1884°E), shown in Fig. 9.2. The water level was measured by water level gauge at Station 1 by Manly Hydraulics Laboratory (https://www.mhl.nsw.gov.au/Station-216410). The water level and current velocity at Station 2 were measured by Acoustic Doppler Current Profiler (ADCP) by Integrated Marine Observation System (IMOS), downloaded from Australian Ocean Data Network (https://portal.aodn.org.au/search). The field wind velocity and direction data were obtained from the Moruya Airport (weather station 69,148: 35.9004°S, 150.1437°E). The hourly significant wave height and wave period data for the duration of the field observations were measured by the Directional Waverider Buoy, located at 35.7031°S, 150.3439°E, obtained through publicly accessible data from Manly Hydraulics Lab (https://mhl.nsw.gov.au/).

3.3   Numerical simulations

3.3.1   The introduction of finite-volume Coastal Ocean Model

The three-dimensional Finite-Volume Coastal Ocean Model (FVCOM) was applied in this study to simulate the hydrodynamics and sediment dynamics in the YRE and BB. FVCOM is a finite-volume, free-surface, and primitive-equations community ocean model developed by Chen et al. (2003). The Smagorinsky turbulent closure schemes (Smagorinsky, 1963) and the modified Mellor and Yamada (1982) level 2.5 (MY-2.5) were applied in this model for horizontal and vertical sediment mixing, respectively. The unstructured triangular grid adopted in FVCOM is suitable for the complicated coastline of coastal estuaries and tidal flats (Chen et al., 2011; Cheng et al., 2020; Xiao et al., 2019; Zhong et al., 2020). The FVCOM-SED sediment module is based on the Community Sediment Transport Model (CSTM), which can simulate unlimited sediment classes for cohesive and noncohesive sediment transport (Ge et al., 2018; Warner et al., 2008) and has been widely applied in coastal areas (Cong et al., 2021; Ge et al., 2015; Niu et al., 2018; Yang et al., 2020; Yellen et al., 2017; Zhong et al., 2020). For detailed description of the sediment transport model, see Warner et al. (2008) and the sediment module introduction in the FVCOM manual (Chen et al., 2012). The governing equations of continuity and sediment transport are given as follows:

$$\frac{\partial Du}{\partial x} + \frac{\partial Dv}{\partial y} + \frac{\partial w}{\partial z} + \frac{\partial \eta}{\partial t} = 0, \tag{9.1}$$

$$\frac{\partial C_i}{\partial t} + \frac{\partial (uC_i)}{\partial x} + \frac{\partial (vC_i)}{\partial y} + \frac{\partial \left(w - w_{s,i}\right)C_i}{\partial z} = \frac{\partial}{\partial x}\left(A_h \frac{\partial C_i}{\partial x}\right) + \frac{\partial}{\partial y}\left(A_h \frac{\partial C_i}{\partial y}\right)$$
$$+ \frac{\partial}{\partial z}\left(K_h \frac{\partial C_i}{\partial z}\right), \tag{9.2}$$

where u and v are the eastward and northward velocities; w is the corresponding vertical velocity component; x, y, and z indicate the east, north, and vertical coordinates, respectively; t is the time; η and h are the sea surface elevation and the mean water depth, respectively; the total water depth is given by $D = \eta + h$; $w_{s,i}$ and C_i are the settling velocity and concentration of sediment i, respectively; K_h and A_h indicate the vertical and horizontal diffusion coefficients. The vertical sediment flux, F, on the seabed due to erosion and deposition processes is given by Eq. (9.3), based on the study of Ariathurai and Krone (1977),

$$F = \begin{cases} E_0\left(\dfrac{|\tau_b|}{\tau_c} - 1\right), |\tau_b| > \tau_c \\[2ex] C_b w_s\left(\dfrac{|\tau_b|}{\tau_d} - 1\right), |\tau_b| < \tau_d \end{cases} \tag{9.3}$$

where E_0 is the erosion coefficient, τ_c and τ_d are the critical stress for re-suspension and deposition, respectively, which can be seen in Table 9.1; C_b indicates the sediment concentration near the bottom layer; τ_b is the shear stress.

The contribution of SSC on the density field is considered in FVCOM-SED, given by Eq. (9.4),

Table 9.1 Model configuration.

Model parameters	Value/method
Model external time step	1.0 second
Horizontal diffusion	Smagorinsky scheme
Vertical eddy viscosity, settling velocity, critical bottom stress for erosion, critical bottom stress for deposition, erosion rate, number of mesh nodes and elements	M-Y 2.5 turbulent closure 1.25×10^{-4} m/s0.08 N/m^{2}0.08 N/m^{2}5.0 $\times 10^{-5}$ kg/m^2/s2988, 5348

From Yang, Gang, Wang, Xiao Hua, Cheng, Zhixin, Zhong, Yi & Oliver, Thomas. (2021). Modeling study on estuarine circulation and its effect on the turbidity maximum zone in the Yalu River Estuary, China. Estuarine, Coastal and Shelf Science, 263. https://doi.org/10.1016/j.ecss.2021.107634

$$\rho = \rho_w + \left(1 - \frac{\rho_w}{\rho_s}\right) C, \tag{9.4}$$

where ρ_w is the density of clear seawater and ρ_s is the density of sediment. The bottom frictional stress in FVCOM is given by Eq. (9.5),

$$\tau_b = \rho C_d |u_b| u_b, \tag{9.5}$$

where u_b is the bottom ($z = z_b$) mean current velocity, C_d is the bottom friction coefficient in a sediment-laden bottom boundary layer, calculated as below:

$$C_d = \max \left\{ \left[\frac{1}{k} \ln \left(\frac{H + z_b}{z_0} \right) \right]^{-2}, 0.0025 \right\}, \tag{9.6}$$

where z_b is the position of the near-bottom sigma layer below the water surface ($\sigma = -0.95$ in this study); z_0 is the bottom roughness (0.001 m), H is the water depth, and k indicates the Von Karman constant (0.4), (Chen et al., 2006b).

3.3.2 Model setup and initial conditions

3.3.2.1 The model setup and validation in the YRE

The model domain of YRE is shown in Fig. 9.4. Surface-water modeling system (SMS) was applied in this study for building the grids, and the coastline and bathymetry data were extracted from sea charts from the Navigation Guarantee Department of the Chinese Navy Headquarters in 2011 (Cheng et al., 2020; GAO et al., 2012). The mesh size ranges from 150 m nearshore to 3000 m at the open boundary with 20

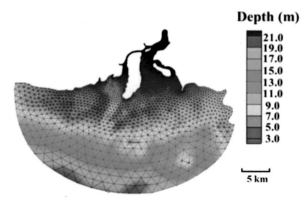

Depth (m)

21.0
19.0
17.0
15.0
13.0
11.0
9.0
7.0
5.0
3.0

5 km

Figure 9.4 Grids of the model domain.
From Yang, Gang, Wang, Xiao Hua, Cheng, Zhixin, Zhong, Yi & Oliver, Thomas. (2021). Modeling study on estuarine circulation and its effect on the turbidity maximum zone in the Yalu River Estuary, China. Estuarine, Coastal and Shelf Science, 263. https://doi.org/10.1016/j.ecss.2021.107634

uniform vertical layers. Wetting-drying processes were included in the coastal areas, with critical water depth 0.05 m. The details of model configuration are shown in Table 9.1. The model was driven by the wind field data from the Climate Forecast System Reanalysis (CFSR) and forced by tides derived from TPXO 7.2 (0.25 degree resolution) at the ocean open boundary. Monthly averaged salinity for the open boundaries and the initial salinity field are taken from the Global Hybrid Coordinate Model (HYCOM). Water discharge and sediment load of the Yalu River (July and August 2009) were taken from Huanggou Station (124.638° E, 40.291°N) located at the upper estuary of the YRE. No sediment input from the open ocean boundary was considered in this model. Only cohesive sediments were considered for all experiments, as most suspended sediments are easily transported and are cohesive in an estuary like the YRE (Brenon & Le Hir, 1999; Cheng et al., 2020). The critical shear stress and erosion rate were assumed to be constant and chosen with different values tested to best fit the field SSC data.

To examine the baroclinic effects on hydrodynamic and sediment dynamics in the YRE, five experiments were conducted in this study and the details can be seen in Table 9.2. All model runs are spun up for 30 days for warm-up purpose.

The measurements from the field trip in August 2009 were used to validate the model results of Case 0. The comparison between field data and model results was shown in Fig. 9.5, and model performance with root mean-square error (RMSE) and correlation coefficient (r) in Table 9.3. Fig. 9.5a–d shows good agreement between the observed and modeled water elevations at Stations Y02 and Y03 during spring and neap tides (Fig. 9.1) (with $r > 0.90$ and RMSE<0.72 m, shown in Table 9.3). The simulated depth-averaged along-channel velocity and salinity also match quite well with the field data (Fig. 9.5 e–l) (for current: $r > 0.85$ and RMSE<0.50 m/second; for salinity: $r > 0.85$ and RMSE$<3.50\%_{oo}$). For the SSC validation, the model has captured the fluctuations and magnitude over time of the measured SSC, with $r > 0.50$. The differences between the measured and modeled SSC at Y03 (RMSE $= 0.34$ kg/m^3, shown in Table 9.3) was also noticed by Cheng et al. (2020): due to the lack of high-frequency river discharge measurements, daily river discharge data were used in this model as the river boundary, which can cause the difference between modeled and field SSC data, and can also lead to the current velocity discrepancy at Y03 between model results and field data.

Table 9.2 Experiments description.

Experiment	Description
Case 0	Forced by wind, salinity, tidal currents, real river sediment load and
Case 1	discharge (~ 700 m^3/second) in July and August 2009Same with Case 0, but barotropic model
Case 2	Same with Case 0, but with constant large river discharge (5500 m^3/second)
Case 3	Same with Case 2, but barotropic model

From Yang, Gang, Wang, Xiao Hua, Cheng, Zhixin, Zhong, Yi & Oliver, Thomas. (2021). Modeling study on estuarine circulation and its effect on the turbidity maximum zone in the Yalu River Estuary, China. Estuarine, Coastal and Shelf Science, 263. https://doi.org/10.1016/j.ecss.2021.107634

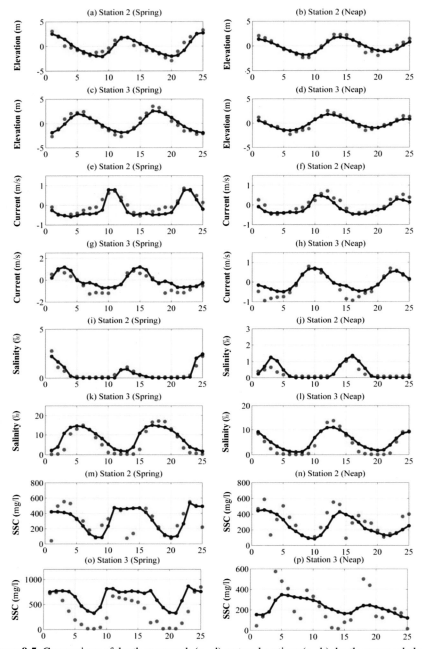

Figure 9.5 Comparison of depth-averaged: (a–d) water elevation; (e–h) depth-averaged along-channel velocity; (i–l) salinity; and (m–p) SSC between the field measurements and FVCOM model results during spring and neap tides in August 2009 at Stations Y02 and Y03.
From Yang, Gang, Wang, Xiao Hua, Cheng, Zhixin, Zhong, Yi & Oliver, Thomas. (2021). Modeling study on estuarine circulation and its effect on the turbidity maximum zone in the Yalu River Estuary, China. Estuarine, Coastal and Shelf Science, 263. https://doi.org/10.1016/j.ecss.2021.107634

Table 9.3 Correlation coefficient (r) and root mean-square error (RMSE) between model results and measurements for water elevation, current, salinity and SSC.

	Y03 (spring)		Y03 (neap)		Y02(Spring)		Y02 (neap)	
	r	RMSE	r	RMSE	r	RMSE	r	RMSE
Elevation (m)	0.97	0.62	0.96	0.47	0.90	0.71	0.95	0.53
Current (m/second)	0.88	0.47	0.95	0.28	0.93	0.22	0.91	0.19
Salinity (‰)	0.92	3.30	0.97	1.99	0.94	0.26	0.88	0.24
SSC (kg/m³)	0.78	0.34	0.55	0.15	0.52	0.15	0.61	0.12

From Yang, Gang, Wang, Xiao Hua, Cheng, Zhixin, Zhong, Yi & Oliver, Thomas. (2021). Modeling study on estuarine circulation and its effect on the turbidity maximum zone in the Yalu River Estuary, China. Estuarine, Coastal and Shelf Science, 263. https://doi.org/10.1016/j.ecss.2021.107634

3.3.2.2 The model setup and validation in the BB

The model domain of BB is shown in Fig. 9.6. The details of model configuration are shown in Table 9.4 and Table 9.5. The model results were validated against the measurements in summer from 1st to 31st January in 2018, and in winter from 1st to 31st July in 2018 with the correlation coefficient (r) and root mean-square error (RMSE) (shown in Fig. 9.7). Fig. 9.7 a−d shows good agreement between the modeled and measured water elevations at Stations 1 and 2 during January and July 2018 (with r > 0.95 and RMSE<0.15 m). The simulated depth-averaged current velocities along the main current axis at Station 2 captured the magnitude and fluctuations over time of the observed current velocity, with r > 0.60 and RMSE<0.20 m/second (Fig. 9.7 e−f). For the wave validation, the model matched well with the field wave height and period data (for wave

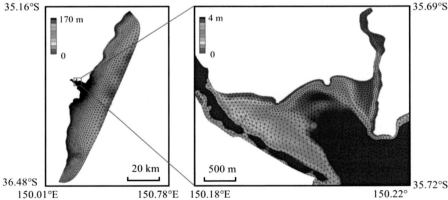

Figure 9.6 Grids of the model domain for Batemans Bay.
From Yang, Gang, Wang, Xiao Hua, Zhong, Yi & Oliver, Thomas S.N. (2022). Modeling study on the sediment dynamics and the formation of the flood-tide delta near Cullendulla Beach in the Batemans Bay, Australia. Marine Geology, 452. https://doi.org/10.1016/j.margeo.2022.106910

Table 9.4 Key parameters of the model configuration in BB.

Model parameters	Value/method
Model external time step	0.5 second
Horizontal diffusion	Smagorinsky scheme (default)
Vertical eddy viscosity, settling velocity, critical erosion stress, critical deposition stress, erosion rate, number of mesh nodes and elements	M-Y 2.5 turbulent closure (default) 1.25×10^{-4} m/s0.08 N/m^{2}0.08 N/ m^{2}5.0 $\times$ 10^{-6} kg/m^2/s8738, 164,199,175, 17,332 (without the square Head)8961, 16,847 (without training wall) 8974, 16,838 (with training wall)

From Yang, G., Wang, X. H., Zhong, Y. & Oliver, T. S. N. (2022). Modeling study on the sediment dynamics and the formation of the flood-tide delta near Cullendulla Beach in the Batemans Bay, Australia. Marine Geology, 452, 106910. https://doi.org/10.1016/j.margeo.2022.106910

Table 9.5 SWAVE parameters.

Parameters	Value and methods
Frequency range (Hz)Frequency bins DirectionDirection binsTriad wave-wave interactionThe lower threshold for Ursell number Quadruplet wave interactionsCoefficient for quadruplet configurationBottom frictionFriction parameterWind inputWhite capping dissipationDepth-induced wave breakingProportionality coefficient of the rate of dissipationRatio of maximum individual wave height over depth	0.05−0.524 Full circle36Lumped Triad Approximation (LTA) *urslim* = 0.01Discrete interaction Approximation (DIA)*lambda* = 0.25Jonswap formulation0.038Third generation; Komen formulationKomen formulationConstant*alpha* = 1.0*gamma* = 0.73

From Yang, G., Wang, X. H., Zhong, Y. & Oliver, T. S. N. (2022). Modeling study on the sediment dynamics and the formation of the flood-tide delta near Cullendulla Beach in the Batemans Bay, Australia. Marine Geology, 452, 106910. https://doi.org/10.1016/j.margeo.2022.106910

height: $r > 0.85$ and RMSE$<$0.45 m; for wave period: $r > 0.60$ and RMSE$<$2.0 seconds), shown in Fig. 9.7 g−j. The differences between the measured and modeled current at station 2 were: (1) due to the lack of higher-resolution bathymetry data as the bathymetry from ETOPO1 Global Relief Model might be not accurate enough to describe the real situation; (2) in addition, the 1-day interval SSH and current circulation data nested into our model was derived from Hycom, the time resolution, and the discrepancy between Hycom modeled and in-situ data can also influence the accuracy of our model. In conclusion, the hydrodynamics model performed well in simulating

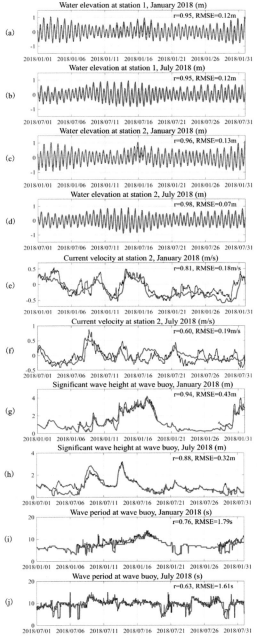

Figure 9.7 Comparison of: (a–d) elevation at Station one and 2; (e–f) depth-averaged streamwise velocity at Station 2; (g–h) significant wave height and (i–j) wave period at the wave buoy in January and July 2018 between the FVCOM model results (*red line*) and field observation (*blue dash line*).
From Yang, G., Wang, X. H., Zhong, Y. & Oliver, T. S. N. (2022). Modeling study on the sediment dynamics and the formation of the flood-tide delta near Cullendulla Beach in the Batemans Bay, Australia. Marine Geology, 452, 106910. https://doi.org/10.1016/j.margeo.2022.106910

the hydrodynamics in BB with high and low RMSE and can be confidently applied for our further study. We note that the current validation for hydrodynamics was based on the stations located outside BB (Station 2 and wave buoy), due to the lack of the measurements inside BB. This is potentially a shortcoming of current work as stations within BB would have been preferable for model validation/calibration in a more nearshore environment.

3.3.2.2.1 The validation for morphology change Given the absence of field sediment and morphology change measurements, and for validating our morphology model results, two high-quality satellite images with low cloud coverage (<15%) on third May and 16th August in 2018 were chosen to retrieve the bathymetry change, during which time several large wave events occurred (Fig. 9.8). This method can provide "indicative" depth changes to inform the numerical modeling.

The chosen two Landsat8 images have been radiometric and atmospheric calibrated by ENVI 5.1. At first, we used the field bathymetry data inside BB (measured by Fugro Pty Ltd on behalf of NSW Office of Environment and Heritage from August to December in 2018, https://datasets.seed.nsw.gov.au/dataset/marine-lidar-topo-bathy-2018) and the reflectance data of Landsat8 image (9:43:57 a.m., on August 16, 2018, local time) to set up an algorithm to retrieve the water depth (4654 for calibrating, 4654 for validation). The algorithm was based on the band ratio model, given by Eqs. (9.7) and (9.8) (Ma et al., 2020), shown as below:

$$H = m1 * x - m0 \tag{9.7}$$

$$x = \frac{\ln(n * \text{Rrs}_{-\text{blue}})}{\ln(n * \text{Rrs}_{\_\text{green}})} \tag{9.8}$$

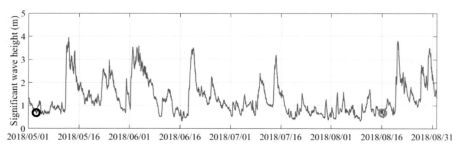

Figure 9.8 Significant wave height at the wave buoy from first May to August 31, 2018 (*black and blue circles* indicate the time for the chosen Landsat eight images on May 3 and 16th August in 2018, respectively).

From Yang, Gang, Wang, Xiao Hua, Zhong, Yi & Oliver, Thomas S.N. (2022). Modeling study on the sediment dynamics and the formation of the flood-tide delta near Cullendulla Beach in the Batemans Bay, Australia. Marine Geology, 452. https://doi.org/10.1016/j.margeo.2022.106910

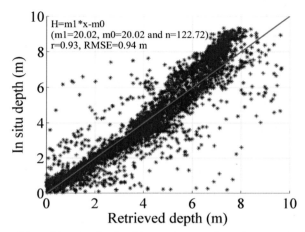

Figure 9.9 The calibrated band ratio algorithm and comparisons between in situ depth and retrieved depth in BB on August 16, 2018.
From Yang, Gang, Wang, Xiao Hua, Zhong, Yi & Oliver, Thomas S.N. (2022). Modeling study on the sediment dynamics and the formation of the flood-tide delta near Cullendulla Beach in the Batemans Bay, Australia. Marine Geology, 452. https://doi.org/10.1016/j.margeo.2022.106910

where H is the water depth derived from the Landsat eight image, Rrs_{-blue} and Rrs_{-green} indicate the reflectance of blue and green band of Landsat 8; the values of $m1$, $m0$ and n can be obtained by minimizing the difference between the retrieved depth H and the field water depth. The retrieving algorithm and the comparison between in situ and retrieved water depth are shown in Fig. 9.9.

The high correlation coefficient (>0.90) and low RMSE (<0.95 m) indicate the algorithm can be applied for retrieving the water depth in BB. Furthermore, the timing for satellite passing BB on third May and August 16, 2018, was different, and the water depth was different due to the tidal elevation change. Therefore, after retrieving the bathymetry, the retrieved bathymetric values of these two images were converted into the high water level data to remove the tidal effects. After the tidal calibration, the two calibrated images (third May and August 16, 2018) and the bathymetry change were shown in Fig. 9.10. The pattern and distribution of the water depth change retrieved from Landsat 8 (Fig. 9.10c) were similar with the our model results in January and July 2018 with wave effects and indicated our model can capture the main process of the morphology change, which can be applied in BB. In addition, although the method of retrieving bathymetry change gives "indicative" depth changes, in-situ calibration using more field morphology evolution data is still required to improve the accuracy and the performance of this method in the future work.

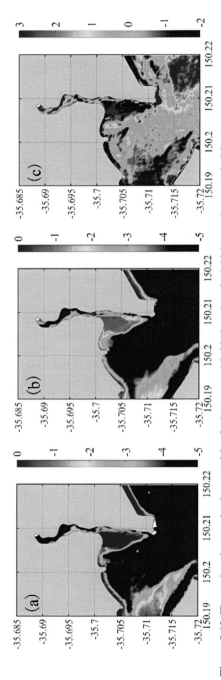

Figure 9.10 The retrieved water depth on (a) May 3, (b) August 16, 2018, and (c) bed thickness change (unit: Meter). From Yang, Gang, Wang, Xiao Hua, Zhong, Yi & Oliver, Thomas S.N. (2022). Modeling study on the sediment dynamics and the formation of the flood-tide delta near Cullendulla Beach in the Batemans Bay, Australia. Marine Geology, 452. https://doi.org/10.1016/j.margeo.2022.106910

4.   Model results and discussion

4.1   Model results and discussion in YRE

4.1.1   Baroclinic effects on estuarine circulation and sediment dynamics in YRE

According to previous studies (Cheng et al., 2020; Yu et al., 2014), the sediment resuspension and tidal pumping can contribute to the TMZ formation in the YRE, based on field data and barotropic models, neglecting the baroclinic effect. In order to understand the importance of baroclinic effects in the YRE, the comparison of hydrodynamic and sediment dynamics between baroclinic and barotropic model runs was investigated in this section (Fig. 9.11).

From baroclinic model (Case 0) results, TMZ was located near the salinity front around 39.84° (Fig. 9.11 2.4 a1, a2, and a7), where the bottom stress was large (Fig. 9.11a5). The large bottom stress can cause sediment resuspension, offering the sediment source for the TMZ. The monthly averaged sediment flux and residual flow show clearly a two-layer structure with landward and seaward flow at bottom and surface layers along section A (Fig. 9.11a3 and a4). Such a structure can transport those resuspended sediments to the upper estuary and contribute to the TMZ formation; this two-layer structure eventually induced a net landward sediment transport (as the SSC was higher near the seabed), until it encountered the seaward directed river transport at upper estuary (at residual flow null point) (Burchard & Baumert, 1998; Festa & Hansen, 1976; Postma, 1967; Yu et al., 2014). The suspended sediments flowing at upper layers were settled due to gravity and trapped in the low layers around lower estuary side of that null point due to the low vertical mixing coefficient under strong stratification (as demonstrated in Fig. 9.12), eventually forming TMZ.

However, this TMZ formation mechanism could not be found from the barotropic model (Case 1). Due to the absence of baroclinic forcing, there was no two-layer structure pattern of the sediment flux and residual flow in Case 1 (Fig. 9.11b3 and b4). TMZ was located at the area where bottom stress was large and strong sediment resuspension occurred, leading to a higher SSC value. Sediment flux showed a divergence pattern at TMZ, where high concentration of sediments can induce SSC gradient, generating a seaward sediment flux downstream of TMZ and a landward flux upstream of TMZ (Yu et al., 2014).

SSC value in Case 0 was lower than that in Case 1 due to the combined influence of bottom stress and vertical stratification: through the comparison of bottom stress in Case 0 and 1, the monthly averaged bottom stress in Case 1 is slightly larger than that in Case 0, making more sediment resuspended into waters; in addition, the stronger stratification (lower vertical mixing coefficients) induced by baroclinic effects can inhibit the vertical mixing (shown in Fig. 9.11a6) and sediment resuspension. These two factors resulted in a lower SSC in Case 0 compared with Case 1 (barotropic run). Although baroclinic effects were considered in Case 0, TMZ was located in the same area in both runs because the river discharge is still relatively low in the

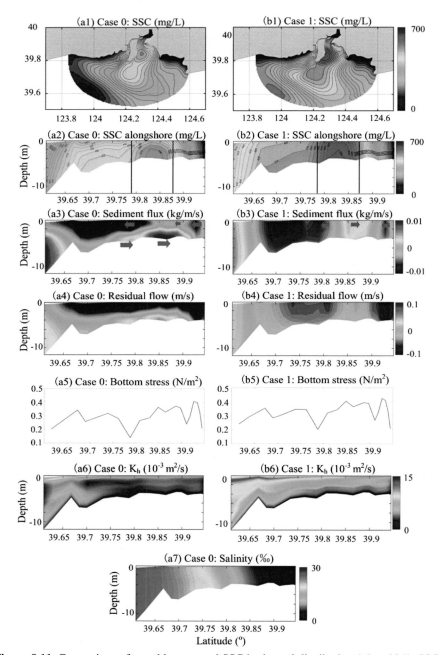

Figure 9.11 Comparison of monthly averaged SSC horizontal distribution (a1 and b1), SSC distribution along section A (a2 and b2), sediment flux (a3 and b3), residual current (a4 and b4), bottom stress (a5 and b5) and vertical mixing coefficient K_h (10^{-3} m^2/second) (a6 and b6) in case 0 and case 1; (a7) salinity distribution in Case 0. (*Red vectors* indicate the sediment flux direction, the *blue line* in a2 and b2 indicated the location of TMZ).

From Yang, Gang, Wang, Xiao Hua, Cheng, Zhixin, Zhong, Yi & Oliver, Thomas. (2021). Modeling study on estuarine circulation and its effect on the turbidity maximum zone in the Yalu River Estuary, China. Estuarine, Coastal and Shelf Science, 263. https://doi.org/10.1016/j.ecss.2021.107634

Lower estuary   **Upper estuary**

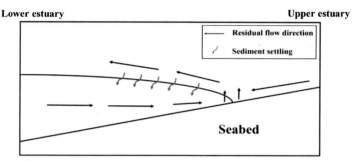

Figure 9.12 Residual flow and sediment trapping process in baroclinic model.
From Yang, Gang, Wang, Xiao Hua, Cheng, Zhixin, Zhong, Yi & Oliver, Thomas. (2021).
Modeling study on estuarine circulation and its effect on the turbidity maximum zone in the Yalu
River Estuary, China. Estuarine, Coastal and Shelf Science, 263. https://doi.org/10.1016/j.ecss.
2021.107634

baroclinic case. In other words, the strong resuspension process can mask the effects of
estuarine circulation as the river discharge influence is not strong enough in this run.

When we increased the river discharge to 5500 m³/second (typical value for flooding events), the salinity front was displaced to the lower estuary (around
39.75°−39.8°N); baroclinic effects on TMZ became stronger and can be clearly
seen (by comparing Case 2 and 3), as shown in Fig. 9.13. The estuarine circulation
during large river discharge occurred at lower estuary (39.6°−39.8°N) and trapped
large amount of sediments around 39.7°−39.8°N (downstream of the flow null point),
contributing to the TMZ formation. However, due to the absence of two-layer structure
estuarine circulation in Case 3, the location of TMZ can be moved seaward to 39.7° N.

Furthermore, in order to distinguish where the baroclinic effects coming from (from
the sediment or salinity), we designed another model test, in which sediment impacts
on water density was excluded (by removing the third term $\left(1 - \frac{\rho_w}{\rho_s}\right)C$ from Eq. (9.4)
in FVCOM). The hydrodynamic and SSC results in this test (Figure not shown) are
nearly same with other experiments considering the effects of sediment-induced density (Case 0 and 2). Therefore, the baroclinic effect in the YRE is mainly caused by
salinity change as the SSC is too low to affect the water density in the YRE.

Results here indicated that the baroclinic effects on hydrodynamics and sediment
transport is important and cannot be neglected in the YRE, and it was mainly induced
by salinity. In addition, model results showed that the along-channel circulation can
transport and trap sediments, influencing the formation of TMZ, especially during
the large river discharge period.

4.1.2 The mechanisms of two-layer estuarine circulation in the YRE

Based on the study of Cheng et al. (2011), along-channel estuarine circulation can be
decomposed into four terms: density-driven flow (UD), nonlinear advection driven
flow (UN), asymmetric tidal mixing-driven flow (UA), and river-induced flow

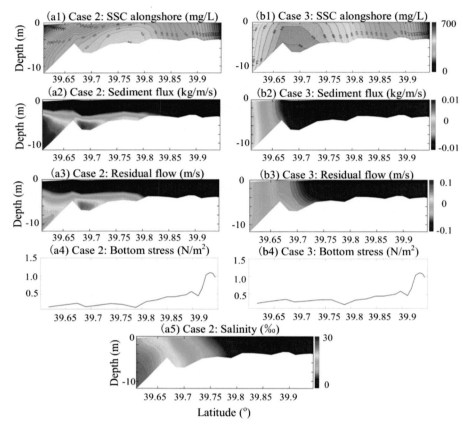

Figure 9.13 Comparison of monthly averaged SSC distribution along section A (a1 and b1), sediment flux (a2 and b2), residual current (a3 and b3), bottom stress (a4 and b4) in Case 2 and Case 3; (a5) salinity distribution in Case 2.

From Yang, Gang, Wang, Xiao Hua, Cheng, Zhixin, Zhong, Yi & Oliver, Thomas. (2021). Modeling study on estuarine circulation and its effect on the turbidity maximum zone in the Yalu River Estuary, China. Estuarine, Coastal and Shelf Science, 263. https://doi.org/10.1016/j.ecss.2021.107634

(UR), by ignoring the lateral processes. This method has been applied in many studies (Song et al., 2013; Xiao et al., 2020; Yan et al., 2019). Based on this method, we decomposed estuarine circulation in the YRE into UD, UA UN, and UR (shown in Figs. 9.14 and 9.15) during low and large river discharge (Case 0 and Case 2). The sum of these four decomposed components of residual currents has a similar along-channel distribution to that directly obtained from modeled residual currents (Figs. 9.14 and Fig. 9.15), which indicates this decomposition method can be suitably applied in the YRE. The details of this decomposition method can be seen in Appendix A in the study of Cheng et al. (2011).

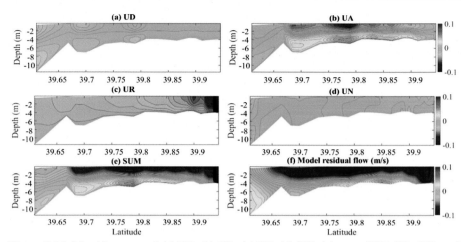

Figure 9.14 Monthly averaged (a) UD; (b) UA; (c) UR; (d) UN; (e) sum of UD, UA, UR, and UN; (f) model residual flow for Case 0 under low river discharge (~ 700 m^3/second), positive: Landward. Unit: m/second.
From Yang, Gang, Wang, Xiao Hua, Cheng, Zhixin, Zhong, Yi & Oliver, Thomas. (2021). Modeling study on estuarine circulation and its effect on the turbidity maximum zone in the Yalu River Estuary, China. Estuarine, Coastal and Shelf Science, 263. https://doi.org/10.1016/j.ecss. 2021.107634

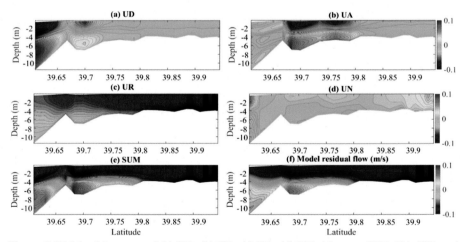

Figure 9.15 Monthly averaged (a) UD; (b) UA; (c) UR; (d) UN; (e) sum of UD, UA, UR, and UN; (f) model residual flow for Case 2 under large river discharge (5500 m^3/second), positive: Landward. Unit: m/second.
From Yang, Gang, Wang, Xiao Hua, Cheng, Zhixin, Zhong, Yi & Oliver, Thomas. (2021). Modeling study on estuarine circulation and its effect on the turbidity maximum zone in the Yalu River Estuary, China. Estuarine, Coastal and Shelf Science, 263. https://doi.org/10.1016/j.ecss. 2021.107634

In the low river discharge case (Fig. 9.14), UA was the dominant factor contributing to the two-layer structure of residual flow, with landward and seaward flow in the bottom and upper layer, which can reach to 39.9° N. Due to the weak baroclinic effects caused by a low river discharge, the magnitude of UD was negligible compared with UA. The seaward river induced flow was strong near the mouth of river, and it decreased to nearly zero downstream with the increasing width of the estuary. UN could be neglected in this case.

However, UD, UA, UR, and UN showed different patterns in large river discharge case (Fig. 9.15). Under strong baroclinic effects, UD became stronger especially at the lower estuary ($\sim$39.65° N) and induced a landward bottom flow to 39.75° N with seaward flow at upper layers. UA also showed a two-layer structure at 39.67°−39.8° N. Seaward UR became stronger compared to the low river discharge case, which can now extend to the lower estuary. UN was landward in the upper estuary, with large value at upper layers. Under the combined effects of these four terms, the two-layer estuarine circulation occurred at the 39.6−39.8° N.

In conclusion, UA plays a dominant role in contributing to the bottom landward residual flow in both cases, which can influence the location of TMZ. The effects of UD are strong only under large river discharge.

4.2 Model results and discussion in BB

4.2.1 Models results in January 2018, summer

In order to understand which factor (wave, tide, and river discharge) dominates the hydrodynamics and sediment dynamics in BB, numerical model runs in two representative seasons were designed (summer and winter) in our study (Table 9.6). The results in January 2018 (summer) are shown in Fig. 9.16.

From the results of Case 1, we noticed that the direction of month-averaged residual flow (RF) at bottom layer (Fig. 9.16a) was from outside the tidal inlet into the inside area, and there are two branches: one was toward the northwest along the training wall and the other one propagated toward the Cullendulla Beach. The magnitude of the sediment flux (SF) was large outside the tidal inlet (>0.04 kg/m/second) and formed a clockwise eddy, shown in Fig. 9.16b. Due to the shelter of the Square Head, the SF decreased dramatically and formed two eddies inside the tidal inlet (<0.03 kg/m/second): one was clockwise along the Surfside Beach and the other was anticlockwise near the Square Head. The SSC at the bottom layer was 0.12−0.18 kg/m^3 in most areas. The monthly averaged significant wave height outside the tidal inlet was higher than 1.0 m in January 2018 and due to the sheltering effects of the Square Head, wave height decreased in the inner inlet, especially in waters near FTD where the wave height was very small (<0.2 m). Furthermore, the decrease of wave height was also caused by the shallow depth of the FTD (<0.1m) and wave breaking. The wave breaking can induce the strong wave orbital velocity (reach to 0.3 m/second) and large bottom stress (>0.4 N/m^2) on the FTD (Gao et al., 2018), leading to the erosion of the FTD for 4.0 cm in Case 1 in January 2018.

Table 9.6 Experiments description.

Experiment	Description
Case 1, Case 2	Forced by wind, river discharge, tidal current and wave in January 2018 Same with Case 1, but without tide
Case 3, Case 4, Case 5, Case 6, Case 7, Case 8, Case 9, Case 10, Case 11, Case 12, Case 13, Case 14, Case 15, Case 16	Same with Case 1, but without river dischargeSame with Case 1, but with large river discharge for Clyde river and Cullendulla creek (200 m^3/second)Same with Case 1, but without wave effectsSame with Case 1, but in July 2018Same with Case 6, but without wave effectsSame with Case 6, but without incoming swell from the open boundarySame with Case 1, but with 5 m flat bottom inside BB (without the FTD)Same with Case 9, but without wave effectsSame with Case 9, but in July 2018Same with Case 1, but with 1 m water depth in the FTD areaSame with Case 1, but with 2 m water depth in the FTD areaSame with Case 1, but with 3 m water depth in the FTD areaSame with Case 1, but with 4 m water depth in the FTD areaSame with Case 9, but without square Head

From Yang, Gang, Wang, Xiao Hua, Zhong, Yi & Oliver, Thomas S.N. (2022). Modeling study on the sediment dynamics and the formation of the flood-tide delta near Cullendulla Beach in the Batemans Bay, Australia. Marine Geology, 452. https://doi.org/10.1016/j.margeo.2022.106910

Fig. 9.17 shows the model results of Case 2 without considering the tidal effects, and we found the similar pattern for RF, SSC, SF, and FTD erosion in Case 2 compared with Case 1. This indicates that the tidal effects in BB are weak and is not the dominant factor controlling the sediment dynamics in our study domain, where the tidal range is less than 2 m.

For exploring the river discharge effects, the river input for Clyde River and Cullendulla Creek was set to 0 m^3/second and 200 m^3/second (peak discharge value for Clyde River during the flooding events in 2018) in Case 3 and 4, respectively. For Case 3, the flow, SSC, and SF patterns were found to be similar with Case 1 (not shown), which was due to low river discharge (<1.0 m^3/second in January 2018); when we increased the river discharge to 200 m^3/second in Case 4 (Fig. 9.18), the RF had a different pattern compared with previous Cases: the RF along the Cullendulla Creek was strong (>0.8 m/second), which can form a southward SF flowing out of the creek mouth. Although the riverine sediment flux can offer sediment source for

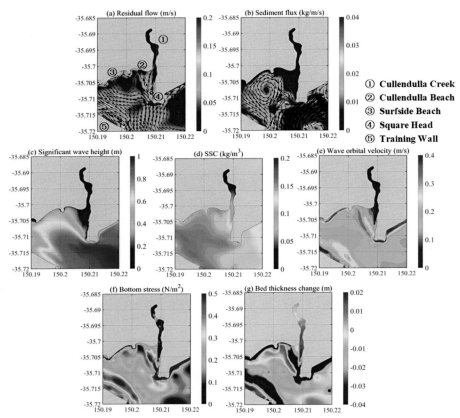

Figure 9.16 Monthly averaged (a) bottom layer residual flow; (b) bottom layer sediment flux; (c) significant wave height; (d) bottom SSC; (e) wave orbital velocity; (f) bottom stress; and (g) bed thickness change in Case 1.
From Yang, G., Wang, X. H., Zhong, Y. & Oliver, T. S. N. (2022). Modeling study on the sediment dynamics and the formation of the flood-tide delta near Cullendulla Beach in the Batemans Bay, Australia. Marine Geology, 452, 106910. https://doi.org/10.1016/j.margeo.2022.106910

the formation of the FTD, due to the strong river discharge and the shallow water depth of the FTD, the bottom stress near the mouth of creek was very large (~ 4 N/m^2), 10 times larger than experiment 1, and can aggravate the erosion of the FTD to 40 cm. We found the SSC dropped to <0.1 kg/m^3 in the waters near the mouth of the Cullendulla Creek, this was caused by the dilution effects of large river discharge.

For Case 5, without considering the wave effects (Fig. 9.19), the patterns for RF, SSC, and SF were totally different from Case 1 with wave effects: the direction for both RF and SF was north-west from the outside to inside of the tidal inlet; the RF became weaker, especially on the FTD; and due to the absence of

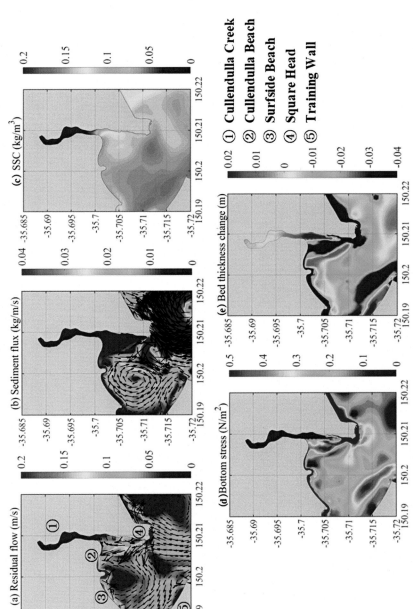

Figure 9.17 Monthly averaged (a) bottom layer residual flow; (b) bottom layer sediment flux; (c) bottom SSC; (d) bottom stress; and (e) bed thickness change in Case 2.

From Yang, Gang, Wang, Xiao Hua, Zhong, Yi & Oliver, Thomas S.N. (2022). Modeling study on the sediment dynamics and the formation of the flood-tide delta near Cullendulla Beach in the Batemans Bay, Australia. Marine Geology, 452. https://doi.org/10.1016/j.margeo.2022.106910

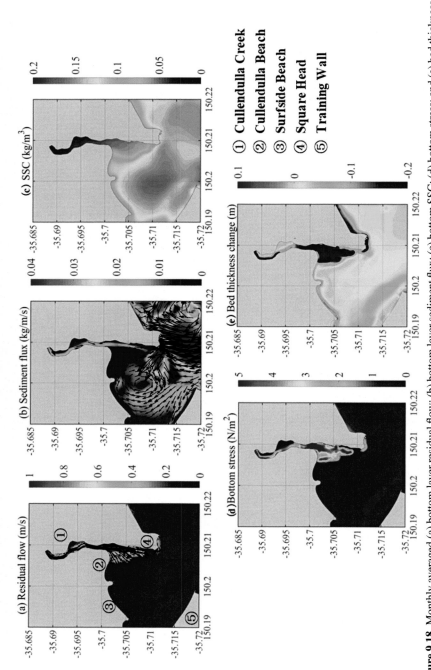

Figure 9.18 Monthly averaged (a) bottom layer residual flow; (b) bottom layer sediment flux; (c) bottom SSC; (d) bottom stress; and (e) bed thickness change in Case 4.

From Yang, G., Wang, X. H., Zhong, Y. & Oliver, T. S. N. (2022). Modeling study on the sediment dynamics and the formation of the flood-tide delta near Cullendulla Beach in the Batemans Bay, Australia. Marine Geology, 452, 106910. https://doi.org/10.1016/j.margeo.2022.106910

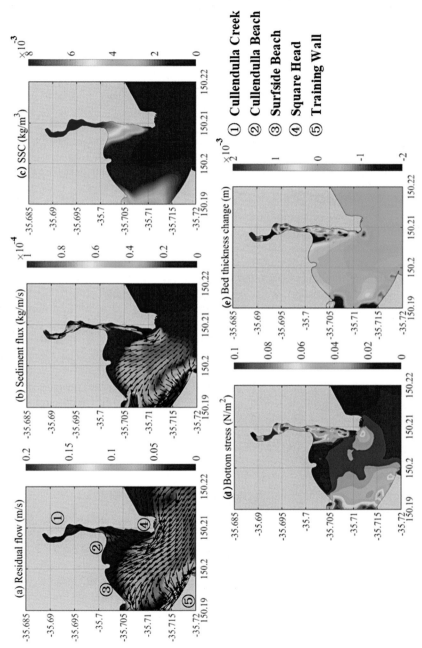

Figure 9.19 Monthly averaged (a) bottom layer residual flow; (b) bottom layer sediment flux; (c) bottom SSC; (d) bottom stress; and (e) bed thickness change in Case 5.
From Yang, Gang, Wang, Xiao Hua, Zhong, Yi & Oliver, Thomas S.N. (2022). Modeling study on the sediment dynamics and the formation of the flood-tide delta near Cullendulla Beach in the Batemans Bay, Australia. Marine Geology, 452. https://doi.org/10.1016/j.margeo.2022.106910

wave effects, the bottom stress on FTD decreased from >0.4 N to < 0.1 N/m^2, the magnitude of SF and SSC also dropped dramatically compared with Case 1, the high value of SSC (~0.01 kg/m^3) near the mouth of Cullendulla Creek was mainly from the riverine sediment input in this case (0.02 kg/m^3); in addition, the bed thickness was only eroded for nearly 2.0 mm, much smaller than Case 1−4 with wave effects.

The difference of hydrodynamic and sediment dynamics between Case 1 and 5 was mainly caused by the wave radiation stress gradient, which was calculated by FVCOM-SWAVE in our study. Wave radiation stress gradient can be regarded as the wave-averaged effects on the current (excess momentum flux carried by the ocean waves), formulated by Longuet-Higgins and Stewart (1964), and implemented by Mellor (2005). Due to the gradient of the radiation stress, the current can be generated in nearshore area (Gao et al., 2018; Mao & Xia, 2020; Song et al., 2020; Symonds et al., 1995). Based on previous studies, the radiation stress method of Mellor (2005) can produce unrealistic flow in nonbreaking wave propagation over a sloped topography, and an alternative approach based on the vortex force formalism that allowed for realistic simulations was developed (Ardhuin et al., 2008; Bennis et al., 2011; Kumar et al., 2012; Uchiyama et al., 2010). However, we would not reopen the debate on radiation stress here, and the objective of this study is to explore the wave effects on the hydrodynamics of BB. Given that water elevation and current calculated by our model have been successfully validated against field measurements, the spurious current is proved to be minimal; in addition, the error from radiation stress formulation is considerably reduced when applying this method for shallow waters with mild slopes like tidal flats (Kumar et al., 2011; Moghimi et al., 2013; Sorourian et al., 2020). Therefore, this method can be applied in this study.

From above model results, we can see that wave effects played the dominant role in controlling the hydrodynamics in BB, which in turn influence the sediment dynamics and FTD erosion. Furthermore, we noticed the FTD was always eroded in Case 1−5, so how was FTD formed, and at what conditions does the FTD begin to accrete? The formation of this FTD requires to be investigated and was discussed in Section 4.2.3, 4.2.4 and 4.2.5.

4.2.2 Model results in July 2018, winter

As introduced in Section 2.2 (Fig. 9.3), the direction of the prevailing wind is northeast and southeast in January and northwest and southwest in July 2018. Due to the sheltering effect of the hills and ranges west of BB (Wang et al., 2021), the monthly averaged wind velocity in July (2.74 m/second) was slightly weaker than that in January (3.65 m/second) The wind field change in different seasons can influence the wave, RF and SF in BB. Therefore, Case 6 was designed to explore the hydrodynamic and sediment dynamics change in July 2018 (winter) in our study domain.

The model results of Case 6 were shown in Fig. 9.20(1) the direction of RF was northwest from the outside to the inner tidal inlet; (2) the SF propagated northward

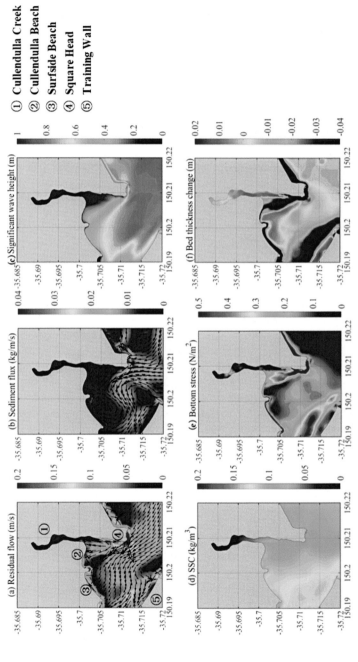

Figure 9.20 Monthly averaged (a) bottom layer residual flow; (b) bottom layer sediment flux; (c) bottom SSC; (d) bottom stress; and (e) bed thickness change in Case 6.

From Yang, Gang, Wang, Xiao Hua, Zhong, Yi & Oliver, Thomas S.N. (2022). Modeling study on the sediment dynamics and the formation of the flood-tide delta near Cullendulla Beach in the Batemans Bay, Australia. Marine Geology, 452. https://doi.org/10.1016/j.margeo.2022.106910

into the inner side of the tidal inlet along the Square Head and turned to westward, merging into the northwest flux along the training wall; the magnitude of SF decreased compared with Case 1 in January; (3) the significant wave height was slightly lower compared with that in January, due to the weaker localized wind velocities in July 2018; however, the overall pattern was similar with that in January, and (4) the bed thickness change still showed an erosion pattern on the FTD for ∼4.0 cm due to the large bottom stress. When we removed the wave effects in Case 7 (not shown), the magnitude of RF, SSC, and SF decreased dramatically, and the FTD still showed a slight erosion pattern, similar with Case 5.

In addition, the pattern and distribution of the water depth change retrieved from Landsat 8 (Fig. 9.10c) were similar with the model results with wave effects (Case 1 and Case 6, shown in Fig. 9.16g and Fig. 9.20f) and support the conclusion that waves are an important driver of erosion of the FTD. Meanwhile, we also noticed that the water depth in the north of the Clyde River channel increased from the Landsat retrieved figure and our model results (Fig. 9.10c and Fig. 9.16g). This is due to the shallow water there and strong bottom stress (Fig. 9.16f), which can erode the seabed there. The siltation in the Clyde River channel was caused by the deep water depth and small bottom stress there (Fig. 9.16f).

From our model results, we noticed that although the wind direction in July is northwestly/southwestly, waves can still propagate into the inner bay with ∼0.8 m significant wave height (Fig. 9.20c), like in January (Fig. 9.16c). Therefore, we produced the simulated wave roses in January and July at chosen inner bay and outer bay stations (red and green dots in Fig. 9.2), as shown in Fig. 9.21. From Fig. 9.21, we can see that the wave directions in the inner bay are from southeast in both months (Fig. 9.21a1 and b1), and the wave directions do not follow the wind direction (Fig. 9.3). Furthermore, when we removed the incoming swell from the open boundary in July 2018 (Case 8), the significant wave height in the inner bay decreased dramatically (not shown). This indicates that the wave in the inner bay is mainly induced by the incoming swell from SSE direction in the outer bay, and due to the bathymetry and coastline orientation, the wave direction changed to southeast in the inner bay.

From above results, we can conclude that wave induced by offshore swells played the dominant factor controlling the hydrodynamics and sediment dynamics in the inner bay, which can always aggravate the erosion of the FTD.

4.2.3 The formation of the FTD

Based on the model results, we noticed that the FTD near Cullendulla Beach was always eroded no matter in summer or winter, with or without wave effects, therefore, the question arises: under what conditions was the FTD formed?

In order to understand how the FTD was formed, a series of model runs were set up where we assumed there is no FTD there. We designed Case 9 to be the same with Case 1 but with flat bottom (5 m) inside the BB to explore the FTD formation in this area. Model results of Case 9 were shown in Fig. 9.22.

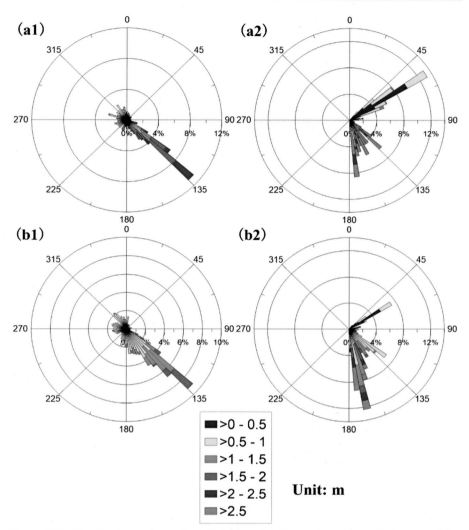

Figure 9.21 Simulated wave roses for (a1) inner and (a2) outer bay stations in January, (b1) inner and (b2) outer bay stations in July 2018.
From Yang, G., Wang, X. H., Zhong, Y. & Oliver, T. S. N. (2022). Modeling study on the sediment dynamics and the formation of the flood-tide delta near Cullendulla Beach in the Batemans Bay, Australia. Marine Geology, 452, 106910. https://doi.org/10.1016/j.margeo. 2022.106910

From the results of Case 9, we can see that: the FTD near Cullendulla Beach (the red circle in Fig. 9.22f) accreted approximately 3.0 cm over the period of 1 month and the direction of monthly averaged RF and SF at bottom layer was from outside toward the inner bay (Fig. 9.22a and b). The SF toward the FTD suggests onshore transport can provide sediment source for the FTD formation. Due

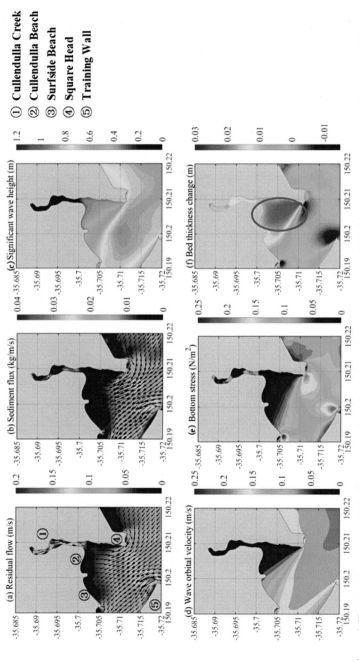

Figure 9.22 Monthly averaged (a) bottom layer residual flow; (b) bottom layer sediment flux; (c) significant wave height; (d) wave orbital velocity; (e) bottom stress; and (f) bed thickness change in Case 9. (*red circle* indicates the location of the FTD). From Yang, Gang, Wang, Xiao Hua, Zhong, Yi & Oliver, Thomas S.N. (2022). Modeling study on the sediment dynamics and the formation of the flood-tide delta near Cullendulla Beach in the Batemans Bay, Australia. Marine Geology, 452. https://doi.org/10.1016/j.margeo.2022.106910

to the sheltering of the Square Head, the wave height in the FTD area was lower than 0.4 m, and the wave orbital velocity there was also very small (<0.05 m/second), which leaded to smaller bottom stress on the FTD area compared with other areas. Therefore, this overall reduction in energy promoted deposition of the FTD. When we removed the wave effects in Case 10 (not shown), there was nearly no bed thickness change, and the magnitude of the SF dropped nearly by 100 times compared with the experiment considering the wave effects, which was mainly caused by the small bottom stress without the wave effects, and the sediment on the seabed was not resuspended. We also designed Case 11 same Case 9 with flat bottom, but in July 2018. The results of Case 11 were similar with Case 9 in January 2018 (not shown). Therefore, sediment deposition can occur on the FTD areas in cases with deeper and flat bathymetry due to onshore sediment transport and a reduction in energy level as a result of the sheltering effects of Square Head.

4.2.4 The equilibrium depth of the FTD near the Cullendulla Beach

Based on the model results in Section 4.2.3, the FTD can be developed in cases with deeper and flat bottom inside the BB. However, in real situations with true bathymetry, the FTD was always eroded due to the shallow water depth and large bottom stress on the FTD. Therefore, the FTD may be a result of a dynamic balance where the FTD is eroded and deepens to a critical value, after which sedimentation would be expected to occur. In order to explore what this critical depth is, four experiments were designed in our study: same with the Case 1, but with 1 m (Case 12), 2 m (Case 13), 3 m (Case 14), and 4 m water depth (Case 15) in the FTD area. Model results can be seen in Fig. 9.23, and the station (red dot) in Fig. 9.23a was chosen to show the 1-month bed thickness change for Cases 12−15.

There was a substantial erosion occurring on the FTD in Case 12, for 10.4 mm at the chosen station (red dot in Fig. 9.23a). When we set the FTD deeper to 2.0 m (Case 13), the FTD began to be slightly developed, with 2.2 mm siltation at the chosen station. Furthermore, we found that the deeper the FTD surface, the more sediment can be silted there, with 16.1 and 22.5 mm siltation at the station for Case 14 and 15, respectively. Therefore, the FTD with a depth of 2 m reached a morphological equilibrium status: if the FTD could be eroded to a depth ~2.0 m, a siltation would occur.

4.2.5 The sheltering effects of the headland

For further explore how the tidal inlet and FTD near Cullendulla Beach is influenced by the hydrodynamic and sediment dynamics inside BB, we designed another experiment (Case 16) where we removed the Square Head. The model results of Case 16 can be seen in Fig. 9.24.

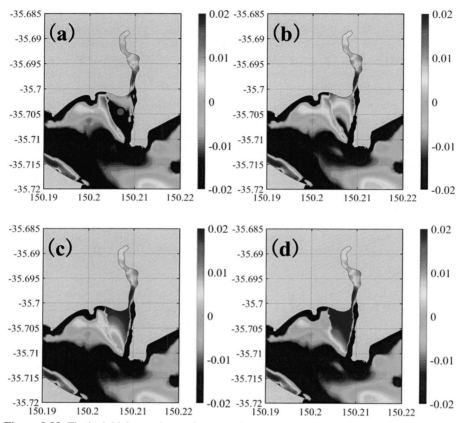

Figure 9.23 The bed thickness change for 1 month in (a) Case 12; (b) Case 13; (c) Case 14; and (d) Case 15 (unit: Meters); the red dot in (a) indicates the location of the chosen station. From Yang, G., Wang, X. H., Zhong, Y. & Oliver, T. S. N. (2022). Modeling study on the sediment dynamics and the formation of the flood-tide delta near Cullendulla Beach in the Batemans Bay, Australia. Marine Geology, 452, 106910. https://doi.org/10.1016/j.margeo. 2022.106910

Without the sheltering effects of Square Head, waves can propagate onto the Cullendulla Beach resulting in a very different pattern of RF and SF. The distribution of monthly averaged bottom stress was relatively uniform in the red rectangular area in Fig. 9.24. Compared with Case 9 with the Square Head (Fig. 9.22f), and the FTD cannot be developed in Case 16. Therefore, the model results suggest that existence of the Square Head plays a critical role in enabling the deposition of the FTD.

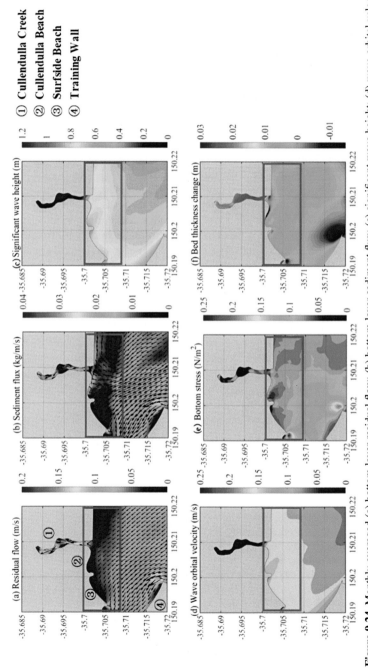

Figure 9.24 Monthly averaged (a) bottom layer residual flow; (b) bottom layer sediment flux; (c) significant wave height; (d) wave orbital velocity; (e) bottom stress; and (f) bed thickness change in Case 16.

From Yang, Gang, Wang, Xiao Hua, Zhong, Yi & Oliver, Thomas S.N. (2022). Modeling study on the sediment dynamics and the formation of the flood-tide delta near Cullendulla Beach in the Batemans Bay, Australia. Marine Geology, 452. https://doi.org/10.1016/j.margeo.2022.106910

5. Conclusions

In summary, this chapter provides a contribution to numerical modeling of suspended-sediment dynamics in estuaries. The sediment model was driven by different forcings (e.g., baroclinic forcing, wave-current interactions, wind, and river discharge), and the model proved its capability in simulating sediment dynamics in any coastal areas by changing the driving forcings. This chapter has provided a better understanding on the effects of estuarine circulation and wave on the sediment dynamics in medium scale estuaries: (1) the two-layer estuarine circulation can influence the sediment transport and contribute to the formation of TMZ, mainly caused by the combined effects of UD, UA, UN, and UR, and (2) the strong gradient of wave radiation stress can generate wave-driven flow and control the sediment dynamics and morphological evolution in coastal areas with shallow water and small tidal range.

The model results on the estuarine circulation in YRE complement knowledge of the along/cross-channel estuarine circulation under different river discharges and their impacts on the TMZ formation, providing novel and more-reliable information about the sediment dynamics in YRE. Unlike YRE, BB is a microtidal estuary with frequent human activities and suffered from strong wave events, which offers us a good opportunity to understand the wave effects on the morphology change in microtidal areas. Modeling results under strong wave events in BB show that waves can dramatically change the hydrodynamic and sediment dynamics patterns even in embayed estuaries. Meanwhile, the model results indicate that FTDs have an equilibrium depth, and waves can not only erode the FTDs but can also supply the sediment source for the formation of FTDs. The mechanisms for FTDs formation and the equilibrium depth can be calculated by the numerical model based on the accurate sediment configuration on the seabed, which offers a way to understand the coastal morphology evolution. In addition, the method using the satellite to retrieve water depth change can also help determine changes in water depth for specified period, such as before and after the strong wave event.

Although the findings in this chapter are site specific, they are of relevance and importance to estuaries under extreme events, which is meaningful and can deliver significant socio-economic benefits to the coastal community and management.

References

Ardhuin, F., Jenkins, A. D., & Belibassakis, K. A. (2008). Comments on "The three-dimensional current and surface wave equations". *Journal of Physical Oceanography, 38*(6), 1340−1350. https://doi.org/10.1175/2007JPO3670.1, 00223670.

Ariathurai, R., & Krone, R. B. (1977). Mathematical modeling of sediment transport in estuaries. *In: Estuarine Processes (Proceedings Third International Estuarine Research Conference, Galveston, U.S.A.: Oct.7-9, 1975), Wiley, M. (Ed.), 2*, 98−106. https://doi.org/10.1016/b978-0-12-751802-2.50015-1. undefined.

Austin, T., Vila-Concejo, A., Short, A., & Ranasinghe, R. (2018). A multi-scale conceptual model of flood-tide delta morphodynamics in micro-tidal estuaries. *Geosciences, 8*(9). https://doi.org/10.3390/geosciences8090324

Bennis, A. C., Ardhuin, F., & Dumas, F. (2011). On the coupling of wave and three-dimensional circulation models: Choice of theoretical framework, practical implementation and adiabatic tests. *Ocean Modelling, 40*(3−4), 260−272. https://doi.org/10.1016/j.ocemod.2011.09.003. 14635003.

Bever, A. J., & MacWilliams, M. L. (2013). Simulating sediment transport processes in San Pablo Bay using coupled hydrodynamic, wave, and sediment transport models. *Marine Geology, 345*, 235−253. https://doi.org/10.1016/j.margeo.2013.06.012, 00253227.

Brand, A., Lacy, J. R., Hsu, K., Hoover, D., Gladding, S., & Stacey, M. T. (2010). Wind-enhanced resuspension in the shallow waters of South San Francisco Bay: Mechanisms and potential implications for cohesive sediment transport. *Journal of Geophysical Research: Oceans, 115*(11). https://doi.org/10.1029/2010JC006172, 21699291.

Brenon, I., & Le Hir, P. (1999). Modelling the turbidity maximum in the seine estuary (France): Identification of formation processes. *Estuarine, Coastal and Shelf Science, 49*(4), 525−544. https://doi.org/10.1006/ecss.1999.0514, 02727714.

Burchard, H., & Baumert, H. (1998). The formation of estuarine turbidity maxima due to density effects in the salt wedge. A hydrodynamic process study. *Journal of Physical Oceanography, 28*(2), 309−321. https://doi.org/10.1175/1520-0485(1998)028<0309:TFOETM>2.0.CO;2. 00223670.

Burchard, H., & Hetland, R. D. (2010). Quantifying the contributions of tidal straining and gravitational circulation to residual circulation in periodically stratified tidal estuaries. *Journal of Physical Oceanography, 40*(6), 1243−1262. https://doi.org/10.1175/2010JPO4270.1, 00223670.

Castelle, B., Turner, I. L., Ruessink, B. G., & Tomlinson, R. B. (2007). Impact of storms on beach erosion: Broadbeach (Gold coast, Australia). *Journal of Coastal Research, 50*, 534−539, 15515036.

Chant, R. J. (2002). Secondary circulation in a region of flow curvature: Relationship with tidal forcing and river discharge. *Journal of Geophysical Research: Oceans, 107*(C9), 14−21.

Chen, C., Beardsley, R. C., Cowles, G., Qi, J., Lai, Z., Gao, G., … Lin, H. (2012). *An unstructured-grid, finite-volume community ocean model: FVCOM user manual.* Cambridge, MA, USA: Sea Grant College Program, Massachusetts Institute of Technology.

Chen, C., Liu, H., & Beardsley, R. C. (2003). An unstructured grid, finite-volume, three-dimensional, primitive equations ocean model: Application to Coastal Ocean and estuaries. *Journal of Atmospheric and Oceanic Technology, 20*(1), 159−186.

Chen, C. S., Beardsley, R. C., & Cowles, G. (2006b). *An unstructured grid, finite volume coastal ocean model: FVCOM user manual, 2nd ed., Rep. SMAST/UMASSD-06-0602, mar. Ecosyst. Dyn. Model. Lab., Univ. Of mass., Dartmouth.*

Chen, C. S., Huang, H., Beardsley, R. C., Xu, Q., Limeburner, R., Cowles, G. W., Sun, Y., Qi, J., & Lin, H. (2011). Tidal dynamics in the Gulf of Maine and New England shelf: An application of FVCOM. *Journal of Geophysical Research: Oceans, 116*. https://doi.org/10.1029/2011JC007054

Chen, Y., He, Q., Shen, J., & Du, J. (2020). The alteration of lateral circulation under the influence of human activities in a multiple channel system, Changjiang Estuary. *Estuarine, Coastal and Shelf Science, 242*, Article 106823.

Cheng, P., Valle-Levinson, A., & de Swart, H. E. (2011). A numerical study of residual circulation induced by asymmetric tidal mixing in tidally dominated estuaries. *Journal of Geophysical Research: Oceans, 116*(C1).

Cheng, Z., Jalon-Rójas, I., Wang, X. H., & Liu, Y. (2020). Impacts of land reclamation on sediment transport and sedimentary environment in a macro-tidal estuary. *Estuarine, Coastal and Shelf Science, 242*, Article 106861.

Cheng, Z., Wang, X. H., Paull, D., & Gao, J. (2016). Application of the geostationary ocean color imager to mapping the diurnal and seasonal variability of surface suspended matter in a macro-tidal estuary. *Remote Sensing, 8*(3), 244.

Cho, J., Song, Y., & Kim, T. I. (2016). Numerical modeling of estuarine circulation in the Geum River Estuary, Korea. *Procedia Engineering, 154*, 982−989. https://doi.org/10.1016/j.proeng.2016.07.586, 18777058.

Christiansen, C., Vølund, G., Lund-Hansen, L. C., & Bartholdy, J. (2006). Wind influence on tidal flat sediment dynamics: Field investigations in the Ho Bugt, Danish Wadden Sea. *Marine Geology, 235*(1−4), 75−86. https://doi.org/10.1016/j.margeo.2006.10.006, 00253227.

Cong, S., Wu, X., Ge, J., Bi, N., Li, Y., Lu, J., & Wang, H. (2021). Impact of Typhoon Chan-hom on sediment dynamics and morphological changes on the East China Sea inner shelf. *Marine Geology, 440*. https://doi.org/10.1016/j.margeo.2021.106578, 00253227.

Dijkstra, Y. M., Schuttelaars, H. M., & Schramkowski, G. P. (2019). A Regime shift from low to high sediment concentrations in a tide-dominated estuary. *Geophysical Research Letters, 46*(8), 4338−4345. https://doi.org/10.1029/2019GL082302, 19448007.

Dufois, F., Verney, R., Le Hir, P., Dumas, F., & Charmasson, S. (2014). Impact of winter storms on sediment erosion in the rhone river prodelta and fate of sediment in the Gulf of Lions (north western mediterranean sea). *Continental Shelf Research, 72*, 57−72. https://doi.org/10.1016/j.csr.2013.11.004, 02784343.

Dyer, K. R. (1997). *Estuaries: A physical introduction.* John Wiley & Sons.

Festa, J. F., & Hansen, D. V. (1976). A two-dimensional numerical model of estuarine circulation: The effects of altering depth and river discharge. *Estuarine and Coastal Marine Science, 4*(3), 309−323. https://doi.org/10.1016/0302-3524(76)90063-3, 03023524.

Gao, G. D., Wang, X. H., & Bao, X. W. (2014). Land reclamation and its impact on tidal dynamics in Jiaozhou Bay, Qingdao, China. *Estuarine, Coastal and Shelf Science, 151*, 285−294. https://doi.org/10.1016/j.ecss.2014.07.017, 02727714.

Gao, G. D., Wang, X. H., Song, D., Bao, X., Yin, B. S., Yang, D. Z., Ding, Y., Li, H., Hou, F., & Ren, Z. (2018). Effects of wave-current interactions on suspended-sediment dynamics during strong wave events in Jiaozhou Bay, Qingdao, China. *Journal of Physical Oceanography, 48*(5), 1053−1078. https://doi.org/10.1175/JPO-D-17-0259.1, 15200485.

Gao, J.-H., Li, J., Wang, H., Bai, F.-L., Cheng, Y., & Wang, Y.-P. (2012). Rapid changes of sediment dynamic processes in Yalu River Estuary under anthropogenic impacts. *International Journal of Sediment Research, 27*(1), 37−49. https://doi.org/10.1016/s1001-6279(12)60014-6, 10016279.

Ge, J., Shen, F., Guo, W., Chen, C., & Ding, P. (2015). Estimation of critical shear stress for erosion in the changjiang estuary: A synergy research of observation, goci sensing and modeling. *Journal of Geophysical Research: Oceans, 120*(12), 8439−8465. https://doi.org/10.1002/2015JC010992, 21699291.

Ge, J., Zhou, Z., Yang, W., Ding, P., Chen, C., Wang, Z. B., & Gu, J. (2018). Formation of concentrated benthic suspension in a time-dependent salt wedge estuary. *Journal of Geophysical Research: Oceans, 123*(11), 8581−8607. https://doi.org/10.1029/2018JC013876. 21699291.

Giddings, S. N., & MacCready, P. (2017). Reverse estuarine circulation due to local and remote wind forcing, enhanced by the presence of along-coast estuaries. *Journal of Geophysical Research: Oceans, 122*(12), 10184−10205. https://doi.org/10.1002/2016jc012479

Goodrich, D. M., & Blumberg, A. F. (1991). The fortnightly mean circulation of chesapeake bay. *Estuarine, Coastal and Shelf Science, 32*(5), 451−462. https://doi.org/10.1016/0272-7714(91)90034-9, 02727714.

Grasso, F., Verney, R., Le Hir, P., Thouvenin, B., Schulz, E., Kervella, Y., Khojasteh Pour Fard, I., Lemoine, J.-P., Dumas, F., & Garnier, V. (2018). Suspended sediment dynamics in the macrotidal seine estuary (France): 1. Numerical modeling of turbidity maximum dynamics. *Journal of Geophysical Research: Oceans, 123*(1), 558−577. https://doi.org/10.1002/2017jc013185

Hansen, D. V., & Rattray, J.,M. (1966). *Gravitational circulation in straits and estuaries.*

Harff, J., Lampe, R., Lemke, W., Lübke, H., Lüth, F., Meyer, M., & Tauber, F. (2005). The Baltic Sea - a model ocean to study interrelations of geosphere, ecosphere, and anthroposphere in the coastal zone. *Journal of Coastal Research, 21*(3), 441−446. https://doi.org/10.2112/04-0217.1

Harris, D. L., Vila-Concejo, A., Austin, T., & Benavente, J. (2020). Multi-scale morphodynamics of an estuarine beach adjacent to a flood-tide delta: Assessing decadal scale erosion. *Estuarine, Coastal and Shelf Science, 241.* https://doi.org/10.1016/j.ecss.2020.106759, 02727714.

Hayes, M. O., & FitzGerald, D. M. (2013). Origin, evolution, and classification of tidal inlets. *Journal of Coastal Research, 69*(10069), 14−33.

Hayes, M. O. (1975). Morphology of sand accumulation in estuaries: An introduction to the symposium. In *Geology and engineering* (pp. 3−22). Academic Press.

Jing, L., & Ridd, P. V. (1997). Modelling of suspended sediment transport in coastal areas under waves and currents. *Estuarine, Coastal and Shelf Science, 45*(1), 1−16.

Kumar, N., Voulgaris, G., & Warner, J. C. (2011). Implementation and modification of a three-dimensional radiation stress formulation for surf zone and rip-current applications. *Coastal Engineering, 58*(12), 1097−1117. https://doi.org/10.1016/j.coastaleng.2011.06.009, 03783839.

Kumar, N., Voulgaris, G., Warner, J. C., & Olabarrieta, M. (2012). Implementation of the vortex force formalism in the coupled ocean-atmosphere-wave-sediment transport (COAWST) modeling system for inner shelf and surf zone applications. *Ocean Modelling, 47*, 65−95. https://doi.org/10.1016/j.ocemod.2012.01.003, 14635003.

Lerczak, J. A., & Rockwell Geyer, W. (2004). Modelling the lateral circulation in straight, stratified estuaries. *Journal of Physical Oceanography, 34*(6), 1410−1428.

Lettmann, K. A., Wolff, J. O., & Badewien, T. H. (2009). Modeling the impact of wind and waves on suspended particulate matter fluxes in the East Frisian Wadden Sea (southern North Sea). *Ocean Dynamics, 59*(2), 239−262. https://doi.org/10.1007/s10236-009-0194-5, 16167228.

Li, L., Ye, T., Wang, X. H., & Xia, Y. (2019). Tracking the multidecadal variability of the surface turbidity maximum zone in Hangzhou Bay, China. *International Journal of Remote Sensing, 40*(24), 9519−9540. https://doi.org/10.1080/01431161.2019.1633701, 13665901.

Liu, X., Zhang, H., Zheng, J., Guo, L., Jia, Y., Bian, C., Li, M., Ma, L., & Zhang, S. (2020). Critical role of wave−seabed interactions in the extensive erosion of Yellow River estuarine sediments. *Marine Geology, 426.*

Longuet-Higgins, M. S., & Stewart, R.w. (1964). Radiation stresses in water waves; a physical discussion, with applications. *Deep Sea Research and Oceanographic Abstracts, 11*(4), 529−562. https://doi.org/10.1016/0011-7471(64)90001-4, 00117471.

Loureiro, C., & Cooper, A. (2018). Temporal variability in winter wave conditions and storminess in the northwest of Ireland. *Irish Geography, 51*(2), 155−170. https://doi.org/10.2014/igj.v51i2.1369, 00750778.

Lu, J., Han, G., Song, D., Oliver, T., Teng, Y., Guo, J., Wu, L., Zhang, C., & Jiang, X. (2021). The cross-shore component in the vertical structure of wave-induced currents and resulting offshore transport. *Journal of Geophysical Research: Oceans, 126*(10). https://doi.org/10.1029/2021JC017311, 21699291.

Ma, Y., Xu, N., Liu, Z., Yang, B., Yang, F., Wang, X. H., & Li, S. (2020). Satellite-derived bathymetry using the ICESat-2 lidar and Sentinel-2 imagery datasets. *Remote Sensing of Environment, 250*, 112047.

Mao, M., & Xia, M. (2020). Monthly and episodic dynamics of summer circulation in lake Michigan. *Journal of Geophysical Research: Oceans, 125*(6). https://doi.org/10.1029/2019JC015932, 21699291.

Mellor, G. (2005). Some consequences of the three-dimensional current and surface wave equations. *Journal of Physical Oceanography, 35*(11), 2291–2298.

Mellor, G., & Yamada, T. (1982). Development of a turbulence closure model for geophysical fluid problems. *Reviews of Geophysics, 20*(4), 851.

Moghimi, S., Klingbeil, K., Gräwe, U., & Burchard, H. (2013). A direct comparison of a depth-dependent Radiation stress formulation and a Vortex force formulation within a three-dimensional coastal ocean model. *Ocean Modelling, 70*, 132–144. https://doi.org/10.1016/j.ocemod.2012.10.002, 14635003.

Neal, C., Leeks, G. J. L., Millward, G. E., Harris, J. R. W., Huthnance, J. M., & Rees, J. G. (2003). Land-ocean interaction: Processes, functioning and environmental management from a UK perspective: An introduction. *Science of the Total Environment, 314–316*, 3–11. https://doi.org/10.1016/S0048-9697(03)00091-3, 00489697.

Niu, Q., Xia, M., Ludsin, S. A., Chu, P. Y., Mason, D. M., & Rutherford, E. S. (2018). High-turbidity events in Western Lake Erie during ice-free cycles: Contributions of river-loaded vs. resuspended sediments. *Limnology & Oceanography, 63*(6), 2545–2562. https://doi.org/10.1002/lno.10959, 19395590.

Olabarrieta, M., Warner, J. C., & Kumar, N. (2011). Wave-current interaction in Willapa bay. *Journal of Geophysical Research: Oceans, 116*(12). https://doi.org/10.1029/2011JC007387, 21699291.

Postma, H. (1967). Sediment transport and sedimentation in the estuarine environment. *American Association of Advanced Sciences, 83*, 158–179.

Pritchard, D. W. (1956). The dynamic structure of a coastal plain estuary. *Journal of Marine Research, 15*(1), 33–42.

Qiu, Z., Xiao, C., Perrie, W., Sun, D., Wang, S., Shen, H., Yang, D., & He, Y. (2017). Using Landsat 8 data to estimate suspended particulate matter in the Yellow River estuary. *Journal of Geophysical Research: Oceans, 122*(1), 276–290. https://doi.org/10.1002/2016JC012412, 21699291.

Ralston, D. K., Talke, S., Geyer, W. R., Al-Zubaidi, H. A. M., & Sommerfield, C. K. (2019). Bigger tides, less flooding: Effects of dredging on barotropic dynamics in a highly modified estuary. *Journal of Geophysical Research: Oceans, 124*(1), 196–211. https://doi.org/10.1029/2018JC014313, 21699291.

Roy, P. S., Cowell, P. J., Ferland, M. A., & Thom, B. G. (1994). Wave-dominated coasts. In *Coastal evolution: Late Quaternary shoreline morphodynamics* (pp. 121–186).

Scully, M. E., Geyer, W. R., & Lerczak, J. A. (2009). The influence of lateral advection on the residual estuarine circulation: A numerical modeling study of the hudson River estuary. *Journal of Physical Oceanography, 39*(1), 107–124. https://doi.org/10.1175/2008JPO3952.1, 00223670.

Smagorinsky, J. (1963). General circulation experiments with the primitive equations. *Monthly Weather Review, 91*(3), 99–164. https://doi.org/10.1175/1520-0493(1963)091<0099:gcewtp>2.3.co;2

Song, D., Wang, X. H., Cao, Z., & Guan, W. (2013). Suspended sediment transport in the deepwater navigation channel, yangtze River Estuary, China, in the dry season 2009: 1. Observations over spring and neap tidal cycles. *Journal of Geophysical Research: Oceans, 118*(10), 5555–5567.

Song, H., Kuang, C., Wang, X. H., & Ma, Z. (2020). Wave-current interactions during extreme weather conditions in southwest of Bohai Bay, China. *Ocean Engineering, 216*, Article 108068.

Sorourian, S., Huang, H., Li, C., Justic, D., & Payandeh, A. R. (2020). Wave dynamics near Barataria Bay tidal inlets during spring—summer time. *Ocean Modelling, 147*. https://doi.org/10.1016/j.ocemod.2019.101553, 14635003.

Symonds, G., Black, K. P., & Young, I. R. (1995). Wave-driven flow over shallow reefs. *Journal of Geophysical Research: Oceans, 100*(C2), 2639—2648. https://doi.org/10.1029/94jc02736

Uchiyama, Y., McWilliams, J. C., & Shchepetkin, A. F. (2010). Wave-current interaction in an oceanic circulation model with a vortex-force formalism: Application to the surf zone. *Ocean Modelling, 34*(1—2), 16—35. https://doi.org/10.1016/j.ocemod.2010.04.002, 14635003.

Uncles, R. J., & Stephens, J. A. (1989). Distributions of suspended sediment at high water in a macrotidal estuary. *Journal of Geophysical Research: Oceans, 94*(C10), 14395—14405.

Valle-Levinson, A., Reyes, C., & Sanay, R. (2003). Effects of bathymetry, friction, and rotation on estuary-ocean exchange. *Journal of Physical Oceanography, 33*(11), 2375—2393. https://doi.org/10.1175/1520-0485(2003)033<2375:EOBFAR>2.0.CO;2

Wang, A., Wang, X. H., & Yang, G. (2021). The effects of wind-driven storm events on partly sheltered estuarine beaches in Batemans bay, New South Wales, Australia. *Journal of Marine Science and Engineering, 9*(3), 314.

Wang, X. H., Pinardi, N., & Malacic, V. (2007). Sediment transport and resuspension due to combined motion of wave and current in the northern adriatic Sea during a bora event in January 2001: A numerical modelling study. *Continental Shelf Research, 27*(5), 613—633.

Wang, X. H., & Pinardi, N. (2002). Modelling the dynamics of sediment transport and resuspension in the northern Adriatic Sea. *Journal of Geophysical Research: Oceans, 107*(C12), 18—21.

Warner, J. C., Sherwood, C. R., Signell, R. P., Harris, C. K., & Arango, H. G. (2008). Development of a three-dimensional, regional, coupled wave, current, and sediment-transport model. *Computers & Geosciences, 34*(10), 1284—1306. https://doi.org/10.1016/j.cageo.2008.02.012, 00983004.

Water Research Laboratory. (2017). *Eurobodalla coastal hazard assessment*. Australia: University of New South Wales. https://www.esc.nsw.gov.au/__data/assets/pdf_file/0020/158132/Eurobodalla-Coastal-Hazard-Assessment-2017.pdf.

Xiao, Z., Wang, X. H., Roughan, M., & Harrison, D. (2019). Numerical modelling of the sydney harbour estuary, New South Wales: Lateral circulation and asymmetric vertical mixing. *Estuarine, Coastal and Shelf Science, 217*, 132—147.

Xiao, Z. Y., Wang, X. H., Song, D., Jalón-Rojas, I., & Harrison, D. (2020). Numerical modelling of suspended-sediment transport in a geographically complex microtidal estuary: Sydney Harbour Estuary, NSW. *Estuarine, Coastal and Shelf Science, 236*, Article 106605.

Xie, D., Pan, C., Wu, X., Gao, S., & Wang, Z. B. (2017). Local human activities overwhelm decreased sediment supply from the Changjiang River: Continued rapid accumulation in the Hangzhou Bay-Qiantang Estuary system. *Marine Geology, 392*, 66—77. https://doi.org/10.1016/j.margeo.2017.08.013, 00253227.

Xu, T., & You, X.y. (2017). Numerical simulation of suspended sediment concentration by 3D coupled wave-current model in the Oujiang River Estuary, China. *Continental Shelf Research, 137*, 13—24. https://doi.org/10.1016/j.csr.2017.01.021, 18736955.

Yan, D., Song, D., & Bao, X. (2019). Spring-neap tidal variation and mechanism analysis of the maximum turbidity in the Pearl River Estuary during flood season. *Journal of Tropical Oceanography, 39*(1), 20—35.

Yang, G., Wang, X. H., Cheng, Z., Zhong, Y., & Oliver, T. (2021). Modelling study on estuarine circulation and its effect on the turbidity maximum zone in the Yalu River Estuary, China. *Estuarine, Coastal and Shelf Science, 263*, Article 107634.

Yang, G., Wang, X. H., Zhong, Y., Cheng, Z., & Andutta, F. P. (2020). Wave effects on sediment dynamics in a macro-tidal estuary: Darwin harbour, Australia during the monsoon season. *Estuarine, Coastal and Shelf Science, 244*, Article 106931.

Yang, G., Wang, X., Ritchie, E. A., Qiao, L., Li, G., & Cheng, Z. (2018). Using 250-M surface reflectance MODIS aqua/terra product to estimate turbidity in a macro-tidal harbour: Darwin harbour, Australia. *Remote Sensing, 10*(7), 997.

Yellen, B., Woodruff, J. D., Ralston, D. K., MacDonald, D. G., & Jones, D. S. (2017). Salt wedge dynamics lead to enhanced sediment trapping within side embayments in high-energy estuaries. *Journal of Geophysical Research: Oceans, 122*(3), 2226–2242. https://doi.org/10.1002/2016JC012595, 21699291.

Yu, Q., Wang, Y., Gao, J., Gao, S., & Flemming, B. (2014). Turbidity maximum formation in a well-mixed macrotidal estuary: The role of tidal pumping. *Journal of Geophysical Research: Oceans, 119*(11), 7705–7724. https://doi.org/10.1002/2014JC010228, 21699291.

Zhong, Y., Qiao, L., Song, D., Ding, Y., Xu, J., Xue, W., & Xue, C. (2020). Impact of cold water mass on suspended sediment transport in the South Yellow Sea. *Marine Geology, 428*. https://doi.org/10.1016/j.margeo.2020.106244, 00253227.

Further reading

Yang, G., Wang, X. H., Zhong, Y., & Oliver, T. S. (2022). Modelling study on the sediment dynamics and the formation of the flood-tide delta near Cullendulla Beach in the Batemans Bay, Australia. *Marine Geology*, Article 106910.

Response of tidal dynamics to the construction of Yangshan Deepwater Harbor, Shanghai, China

Wenyun Guo [1], Wei Mao [1], Jianzhong Ge [2], Xiao Hua Wang [3] and Pingxing Ding [2]
[1]College of Ocean Science and Engineering, Shanghai Maritime University, Shanghai, China; [2]State Key Laboratory of Estuarine and Coastal Research, East China Normal University, Shanghai, China; [3]The Sino-Australian Research Consortium for Coastal Management, School of Science, University of New South Wales, Canberra, ACT, Australia

1. Introduction

Coastal zones and estuaries are characterized by abundant natural resources. Thus, coastal constructions, such as reclamations and harbor constructions, are conducted to satisfy the rapid socio-economic development. The constructions can modify local hydro- and sediment dynamics. And it has been realized that the cumulative impacts of constructions may lead to irreversible influence on the natural environments and threaten the dynamical and ecological functions of original coastal systems (Winterwerp et al., 2013). For example, construction of a fill-type causeway in Florida reduced the tidal prism at inlets and ultimately resulted in closure of Blind Pass and Dunedin Pass (Davis & Barnard, 2003). Sequential reclamations in the Jiaozhou Bay had switched the bay from erosion to an accretion pattern (Gao et al., 2018). Thus, understanding response of hydro- and sediment dynamics to these constructions is of great scientific importance and application values.

The Caofeidian Sea, which located in the northwest of the Bohai Sea, China, is characterized by a combination of natural deep channels and vast tidal flats. This geographic setting makes it to be an ideal coast to construct a deep-water harbor. Thus, significant reclamations are carried out since 2004. Lu et al. (2009) predicted that the annual siltation in the harbor basin and channel is 0.35−1.31 m adopting a 2D numerical model. Kuang et al. (2012) indicated that the reclamations will lead to reduction in suspended sediment concentration and potentially increasing landslide risks due to steep slope and sedimentation in Laolonggou creek. These reclamations induced significant reduction of its sediment storage capability, resulting in continuous loss of sediment volume of tidal flat (Liang et al., 2018). The reclamations induce adjustments in hydro- and sediment dynamics through different mechanisms. In the channels of the western uninterrupted coast, reclamations induce morphological evolution mainly through damping the long-shore sediment transport. In the eastern inlet-

Current Trends in Estuarine and Coastal Dynamics. https://doi.org/10.1016/B978-0-443-21728-9.00010-7

interrupted coast, tidal asymmetry plays an important role in the response of topographies in the channels (Liang et al., 2018).

The Yangshan Deepwater Harbor (YDH) was constructed on the Qiqi Archipelago, which is located in the conjunction of the Changjiang Estuary and the Hangzhou Bay, China. Three important passages are closed by the construction of the YDH. After the closure of each passage, remarkable siltation zones were developed on both sides of the closing dam (Ying et al., 2012). However, the closure of the Jiangjunmao-Dazhitou passage developed extensive accretion in the central and southern of the harbor basin. The area reached 12.4 km^2 (Ying et al., 2012; Yu & Zhang, 2006). Dynamics responsible for this extensive siltation has been investigated by several researchers. Tianjin Academy of Water Transport Engineering of the Ministry of Transport (2006) regards that this siltation is mainly related to the perturbance from dredging of harbor basins on sediment transport and the scoring of passages in the southern-islands chain, and the siltation is unsustainable. Liu (2008) believed that the formation of this extensive siltation can mainly attribute to the less difference between flood and ebb velocities and will be on-going. Ding and Chen (2007) regarded that there are two factors that contribute to the extensive siltation: one is the leakage of filling earth for reclamation; the other is that this region is located in a transition zone where in the south net sediment transport is in ebb direction and in the north net sediment transport is in flood direction, and it favors sediment siltation. At present, there is no consensus on the dynamics responsible to this extensive siltation.

Though there are many studies on the responses of local hydro- and sediment dynamics to the construction of the YDH, little of them focus on the tidal dynamics. In fact, tidal dynamics is the most important physical setting in coastal waters, and it dominates the hydrodynamics in the YDH sea area. Thorough comprehension of tidal dynamics in the YDH and its response to the constructions may conduce to the understanding of sediment transport and topographic changes during constructions.

Tidal waves are strongly affected by bottom friction, local geometries, and river discharges when they travel into coastal zones. The inequality between the propagation speed of high tide and that of low tide in shallow waters will lead to transformation in tidal waves, developing tidal asymmetry. Tidal asymmetry is recognized as one of the most important processes responsible for residual sediment transport and associated large-scale morphological changes in coastal waters (de Swart & Zimmerman, 2009). There are three types of tidal asymmetry: (1) unequal rising and falling tidal durations of vertical tides (called tidal duration asymmetry); (2) uneven peak ebb and flood velocities (called peak current asymmetry); and (3) unequal high water and low water slack durations in tidal currents (called slack water asymmetry), all of which are commonly used in the analysis of sediment transport (Gong et al., 2016; Guo et al., 2019). A tide with shorter rising duration, stronger peak flood currents, or longer high water slack is regarded as a flood-dominated tide. The flood dominance is in the perspective of sediment transport that this type of tide always causes flood-directed residual sediment transport. Tidal asymmetry can be regarded as a result of interactions among tidal constituents with specific frequency relationship. In most semi-diurnal tidal regimes, tidal duration asymmetry mainly results from the nonlinear interaction between M_2 and its first overtide M_4 (van Maren & Winterwerp, 2013) and the interaction of M_2, S_2, and their compound tide MS_4 (Song et al., 2011; Speer et al., 1991). Tidal duration asymmetry is a common phenomenon in the oceans and seas. Even without overtides or compound tides, tidal duration asymmetry can be induced by the interactions among principle tides, such as the interaction of K_1, O_1, and M_2 (Hoitink et al., 2003; Song et al., 2011).

Without doubt, coastal constructions can impact tidal asymmetry. The research of (Bolle et al., 2010) on Western Scheldt showed that dredging in tidal channels lead to decrease of flood dominance and sedimentation in the tidal channels result in greater flood dominance. Similar conclusion is also found in the research of (Colby et al., 2010) on basis of observations at the River Murray Mouth. Li et al. (2012) suggested that tidal duration asymmetry would increase by 100% if the mangrove areas were removed in Darwin Harbor, Australia; if the tidal flats were also removed, the increase would be up to 120%. Along with the land reclamation in Jiaozhou Bay, China, tidal duration asymmetry induced by interaction between M_2 and M_4 has increased significantly (Gao et al., 2014). Coastal constructions can even lead to qualitative transition between flood dominance and ebb dominance. Such as the reclamations along the west coast of Korea, especially the construction of Siwha dike, they have changed the coast from ebb dominance to flood dominance (Suh et al., 2014).

The primary objective of this study is to investigate the responses of tidal constants and tidal asymmetry to the construction of the YDH and explore the dynamic mechanisms predominant the responses. Tidal elevations during the constructions, including a longtime tidal gauge station (Xiaoyangshan) and some temporary tidal gauges are gathered for our investigation. A high-resolution numerical model is established to explore the spatial difference in tidal responses and the dynamic mechanisms governing these changes in tides.

2. Yangshan Deepwater Harbor

2.1 Geographic setting

The YDH was constructed at the Qiqu Archipelago, which is about 32 km to the southeast of Shanghai (Fig. 10.1). A famous cross-sea bridge, named Donghai Bridge, connects the YDH to Shanghai. There were 69 islands with total area of 10.72 km^2 distributed in the Qiqu Archipelago. The islands and passages make the coastlines and bathymetries extremely complicated in this sea area, resulting in intricate hydro- and sediment dynamics.

The islands form two chains: the northwest-southeast trending northern-island chain, where Xiaoyangshan Island is located, and the west-east trending southern-island chain, where Dayangshan Island belongs. The angle between the two island chains was about 40 degrees. Two passages with completely different topographical characteristics were formed between the two island chains: the narrow (~ 1 km) but much deeper (maximum depth exceeding 85 m) East Entrance and the wide (~ 8 km) and shallow (averaged depth of ~ 10 m) West Entrance. The East Entrance, West Entrance, and the two island chains enclosed an expansive well-sheltered sea

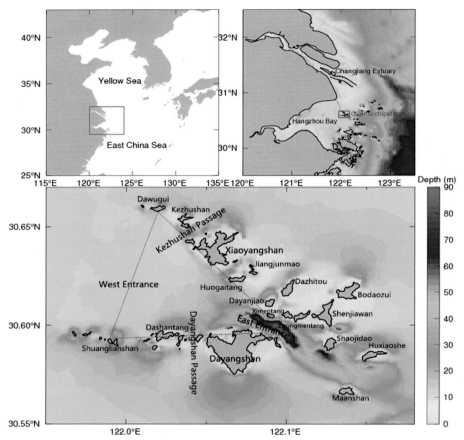

Figure 10.1 Map of the Qiqu Archipelago where the Yangshan Deepwater Harbor (YDH) constructed. The names of the islands and passages are denoted. Dot-dashed polygon indicates the Inner Harbor Area (IHA). The color indicates the depth.

area with acreage of 35.9 km^2. This sea area is referred as Inner Harbor Area (IHA). Due to the sheltering effect, the annual-averaged wave height in the IHA is only 0.47 m, forming an ideal sea area berthing ships.

The average tidal range is ~2.7 m in the YDH sea area, and the typical spring tidal range (recurring 10% of time) is 3.78 m. This sea area belongs to irregular semidiurnal tidal regime, with prominent semiannual tidal constituents M_2 and S_2, following by annual tidal constituents K_1 and O_1. Tidal current is strong in this sea area. The maximum current velocity can reach over 3 m/second, and the average current velocity is about 1.0 m/second. The main direction of flood tide is west-northwesterly and that of ebb tide is east-southeasterly. The waves in the IHA are weak forced by the sheltering effect. It is observed that averaged wave height is only about 0.47 m. Thus, tidal current dominates the hydrodynamics in the IHA.

2.2 Wind characteristics

The YDH sea area belongs to a humid subtropical monsoon climate. Northerly winds prevail in winter, while southeasterly winds reign in summer. Fig. 10.2 shows the wind rose at Xiaoyangshan Island for the years from 2002 to 2012. The data are retrieved from the 6-hourly ERA-5 reanalysis product of European Center for Medium-range Weather Forecasts (ECMWF). It is obvious that the westerly winds are of rare occurrence. There is no clear preferential wind direction for wind speeds <4 m/second, but winds with larger speeds (>6 m/second) are usually from North-northwest to North-east during winter and from South to Southeast during summer. Strong winds with speed >10 m/second always occurring in the northerly wind.

The annual averaged wind speed is greater than 5.7 m/second. The mean wind speed is very different in wind directions, the north-northwesterly wind have maximum mean wind speed of over 7 m/second, but the mean speed of west-southwesterly wind is smaller than 4.0 m/second. However, since the frequent typhoons during summer or autumn seasons, the yearly maximum wind speed always not occur in the winter season with strong northly winds.

2.3 Construction of the YDH

On the condition of fast development of international trades and rapid increase in cargo transportations, China decided to construct a new deepwater harbor for the Shanghai Port, thus the YDH construction project emerged. The constructions

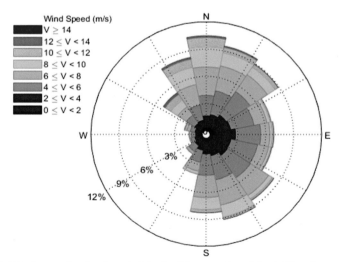

Figure 10.2 Wind rose plot for the wind in the Xiaoyangshan Island. The plot represents the probability of occurrence (in %) per wind direction and wind speed range. They are based on the reanalysis data from ERA-5 with a temporal resolution of 6 hours. Colors denote wind speed in m/second.

started in 2002 and finished its third phase in 2008, and the fourth phase was constructed during 2009–2018 (Fig. 10.3 and Table 10.1). In the former three phases, at least one channel was closed during each phase. Phase I closed the Xiaoyangshan-Huogaitang passage in 2003 and formed a quay with length of 1600 m, phase II closed the Dawugui-Kezhushan passage in 2004 but constructed a quay on the south of the Xiaoyangshan Island, which length is 1120 m, phase III closed the Jiangjunmao-Dazhitou passage in 2005 and formed a quay with length of 1860 m. The close of passages extremely altered the coastlines of the YDH sea area, which had induced dramatic adjustment in hydro- and sediment dynamics, and thus geomorphological changes (Ying et al., 2011). Our investigation mainly focuses on the influence of passage-closures on tidal dynamics; thus, the fourth phase is excluded in this study.

To satisfy the land demand of the YDH, extensive reclamation works were conducted in each phase. The reclamation in phase I had used 25.0 million m^3 earths and formed harbor land with area of 1.3 km^2. In phase II, the reclamation works used 4.0 million m^3 earths and constructed 0.89 km^2 harbor land area. Phase III reclamation produced 5.9 km^2 land area for the YDH. These reclamations offer much land for the operation of the YDH. On the north of the harbor, reclamations were continued for the landward sediment transport there.

The harbor came into service in October 2005 after the dredging of main navigation channel and the dredging of the phase I harbor basin. The phase II harbor basin was then dredged in the next year. At present, the YDH can keep in a good navigation condition with tidal current action and little regular dredging.

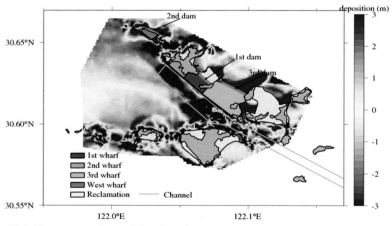

Figure 10.3 The constructions of the Yangshan Deepwater Harbor (YDH), including the passages-closure, land reclamation and dredging works. Colors indicate changes in topography from 2002 to 2008.

Table 10.1 Timeline of the Yangshan Deepwater Harbor construction Projects.

Time	Main constructions
2002.04−2004.06	Phase I: close Xiaoyangshan-Huogaitang passage and land reclamation
2003.03−2005.10	Phase II: close Dawugui-Kezhushan passage and land reclamation
2004.05−2008.01	Phase III: close Jiangjunmao-Dazhitou passage and land reclamation
2009.01−2018.12	phase IV: land reclamation and diversion dikes
2004.12−2005.11	Dredging (16.5 m) of main navigation channel
2005.05−2005.10	Dredging (16.0 m) of harbor basin of phase I
2006.05−2006.10	Dredging (16.0 m) of harbor basin of phase II

3. Observed tidal response to the constructions

3.1 Tidal elevation and tidal range

The tidal elevations at Xiaoyangshan station exhibit strong variabilities. As an example, the maximum tidal level is 6.38 m, but the minimum is only 2.11 m during August 2002 (Fig. 10.4a). The depth can be over 20.0 m during high tides, giving the YDH the ability to berth the globally largest container ships to date.

A prominent spring-neap cycle can be found in the tides of the YDH. During August 2002, the maximum tidal range is 3.76 m, and the minimum tidal range is 1.47 m (Fig. 10.4b). The large difference of tidal ranges in spring and neap tides implies that tidal dynamics and sediment transport may be quite different during spring and neap tides. As shown in Fig. 10.4a, the tides in the YDH are dominated by

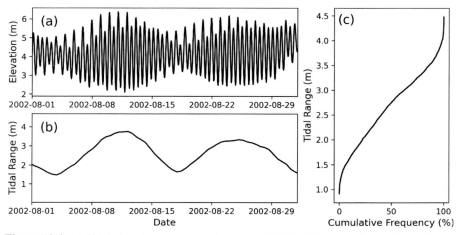

Figure 10.4 (a) Tidal elevation process during August 2002 at Xiaoyangshan station; (b) the daily tidal range time-series during August 2002; and (c) the cumulative frequency curve of tidal ranges at Xiaoyangshan station during 2001−05.

semidiurnal tides, and the adjacent two high tides are obviously unequal. Such diurnal inequality can mainly be attributed to the angle between equatorial plane and the Moon, and thus diurnal tidal constituents are generated.

Fig. 10.4c presents the cumulative frequency curve of tidal ranges basing on observed tidal levels during 2001–05. It is obvious that the minimum tidal range is less than 1.0 m, and the maximum tidal range can nearly reach 4.5 m, and the mean tidal range is 2.69 m. The statistics imply that the YDH sea area belongs to meso-tidal coasts.

Water levels in the YDH sea area are intensively impacted by the water discharge from the Changjiang river. As Fig. 10.5 shows, an obvious annual cycle can be found in the 30-days mean tidal elevations at Xiaoyangshan station, with a varying range of about 0.4 m and reach peak values around September. This annual cycle can mainly be attributed to the seasonality in the river discharge of Changjiang, since the discharge is much higher during June–September (Fig. 10.5).

Tidal range is an important parameter denoting the intensity of tidal movement. Our statistical analysis of the monthly tidal ranges at Xiaoyangshan station is drawn in Fig. 10.6. The results demonstrate that the tidal range in the YDH sea area owing an obvious semiannual cycle, general reaching its maximum in March and September, and its minimum in June and December. The varying range is over 0.1 m. Such semi-annual cycle can be explained by that only twice a year Sun, Earth, and Moon are in a line and align approximately with the Moon's perigee.

A slightly increasing tendency is presented by Fig. 10.6. The yearly averaged tidal range was around 2.69 m before 2003 and abruptly increased to 2.74 m in 2004; then

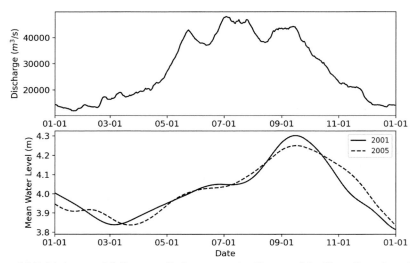

Figure 10.5 (a) Averaged daily water discharge at station Datong of the Changjiang river during 2001–05; (b) 30-days moving smoothed water levels at Xiaoyangshan station during 2001–05.

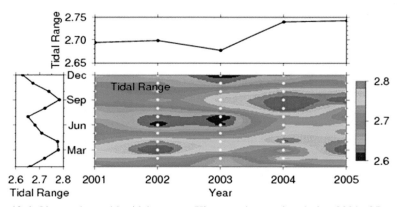

Figure 10.6 Observed monthly tidal range at Xiaoyangshan station during 2001−05.

the tidal range kept 2.74 m in 2005. This increase in tidal range may highly relate to the constructions of the YDH.

3.2 Harmonic analysis

The construction of the YDH had strongly altered the coastlines and topographies in the sea area. Adjustment of tides is also expected since the importance of topographies in tidal dynamics. To investigate the adjustment of tides during the constructions, we had gathered observed tide elevation data from 2001 to 2007 at Xiaoyangshan station. There were at least successive 1-month tidal elevations in each year. A widely used T_TIDE package (Pawlowicz et al., 2002) was adopted to retrieve the tidal constituents. Results are displayed in Table 10.2.

The amplitudes of the semidiurnal tidal constituents M_2 and S_2 had increased after the constructions of phases I−III, with an increment of 6 cm in M_2 and 2 cm in S_2. The increase mainly occurred from 2003 to 2004, just after the close of Xiaoyangshan-Huogaitang passage during phase I. The diurnal tide K_1 and O_1 also increased slightly

Table 10.2 The observed tidal amplitude (m) at Xiaoyangshan station. $F = (a_{O1} + a_{K1})/(a_{M2} + a_{S2})$ is the form number, and $G = (a_{M4} + a_{MS4})/(a_{M2} + a_{S2})$.

Year	K_1	O_1	M_2	S_2	M_4	MS_4	F	G
2001	0.29	0.18	1.25	0.50	0.05	0.04	0.27	0.05
2002	0.30	0.19	1.25	0.50	0.06	0.05	0.28	0.06
2003	0.30	0.19	1.25	0.50	0.07	0.05	0.28	0.07
2004	0.31	0.19	1.29	0.51	0.08	0.06	0.28	0.08
2005	0.31	0.20	1.30	0.52	0.08	0.07	0.28	0.08
2006	0.31	0.20	1.31	0.52	0.08	0.07	0.28	0.08
2007	0.31	0.20	1.31	0.52	0.08	0.07	0.28	0.08

(2 cm) from 2001 to 2007. To the shallow-water tides M_4 and MS_4, their amplitudes were increased slightly by 0.03 m. Though the increasing magnitudes are small, comparing to their small values (0.05 and 0.04 m) before project, their amplitudes were increased by over 50% from 2001 to 2007, showing significant changes in shallow-water tides. Since the important roles of them in tidal duration asymmetries as most semi-diurnal tidal regimes show, tidal asymmetries are expected to be adjusted significantly.

Adopting the tidal constants, the tidal form number can be calculated as the ratio of semidiurnal tidal amplitudes to diurnal tidal amplitudes:

$$F = (a_{O1} + a_{K1}) / (a_{M2} + a_{S2}) \qquad (10.1)$$

where F is the tidal form number, and a_{O1}, a_{K1}, a_{M2}, and a_{S2} are the amplitudes of the tidal constituents of O_1, K_1, M_2, and S_2, respectively.

Tidal form is classified as regular semidiurnal for $0.0 < F \leq 0.25$; irregular semi-diurnal when $0.25 < F \leq 1.5$; irregular diurnal when $1.5 < F \leq 3.0$; and regular diurnal when $F > 3.0$. As Table 10.1 shows, the tidal form number at YDH sea area is about 0.28, indicating that the tidal form is irregular semidiurnal and very close to regular semidiurnal tidal form. The constructions induced synchronous reinforcement in semidiurnal and diurnal tidal constituents. This synchronous reinforcement resulted in little variations in the tidal form number, only changed F from 0.27 to 0.28.

The topography could response to the constructions very rapidly. After the closure of each passages, two distinct siltation zones were formatted just aside the passage-closing dam (Ying et al., 2012). The constructions and siltation had resulted in significant amplification in shallow-water tides M_4 and MS_4 from 2001 to 2007. The M_4 amplitude had increased from 5 to 8 cm, and MS_4 amplitude had increased from 4 to 7 cm. Though the absolute increment of shallow-water tide amplitudes were smaller than that of M_2 amplitudes, the significant increase in the relative values made the tidal number G (defined to $(a_{M4}+a_{MS4})/(a_{M2}+a_{S2})$ in this study) increase from 0.05 to 0.08, which implying prominent changes in tidal asymmetry—a crucial dynamic mechanism affecting the net transport of sediment. However, this significant augment in G may be site-specific, since the extreme complexity in topographic responses to constructions of the YDH (Ying et al., 2012).

No evident changes in tidal phases were shown by the observed tidal elevations in the Xiaoyangshan station, so the tidal phases are not listed in Table 10.2.

3.3 Tidal asymmetry

To further investigate the response of tidal asymmetry to the constructions, we had gathered hourly tidal elevations during consecutive 5 years from 2001 to 2005 at the Xiaoyangshan station.

There are various methods that can be employed to quantify tidal asymmetry, including harmonic method and statistical method. The harmonic method has been widely used for a long time, which assessing tidal asymmetry based on the phase differences of the tidal constituents that interact and create tidal wave deformation. The statistical methods evaluate tidal asymmetry by calculations of probability distribution function of tidal heights and tidal durations and the skewness of the time derivative of tidal water levels or the transformed skewness of tidal water levels. A review of these methods can be found in (Guo et al., 2019). In this study, we adopted a skewness-based method proposed by (Song et al., 2011) to quantify the response of tidal asymmetry to the constructions of the YDH. Two duration skewness, that is, the skewness on basis of hourly water levels and the skewness on basis of tidal constants are evaluated. Their calculating formulas are

$$\gamma_1(t) = \frac{\mu_3(t)}{\mu_2^{3/2}(t)} = \frac{\frac{1}{T}\int_{t-T/2}^{t+T/2} \zeta^3(\tau)d\tau}{\left[\frac{1}{T}\int_{t-T/2}^{t+T/2} \zeta^2(\tau)d\tau\right]^{3/2}} \tag{10.2}$$

$$\gamma_N \approx$$

$$\frac{\sum_{2\omega_i=\omega_j}\frac{3}{4}a_i^2\omega_i^2 a_j\omega_j\ sin\ (2\varphi_i - \varphi_j) + \sum_{\omega_i+\omega_j=\omega_k}\frac{3}{2}a_i\omega_i a_j\omega_j a_k\omega_k\ sin\ (\varphi_i + \varphi_j - \varphi_k)}{\left(\frac{1}{2}\sum_{i=1}^{N} a_i^2\omega_i^2\right)^{3/2}}$$

$$\tag{10.3}$$

respectively, where μ_2 is the second moment about zero, μ_3 is the third moment about zero, $\zeta(\tau)$ is the tidal elevation time derivative at time τ, and T is the tidal cycle; a_n, ω_n, and φ_n are the amplitude, frequency, and phase of tidal constituent n, respectively. Results are shown in Fig. 10.7.

The blue line γ in Fig. 10.7 represents the calculated duration skewness adopting Eq. (10.2) with integral time period $T = 1$ year. The red line γ_N is the duration skewness calculated by Eq. (10.3). The small differences and consistent trends

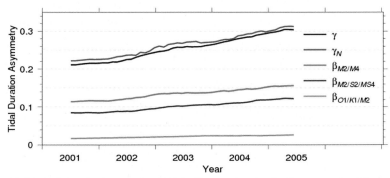

Figure 10.7 Tidal duration asymmetry and the three main contributions at station Xiaoyangshan based on the observed sea-level data from 2001 to 2005.

between them suggest that the method developed by (Song et al., 2011) is highly reliable. The positive values of γ and γ_N imply that tides at Xiaoyangshan station are flood-dominant. From 2001 to 2005, the value of γ increased continuously, indicating increasing ebb tide durations and enhancing flood-dominant tidal asymmetry at Xiaoyangshan station. This result is consistent with the analysis of the changes in flood-ebb durations on basis of observations from (Yu et al., 2013, p. 163).

From 2001 to 2002, γ changed little before the constructions. However, from 2002 to 2003, the value of γ suddenly increased by 0.03, implying significant impact from the constructions of the YDH. The following constructions could continuously increase the tidal duration asymmetry at Xiaoyangshan station, but the increase rate was much slower than that of the phase I project. This might attribute to the farther distance of phase II and phase III projects from the Xiaoyangshan station.

According to (Song et al., 2011), the couples or triplets of interacting tidal constituents satisfying the frequency relationship of $2\omega_1 = \omega_2$ or $\omega_1 + \omega_2 = \omega_3$, respectively, can generate persistent tidal asymmetry. The contribution of each tidal interactions can be separated from Eq. (10.3), and reads

$$\beta_2 = \frac{\frac{3}{4}a_1^2\omega_1^2 a_2\omega_2 \, sin\,(2\varphi_1 - \varphi_2)}{\left(\frac{1}{2}\sum_{i=1}^{N}a_i^2\omega_i^2\right)^{3/2}}, 2\omega_1 = \omega_2 \tag{10.4}$$

$$\beta_3 = \frac{\frac{3}{2}a_1\omega_1 a_2\omega_2 a_3\omega_3 \, sin\,(\varphi_1 + \varphi_2 - \varphi_3)}{\left(\frac{1}{2}\sum_{i=1}^{N}a_i^2\omega_i^2\right)^{3/2}}, \omega_1 + \omega_2 = \omega_3 \tag{10.5}$$

Fig. 10.7 also show the main three tidal interactions that contribute to tidal asymmetry at Xiaoyangshan station. It is obvious that the M_2/M_4 and $M_2/S_2/MS_4$ interactions dominate the tidal duration asymmetry at this station. Their contribution to the total skewness is nearly 90%. From 2001 to 2005, the increase in tidal skewness could mainly be attributed to the enhanced M_2/M_4 and $M_2/S_2/MS_4$ interactions. Combining with the significant amplification of shallow-water tidal amplitudes, we can conclude that the significant enhancement in shallow-water tides is the main cause for the increase of tidal duration asymmetry at the Xiaoyangshan station.

The tides are chaotic in the YDH sea area relating to the complex topography. We gathered successive 15-days tidal elevations during May 2008 at nine stations. Their tidal skewness are calculated, and the results are displayed in Fig. 10.8. It is obvious that the tidal skewness is in a highly spatial variation. Skewness is high in the west and relatively small in the east. The maximum skewness in 2008 is 0.27, occurred at

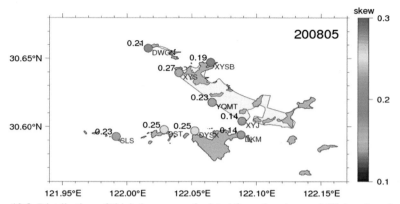

Figure 10.8 Distribution of tidal skewness calculated from the observed 15-days hourly sealevels from during May 2008.

Xiaoyangshan station, while the minimum skewness is only 0.14, occurred in the East Entrance.

4. FVCOM model study

4.1 Numerical model

Finite-Volume Coastal Ocean model (FVCOM) is a comprehensive ocean model integrating hydrodynamics, sediment transport and geomorphic evolution, water quality, ecology, and sea ice (Chen et al., 2003, 2013). The model solves the three-dimensional primitive equations adopting finite-volume approach, which combines the geometric flexibility of finite-element methods and the computational efficiency of finite-difference methods, can provide a much better representation of mass, momentum, salt, and heat conservation in coastal and estuarine regions with complex geometry. The equations are closed by Mellor-Yamada 2.5 order vertical turbulence closure model and Smagorinsky horizontal turbulence closure model. Multiple vertical coordinate systems, such as z-coordinate, σ-coordinate, and s-coordinate, are offered to fit the irregular bottom terrain, and nonoverlapping unstructured triangular mesh is employed to discretize the horizontal computational domain. The unstructured grid can fit the shoreline well and is convenient for local mesh refinement. The model uses the dry-wet boundary technique to deal with the dynamic inundation process of tidal flats. For the computing efficiency, a mode splitting method, which is successfully used in POM and ROMS, is adopted to divide the currents into external and internal modes that can be computed using two distinct time steps.

FVCOM calculates the bottom friction using the quadratic drag law:

$$\left(\tau_x, \tau_y\right) = C_d \sqrt{u^2 + v^2}\,(u, v) \tag{10.6}$$

The drag coefficient for bottom friction is determined by matching a logarithmic bottom layer to the model at a height z_{ab} above the bottom, namely:

$$C_d = max\left(\frac{\kappa^2}{ln(z/z_{zb})^2}, 0.0025\right) \tag{10.7}$$

where $\kappa = 0.4$ is the von Karman constant, z is the distance from the seabed to the position of u and v, and z_0 is the bottom roughness parameter.

4.2   Model configuration

The YDH sea area is located at the water and sediment exchange area between the Changjiang Estuary and the Hangzhou Bay (Fig. 10.1). Model study of water, sediment, and even ecology in the YDH sea area needs to consider the combined effect of the Changjiang Estuary and the Hangzhou Bay. The irregular coastlines and complex topographies in the YDH sea area prompt the requirement of very fine spatial resolution to express the coastlines and topographies in a high accuracy in numerical models. And the time step should be short enough, limited by the CFL condition. Considering the computation efficiency of our numerical model, a model nesting technique is used in our model. As shown in Fig. 10.9, the YDH model domain with a high spatial resolution of 50 m is nested to the well-validated Changjiang Estuary FVCOM (CJE-FVCOM) model developed by Ding Pingxing's group at the East Normal University of China, which has spatial resolution of 200 m (Ge et al., 2013, 2022; Zhou et al., 2019). There are 10 vertical sigma layers specified in the CJE-YDH nesting FVCOM model system.

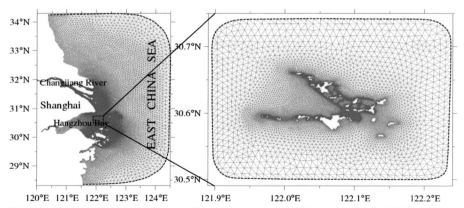

Figure 10.9 The Changjiang Estuary-Yangshan deepwater harbor nesting FVCOM model mesh used in this study. The mesh is refined on the boundaries of constructions.

In the CJE-FVCOM model, the upstream boundary of the Changjiang Estuary reaches Datong, which is about 630 km upstream from the river mouth. There locates a representative hydrologic station which measures water and sediment flux is regarded as the Changjiang river into the sea. The upper boundary of the Qiantang River reaches Laoyancang. The model domain covers an extensive sea area stretching to 124.5 degrees E to the east, 34.5 degrees N to the north and 28.5 degrees N to the south, embracing the Changjiang Estuary, the Hangzhou Bay and the Zhoushan Archipelago. More details about this model can refer to (Ge et al., 2013).

The domain of YDH-FVCOM model covers all islands in the Qiqu Archipelago. The maximum resolution of the model mesh at the open boundary is 1000 m. The mesh resolution is refined to 50 m on the coastlines and in the passages, providing accurate geometric fitting for the irregular coastlines and steep topographies. Moreover, after each phase of the project, new boundaries are formed. The original mesh considers these boundaries and set a high resolution (50 m) on these boundaries. Therefore, a mesh after any project phases can be obtained through removing the mesh cells where the sea had become land after this phase without changing other mesh cells. This scheme not only provides great facilitates for this study but also eliminates imparity of simulated physical processes caused by mesh discrepancy.

Before the constructions, the model mesh has 33,781 triangles and 17,647 nodes. After the constructions, the triangle number decreased to 27,147, and the node number decreased to 14,352. The current velocities in the passages are extremely strong, thus the time step of the YDH-FVCOM model is set to a small value as 0.2 seconds. In addition, considering the significant influence of bottom friction on tidal waves, the model adopts a spatially varying bottom roughness parameter on basis of measured seabed sediment particle size and other data. The bottom roughness parameter is overall large in the passages and relatively small in the open seas. More details about the YDH-FVCOM model can be referred to (Guo, 2017; Guo et al., 2018).

The open sea boundary is driven by 13 tidal constituents: M_2, S_2, K_1, O_1, N_2, K_2, P_1, Q_1, M_f, M_m, M_4, MS_4, and MN_4. They are from the latest OSU TOPEX/POSEIDON (TPXO9, https://www.tpxo.net/global/tpxo9-atlas) global ocean tide model after slight adjustment. The model is cold-started, that is, the initial tidal elevations and currents are set to 0. Daily water discharge at Datong and monthly water discharge of the Qiantang River are specified on the river boundary. The model results are much reliable after 10-days spin-up.

The CJE-YDH numerical model has been well validated in (Guo, 2017; Guo et al., 2018). Readers can refer to these works for more details.

4.3 Numerical experiments

Due to the continuity of the YDH constructions during 2002—08, it is hard to find a time point to distinguish two different engineering stages. In view of the crucial impact of the passage-closing works on the local hydrodynamics, the numerical experiments are set mainly in according with the time of that passages closed. There are no passages closed in phase IV; thus, phase IV is not considered in this numerical study.

Table 10.3 Description of the nine numerical experiments in this study.

Exps	Depth	Project setting
CR	November 1998	—
A3	September 2003	Phase I passage-close and wharf
A4	May 2004	Add phase II passage-close on A3
A5	April 2005	Add phase III passage-close and phase II wharf on A4
A8	April 2008	All projects of phases I ~ III
B3	November 1998	Same with A3
B4	November 1998	Same with A4
B5	November 1998	Same with A5
B8	November 1998	Same with A8

On the other hand, since the amplitudes of the shallow water tides are inversely proportional to the water depth in theory (Pugh & Woodworth, 2014), the topographic changes caused by the constructions are also an important factor affecting the tidal dynamics in the sea area. For this purpose, nine numerical experiments were designed, as shown in Table 10.3. The numerical experiments CR, A3, A4, A5, and A8 represent the scenarios before the constructions, after the phase I passage-closing works, after the phase II passage-closing works, phase III passage-closing works, and after all the constructions of phases I ~ III. B3 to B8 then set the topography the same as the CR to discard the influence of topographic adjustment on tides, distinguishing the influence of engineering constructions itself on tides. The topographic data used in this study is from the Third Institute of Waterway Design, Ministry of Communications, China. As only one extensive topographic survey was carried out before the project in November 1998, we can only use these data to assess the preproject topography. This approximation should be reasonable on the condition that this sea area is on a slight deposition state naturally, with a siltation rate of about 2.3 cm per year (Chen, 2000). All the numerical experiments in Table 10.3 were simulated for 40 days using the same driving forces with 10-days spin-up.

4.4 Tidal constants

According to the analysis of Xiaoyangshan tidal elevations, the amplitudes of M_2 and M_4 tidal constituents increase greatly along with the constructions of YDH. In addition, their interaction dominates the tidal duration asymmetry in the YDH sea area. Therefore, we mainly focus on the harmonic constants of the two tidal constituents. As shown in Fig. 10.10, M_2 tidal wave propagates westward in the YDH sea area. Its amplitude magnifies gradually during its propagation. Mainly due to the shallow-water effect, the M_2 amplitudes is relatively large in the shallow north. Correspondingly, the M_4 amplitude is also relatively large in the north with strong shallow-water effect. The irregular coastlines and complex topographies around the Qiqu Archipelago introduce chaos into the spatial distribution of M_2 and M_4 amplitudes.

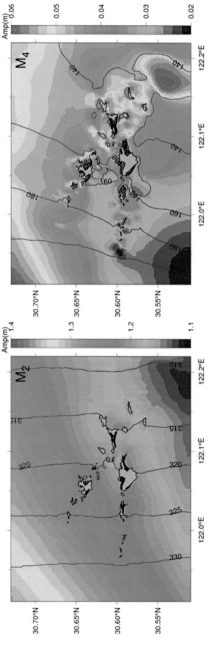

Figure 10.10 Distribution of (*left*) M2 amplitude, and (*right*) M4 amplitude of Exp CR. *Black* lines display the corresponding co-phase lines.

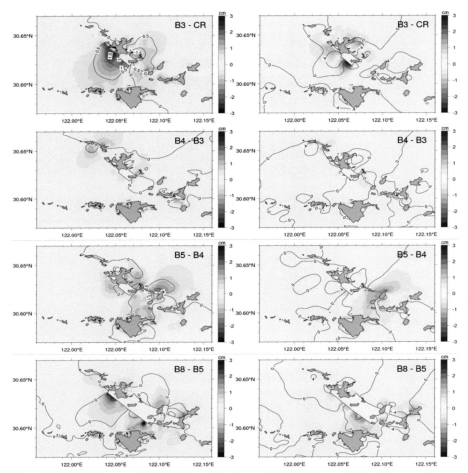

Figure 10.11 Differences of (*left*) M_2 amplitudes and (*right*) M_4 amplitudes between the adjacent experiments. Lines indicate the corresponding phase differences.

The influence of the constructions of the YDH on tidal dynamics is quite different in space (Fig. 10.11). Phase I of the YDH construction project had closed the Xiaoyangshan-Huogaitang passage, blocking tidal propagation through this passage. This results in significant reduction of M_2 amplitudes in the lee, accompanying by delayed M_2 phases. Meanwhile, in the stoss side, M_2 amplitudes are amplified accompanying by advanced phases. Similar result is found in the M_4 tide. This phenomenon (reduction in the lee and amplification in the stoss) is also found after the close of Dawugui-Kezhushan passage in phase II and the close of Jiangjunmao-Dazhitou passage in phase III. While, mainly due to the nearly parallel relationship between the direction of the dam and the tidal propagating direction, the influence of passage-close in phase II is much smaller than that caused by that in phase I and phase III, no matter in amplitude-reducing areas or in amplitude-reducing values. The bathymetrical changes superimpose some disturbances on this change pattern but not shift it (not shown). Similar phenomenon of "amplification in the stoss" is also found in the west coasts of Korea. Major tidal constituents in the seaside are amplified and advanced after construction of a dyke and two seawalls in the Mokpo Coastal Zone (Byun et al., 2004; Suh et al., 2014).

Fig. 10.11 also show that the close of Jiangjunmao-Dazhitou passage in phase III lead to significant reduction in M_2 and M_4 amplitudes around the East Entrance. The reducing magnitudes are 0.9 and 0.6 cm, respectively. The reduction in M_2 amplitudes is even stronger than that in the lee of the dam, where the reduction is only by 0.7 cm. The reclamations in phase III further dramatically reduce the M_2 amplitudes in the East Entrance by about 2 cm but notably amplifies the M_2 amplitudes in the harbor basin of phase II. Similar phenomenon is not found in M_4 amplitudes. Observations of currents (figure not shown) show that the flood direction has been transformed to be more northly to be parallel to the new coastline. We believe that this is a plausible interpretation for the M_2 amplification in the harbor basin of phase II.

The constructions of phases I $\sim$ III lead to extensive reduction of M_2 amplitudes in the IHA, especially that around the East Entrance (Fig. 10.12). However, for the M_4 tide, its amplitudes are not reduced in the whole IHA, only in the East Entrance and the harbor basin of phase III. As compared in Fig. 10.11, bathymetrical changes play a role in compensation to the reduction of tidal amplitudes and result in that M_2 and M_4 amplitudes are reduced only around the East Entrance. The topographic adjustment transforms the response of M_2 tidal constituent around the Kezhushan passage to remarkable increase in amplitudes. The amplification of M_2 and M_4 tide around Xiaoyangshan station is mainly due to the changes in bathymetries induced by the constructions, further confirming our deduction in above that the results of Xiaoyangshan station is site-specific.

The constructions also lead to changes in tidal phases. The M_4 phases advance about 4 degrees in the East Entrance and lag 4 degrees in the harbor basin of phase II. In overall, the influence of constructions on tides can be restricted in a scope of 4 km around the specific construction. But the induced bathymetric adjustment can notably expanse the influence scope and complicate the tidal responses in space.

4.5 Tidal choking in the East Entrance

Tidal choking refers a geometric feature in a narrowed channel. Previous studies indicated that when tides travel into a narrow, frictional channel, their amplitudes are damped and phases are delayed (Keulegan, 1967; Kjerfve & Knoppers, 1991). This tidal choking phenomenon is mainly found in semi-enclosed seas such as lagoons and bays that are connected to the open ocean by narrow and shallow channels (Byun et al., 2004). In a choked system, the narrow channel acts as a low-pass filter, damps more in high-frequency tides but less in low-frequency (Keulegan, 1967; Mac-Mahan et al., 2014; Moody, 1988, pp. 137−156). There are many simplified theoretical models that describe the dependence of tidal choking effect on the geometries and physics of the narrow channel and tidal frequencies (Hill, 1994; Keulegan, 1967;

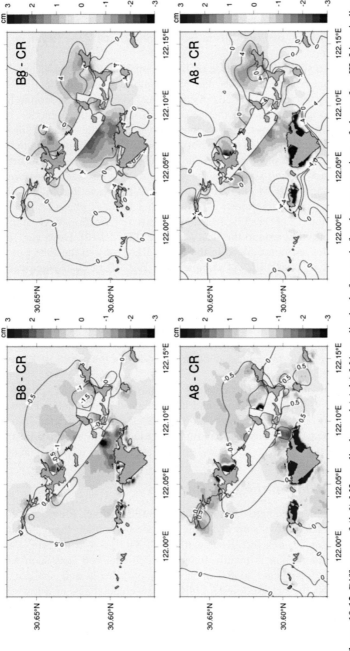

Figure 10.12 Difference of (*left*) M2 amplitudes and (*right*) M4 amplitudes before and after the constructions of phases I ~ III. Lines indicate the corresponding phase difference.

Stigebrandt, 1980). Substantially, there are two main mechanisms responsible for the amplitude reduction and phase lag caused by the tidal choking effect: one is the stronger frictions relating to the rapid current, which reduces the total tidal energy, and the other is the conversion between tidal potential energy and tidal kinetic energy, as defined by Bernoulli's equation.

Many coastal constructions, particularly the passage closure works or dredging works, can reduce or enhance the tidal choking effect. For example, the construction of a dyke and two seawalls on the coast of Mokpo, South Korea, changed this region from a tidal-choked system to a nonchoked one (Byun et al., 2004). The numerical investigation of Li et al. (2011) shows that the removal of the arms in Darwin Harbor, Australia, will lead to disappearance of the tidal choking effect.

The significant reduction in the amplitudes of M_2 and M_4 near the East Entrance of YDH is most likely related to the enhanced tidal choking. For confirm this hypothesis, a longitudinal section through the East Entrance was selected for analysis (Fig. 10.13). Bathymetric data on this longitudinal section show that the depth in the East Entrance can nearly reach 90 m before the constructions. There was a small hill on the east of the

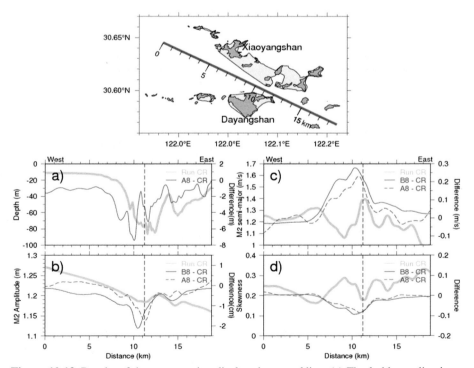

Figure 10.13 Results of the cross-section displayed as a *red* line. (a) The *bold gray* line indicates the depth used in Exp CR, the red line denotes the difference between the depth used in Exp A8 and CR. (b), (c), and (d) are similar with (a), but for M_2 amplitude, the semi-major axis of M_2 current, and the duration asymmetry respectively.

East Entrance at 14 km, about 20 m higher than the adjacent seabed. From 2001 to 2003, the topography changes little. After 2004, the impact of the constructions on the topography has gradually emerged. In general, the East Entrance is in a trend of deepening. The change pattern presents a double-valleys structure, that is, the water depth in the middle of the East Entrance deepens slowly, while the surrounding area deepens faster. From 2001 to 2008, the constructions of phases I ~ III have led to deepening of over 3.0 m in the East Entrance, and the East Entrance is still in a state of scouring.

Before the constructions, there is a slight decrease in M_2 amplitude as the tide travels from east to west through the East Entrance (the thick gray line in Fig. 10.13b), and then the amplitude increases continuously due to the shallow-water effect. The M_2 phase also lags on the position of M_2 amplitude decrease. These features imply that the East Entrance is slightly choked before the constructions.

The choking intensity can be estimated by the ratio of the tidal range inside and outside the passage (Glenne & Simensen, 1963; Hill, 1994; Rydberg & Wickbom, 1996). Because the tidal range ratio of the East Entrance does not change significantly, we use the amplitude difference of M_2 tidal constituent to quantify the choking intensity. The M_2 amplitudes at 10 and 13 km along the selected section are employed to mark the M_2 amplitudes inside and outside the East Entrance. The two positions have the same water depth (approx. 60 m) before the constructions. Results are presented in Table 10.4. Greater reduction in M_2 amplitude accompanying with more M_2 phase lag implies stronger choking effect.

Comparing the results of numerical experiments B3 and CR, the passage-closure work in phase I works reduced the M_2 amplitude drop but have little impact on the M_2 phase lag. The results for B3 and B4 are nearly equal, suggesting that the passage-closure work in phase II can throw little impact on the tidal choking at the East Entrance. The results of experiment B5 show that a significant enlargement occurs in the M_2 amplitude drop compared to B4, accompanied by a larger phase lag, implying remarkable enhancement of tidal choking at the East Entrance by the phase passage-closure work. According to the results of experiment B8, the land reclamation in phase III also thrown significant impacts on the tidal choking at the East Entrance.

The bathymetries at the East Entrance of experiments group A are deeper than that of experiments group B. Table 10.4 shows that the deeper topography of the choking channel will attenuate the choking effect (compare A4 and B4, A8, and B8). However, experiment A5 exhibits stronger choking effect than experiment B5. This is mainly related to the narrowing of the East Entrance induced by the land reclamation works in the southern-islands chain during 2004−05, particularly that at the Dayangshan Island. Observed data show that the cross-sectional area of the East Entrance decreases by 7% during this period, though it is deeper.

Stronger tidal choking means that there will be a greater difference in water levels between the inside and outside of the passage. Limited by momentum conversation, water current in the choked channel will be stronger accordingly (Stigebrandt, 1999). The stronger current is also able to be noticed at the East Entrance (Fig. 10.13c). Before the constructions, the main semi-axis of M_2 tidal current at the

Table 10.4 The difference of M_2 amplitude and phase, and the difference of M_4 amplitude and phase between the position of 10 km and of 13 km at the cross-section displayed as in Fig. 10.13.

	Difference	CR	A3	A4	A5	A8	B3	B4	B5	B8
M_2	Amplitude (cm)	−0.8	−0.3	−0.1	−2.2	−1.4	−0.3	−0.3	−0.9	−2.0
	Phase (°)	3.1	3.0	2.8	3.8	3.3	3.1	3.0	3.4	3.6
M_4	Amplitude (cm)	−0.4	−0.4	−0.8	−1.0	−1.3	−0.4	−0.5	−0.8	−1.5
	Phase (°)	4.9	5.4	2.8	0.7	−8.0	4.8	5.3	−0.3	−14.1

East Entrance was 0.3 m/second larger than that at 13 km along the cross section, where outside the East Entrance. This value was remained after the passage closure works in phase I and phase II but significantly increased by about 0.1 m/second after the passage closure work in phase III, and further enlarged by the land reclamation in phase III. A comparison of the dashed and solid lines in Fig. 10.13c indicates that the reduction in the cross-sectional area of the East Entrance contributes to the strength of water current. The results shown in this study that deeper topography in the East Entrance tend to reduce the choking effect, which is consistent with the results of multiple theorical models of tidal choking (Hill, 1994; Keulegan, 1967; Stigebrandt, 1980). In overall, the construction of the YDH enhances the choking effect of the East Entrance and enhances the water currents in the East Entrance, but the deeper bathymetries can partly attenuate the stronger choking effect.

Tidal choking cause greater amplitude reduction for high-frequency tides (MacMahan et al., 2014; Moody, 1988, pp. 137−156). In the East Entrance, the decrease of M_2 amplitude is very small, so can we use M_4 tide instead of M_2 to quantify the choking intensity? Table 10.4 also presents the amplitude reduction and phase lags of M_4 inside and outside the East Entrance. The results show that for short tidal passages such as the East Entrance, the amplitude differences of M_4 can also be used to quantify the choking intensity without topographic changes. And its amplitude difference is essentially the same as that of M_2, but the relative changes is much more significant considering that the M_4 amplitude is an order of magnitude smaller than that of M_2. However, when the water depth changes, due to the sensitivity of shallow-water tide to water depth, M_4 tide is harder to be applied in quantifying the choking intensity. Moreover, there is no objective rule in the responses of M_4 phase difference to bathymetric changes. Sometimes, the M_4 phase in the inlet can be smaller than that outside the inlet.

It should be noticed that the variations of tidal constants at the East Entrance is quite different from that observed at the Xiaoyangshan station: the construction of the YDH enhanced the M_2 amplitude at the Xiaoyangshan station but reduced the amplitude at the East Entrance. Different dynamic mechanisms dominate the response of tides to the constructions in different regions. According to the results in Fig. 10.10, the increase in amplitude at Xiaoyangshan station is mainly due to topographic adjustment, while at the East Entrance, the decrease in amplitude is mainly related to stronger tidal choking.

4.6 Tidal asymmetry

Fig. 10.14 exhibits the distribution of skewness of the time derivative of water level in the YDH sea area before the constructions. The skewness is greater than zero in most sea areas, indicating that the ebb time in this sea area is longer. In the IHA, the skewness is nearly uniform in space, except for that in the center is slightly smaller. Due to the shallow-water effect, tidal skewness on the northwest of the YDH is extremely large. It is believed that shallow waters are favored to flood-dominated tidal asymmetry by many studies (Dronkers, 1986; Friedrichs & Madsen, 1992; Speer et al., 1991). Our numerical results for the YDH sea area are consistent with their conclusion. However, tidal asymmetry is also extremely strong in a deep trough on the southeast of the sea area, which can mainly attribute to the development of shallow-water tides under the circumstances of strongly changing topography of this local sea area (Fig. 10.10).

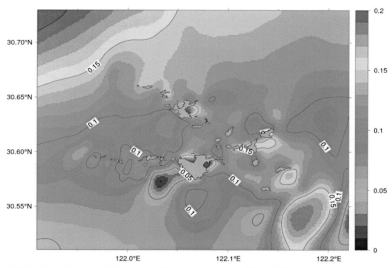

Figure 10.14 Distribution of tidal duration asymmetry of Exp CR. Large values mean longer falling duration.

On the south of Dashantang of the southern-islands chain, there is an area of very small tidal skewness.

The response of tidal asymmetry shows significant spatial differences along with the advance of constructions (Fig. 10.15). After the passage-close work in Phase I, tidal asymmetry significantly reduced tidal asymmetry in the lee of the constructed dam, mainly relating to the reduction in the amplitudes of M_2 and M_4 tides. On the contrary, in the stoss of the constructed dam, tidal asymmetry increases significantly forced by increased amplitudes of M_2 and M_4 tides. Fig. 10.13 also indicate that the close of Xiaoyangshan-Huogaitang passage has increased the tidal skewness around the Xiaoyangshan Station, which is consistent with the observed results.

The Dawugui-Kezhushan passage has been closed in Phase II. No remarkable changes of tidal constants are retrieved from our numerical results, also no notable changes in tidal asymmetry are observed. During the period of 2001−04, though the topographic adjustments can impact tidal asymmetry, their influence is mainly restricted in the south of the southern-islands chain, changes of tidal asymmetries are primarily induced by the direct influence of constructions.

The close of Jiangjunmao-Dazhitou passage in phase III not only gives significant impacts on the tidal asymmetry on both sides of the closing dam but also reduces the tidal asymmetries around the East Entrance. The comparison reveals that from 2004 to 2005, changes in tidal asymmetries in the central and western parts of the IHA is primarily generated by the adjustment of topography. Whereas, around the East Entrance, the direct influence of constructions dominates the changes in tidal asymmetries. During this period, changes in tidal asymmetry around Xiaoyangshan station are also dominated by topographic adjustment.

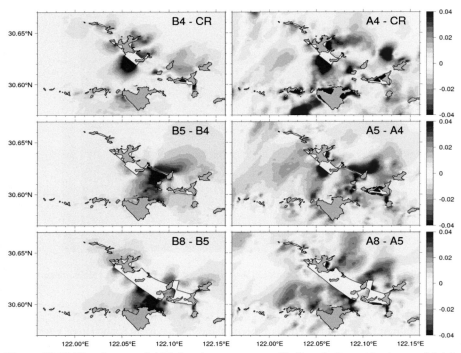

Figure 10.15 The changes of tidal duration asymmetry. (*Left*) no bathymetric change, (*right*) considering bathymetric change.

The land reclamation in Phase III extended the length of the East Entrance by 2 km, which significantly enhanced the tidal choking effect in this passage, and decreased the amplitudes of M_2 and M_4 tide, reducing the tidal asymmetries. From 2005 to 2008, the direct impact of the constructions on tidal asymmetries in the IHA mainly occurs in the East Entrance, with a slight increase in the Kezhushan passage. In the south-western part of the IHA, the topographic adjustment causes extensive reduction in tidal asymmetries.

In overall, the spatial distribution of changes in tidal asymmetries is very close to the distribution of amplitude changes of shallow-water tide M_4 (Figs. 10.13 and 10.14). This implies the importance of shallow-water tides in the tidal asymmetries of the sea area, which is consistent with that the tidal asymmetry is dominated by M_2/M_4 and $M_2/S_2/MS_4$ as shown by the observed tidal elevations at Xiaoyangshan station. As shown by Fig. 10.13, the direct impact of the constructions is mainly restricted in a scope within 4 km of the specific construction. But the resulting topographic adjustments, in turn, carry these impacts to a larger extension and complicate the spatial distribution of changes in tidal asymmetry. To the north of the IHA, there is a widespread increase in tidal asymmetry, and the scope is nearly consistent with the deposition zone to the north of the IHA.

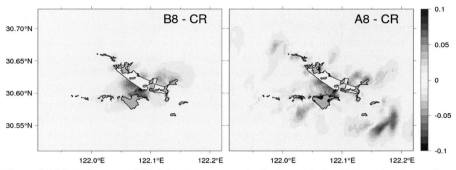

Figure 10.16 Difference of tidal duration asymmetry before and after the constructions of phases I ~ III, (*left*) no bathymetric changes, (*right*) considering bathymetric changes.

The YDH project has reduced the tidal asymmetry around the East Entrance by approximately 20%, mainly due to the passage-close and land reclamation works in Phase III. This is consistent with the observed decrease of ebb duration in this sea area (Ding & Chen, 2007). This extension of reduced tidal asymmetry can spread into the IHA of 3 km, reaching the harbor basin of Phase I. Fig. 10.16 also displays that the response of tidal asymmetry around Xiaoyangshan station is quite different with that in most parts of the IHA. Calculations show that from 2001 to 2008, the tidal asymmetry in the IHA decreased by 15% (from 0.12 to 0.10), which is completely converse to the increase of tidal asymmetry at Xiaoyangshan station according to the observations.

According to Eqs. (10.3) and (10.4), we can calculate the spatial distribution of the contribution of M_2/M_4 interaction and $M_2/S_2/MS_4$ interaction to the tidal duration asymmetry. Results are displayed in Fig. 10.17. The latter contributes roughly half

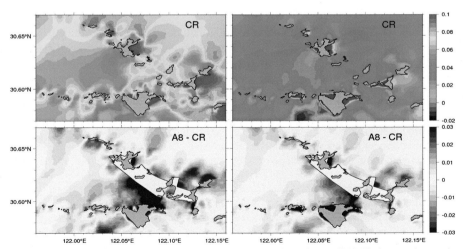

Figure 10.17 Contribution of (*left*) M_2/M_4 interaction and (*right*) $M_2/S_2/MS_4$ interaction to the tidal duration asymmetry and their difference before and after Yangshan Harbor's construction.

as much as the former, and this relationship is largely consistent spatially. Their response to the construction of YDH is generally synchronized.

Because tidal choking induces more amplitude reduction on the high-frequency tides, the stronger tidal choking will lead to relatively more reduction of M_4 amplitude than that of M_2 tide, resulting in a decrease in the amplitude ratio of M_4 to M_2. This implies a decrease in tidal asymmetry. Changes of tidal asymmetries in the East Entrance of the YDH obviously reflect this dynamic mechanism with the constructions. Tidal choking in the East Entrance is significantly enhanced by the passage closure and land reclamation works in phase III, leading to a reduction in the amplitude of the M_2 and M_4 tides in the area, with more significant reduction in the M_4 tide. As a result, tidal asymmetry is reduced significantly. No matter the M_2/M_4 interaction or the $M_2/S_2/MS_4$ interaction, they both embody this important mechanism of tidal responses (Fig. 10.14).

The shallow-water effect is an important mechanism responsible for the formation of shallow-water tides, influence of changes in bathymetries on tidal asymmetry thus cannot be ignored. The deepening of the water depth in the East Entrance will weaken the tidal choking, thereby enhancing the tidal asymmetry; however, it may also lead to weaker shallow-water effect and reduce tidal asymmetry. A dynamic equilibrium between tidal choking and water depth also seems to be emerged by the constructions. The constructions enhance the tidal choking in the East Entrance, thereby increasing the current velocities, and the bottom shear stress increases accordingly, which causes seabed scouring. The deeper bathymetries in turn weaken the tidal choking effect. Thus, a dynamic equilibrium is reached between. However, this equilibrium mechanism needs to be confirmed by further studies of sediment dynamics.

5. Summary

We investigate response of tidal constants, tidal choking, and tidal asymmetry to the construction of the YDH, especially the passage closure works, in this study. Observed tidal elevations and a high-resolution numerical model system based on FVCOM are employed for this study. Results show that the tidal responses are highly varying in space, and some meaningful conclusions are captured by our careful analysis.

1. The passages closure works decrease tidal amplitudes in the lee, but increase the tidal amplitudes in the stoss.

    The closure of passages block tides travel through these passages and leads to decrease in tidal amplitudes in the lee, accompanying by lags in tidal phases. By contrast, in the stoss, tides are increased in amplitudes and advanced in phases. This mechanism is embodied by the closures of all the three important passages during the construction of the YDH. While, the dam closing Dawugui-Kezhushan passage in phase II is in a direction of nearly parallel to the direction of tidal propagation, thus this passage closure work leads to little changes in tides. The Jiangjunmao-Dazhitou passage is the widest one among the closed passages and transport much tidal energies before the constructions. The closure of it brings the most extensive response of tides. The amplified tides in the stoss lead to strengthen tidal duration asymmetries, and the weakened tides in the lee also result in weakened tidal duration asymmetries. This common response is conductive to researches in similar coastal and marine constructions.

2. Shallow-water tides dominate the tidal asymmetries in the YDH and its response to the constructions.

   We employ a skewness method to quantify tidal duration asymmetry in this study. Observations and numerical model results show that tidal skewness in the YDH is positive, implying flood-dominated tidal asymmetry (longer ebb duration). The couple interaction M_2/M_4 and triplet interaction $M_2/S_2/MS_4$ predominates the tidal asymmetry in the YDH, and the former contributes twice of the later. Response of tidal asymmetry to the constructions is mainly related to the development of shallow-water tides M_4 and MS_4.

3. Stronger tidal choking decreases tidal asymmetry

   Dynamic mechanisms responsible for the tidal asymmetries and its response to the constructions of the YDH is quite different in space. Both observations and numerical model results show that the tidal skewness in the East Entrance is obviously smaller than that at the Xiaoyangshan station. Our numerical results reveal that this difference is mainly related to the weak shallow-water tides induced by strong tidal choking in the East Entrance.

   In the central and western part of the IHA, response of tidal amplitudes and tidal asymmetries is dominated by adjustment in bathymetries. The siltation in this region can amplify shallow-water tides and enhance tidal asymmetries accordingly. Whereas, in the East Entrance, the constructions strengthen the choking effect in this passage, which is not favored to the development of high-frequency shallow-water tides, and reduce the tidal asymmetries here. This result implies that conclusions only on basis of observations with only one station tend to be unsound in a complex multichannels system.

References

Bolle, A., Bing Wang, Z., Amos, C., & De Ronde, J. (2010). The influence of changes in tidal asymmetry on residual sediment transport in the Western Scheldt. *Continental Shelf Research, 30*(8), 871−882. https://doi.org/10.1016/j.csr.2010.03.001

Byun, D. S., Wang, X. H., & Holloway, P. E. (2004). Tidal characteristic adjustment due to dyke and seawall construction in the Mokpo Coastal Zone, Korea. *Estuarine, Coastal and Shelf Science, 59*(2), 185−196. https://doi.org/10.1016/j.ecss.2003.08.007

Chen, C., Beardsley, R. C., & Cowles, G. (2013). *An unstructured grid,: FVCOM user manual,* 404 pp., New Bedford.

Chen, C., Liu, H., & Beardsley, R. C. (2003). An unstructured grid, finite-volume, three-dimensional, primitive equations ocean model: Application to coastal ocean and estuaries. *Journal of Atmospheric and Oceanic Technology, 20*(1), 159−186. https://doi.org/10.1175/1520-0426(2003)020<0159:AUGFVT>2.0.CO;2

Chen, S. (2000). Erosion and accretion characteristics and their causes in the Qiqu Archipelago in the recent century. *Marine Science Bulletin, 19*(1), 58−67.

Colby, L. H., Maycock, S. D., Nelligan, F. A., Pocock, H. J., & Walker, D. J. (2010). An investigation into the effect of dredging on tidal asymmetry at the River Murray mouth. *Journal of Coastal Research, 26*(5), 843−850. https://doi.org/10.2112/08-1143.1

Davis, R. A., & Barnard, P. (2003). Morphodynamics of the barrier-inlet system, west-central Florida. *Marine Geology, 200*(1−4), 77−101. https://doi.org/10.1016/S0025-3227(03)00178-6

de Swart, H. E., & Zimmerman, J. T. F. (2009). Morphodynamics of tidal inlet systems. *Annual Review of Fluid Mechanics, 41*(1), 203–229. https://doi.org/10.1146/annurev.fluid. 010908.165159

Ding, P., & Chen, S. (2007). *Analysis of changes in water and sediment environment and erosion and deposition before and after Yangshan Port project.* State Key Laboratory of Estuarine and Coastal Research (In Chinese).

Dronkers, J. (1986). Tidal asymmetry and estuarine morphology. *Netherlands Journal of Sea Research, 20*(2–3), 117–131. https://doi.org/10.1016/0077-7579(86)90036-0

Friedrichs, C. T., & Madsen, O. S. (1992). Nonlinear diffusion of the tidal signal in frictionally dominated embayments. *Journal of Geophysical Research, 97*(C4). https://doi.org/ 10.1029/92jc00354

Gao, G. D., Wang, X. H., & Bao, X. W. (2014). Land reclamation and its impact on tidal dynamics in Jiaozhou Bay, Qingdao, China. *Estuarine, Coastal and Shelf Science, 151,* 285–294. https://doi.org/10.1016/j.ecss.2014.07.017

Gao, G. D., Wang, X. H., Bao, X. W., Song, D., Lin, X. P., & Qiao, L. L. (2018). The impacts of land reclamation on suspended-sediment dynamics in Jiaozhou Bay, Qingdao, China. *Estuarine, Coastal and Shelf Science, 206,* 61–75. https://doi.org/10.1016/j.ecss.2017. 01.012

Ge, J., Ding, P., Chen, C., Hu, S., Fu, G., & Wu, L. (2013). An integrated East China Sea-Changjiang Estuary model system with aim at resolving multi-scale regional-shelf-estuarine dynamics. *Ocean Dynamics, 63*(8), 881–900. https://doi.org/10.1007/s10236-013-0631-3

Ge, J., Lu, J., Zhang, J., Chen, C., Liu, A., & Ding, P. (2022). Saltwater intrusion-induced flow reversal in the Changjiang estuary. *Journal of Geophysical Research: Oceans, 127*(11). https://doi.org/10.1029/2021JC018270

Glenne, B., & Simensen, T. (1963). Tidal current choking in the landlocked fjord of nordå-syatnet. *Sarsia, 11*(1), 43–73. https://doi.org/10.1080/00364827.1963.10410284

Gong, W., Schuttelaars, H., & Zhang, H. (2016). Tidal asymmetry in a funnel-shaped estuary with mixed semidiurnal tides. *Ocean Dynamics, 66*(5), 637–658. https://doi.org/10.1007/ s10236-016-0943-1

Guo, L., Wang, Z. B., Townend, I., & He, Q. (2019). Quantification of tidal asymmetry and its nonstationary variations. *Journal of Geophysical Research: Oceans, 124*(1), 773–787. https://doi.org/10.1029/2018JC014372

Guo, W. (2017). *The time-varying characteristics of tidal duration asymmetry and its response to project.*

Guo, W., Wang, X. H., Ding, P., Ge, J., & Song, D. (2018). A system shift in tidal choking due to the construction of Yangshan Harbour, Shanghai, China. *Estuarine, Coastal and Shelf Science, 206,* 49–60. https://doi.org/10.1016/j.ecss.2017.03.017

Hill, A. E. (1994). Fortnightly tides in a lagoon with variable choking. *Estuarine, Coastal and Shelf Science, 38*(4), 423–434. https://doi.org/10.1006/ecss.1994.1029

Hoitink, A. J. F., Hoekstra, P., & Van Maren, D. S. (2003). Flow asymmetry associated with astronomical tides: Implications for the residual transport of sediment. *Journal of Geophysical Research: Oceans, 108*(C10), 1–38. https://doi.org/10.1029/2002jc001539

Keulegan, G. H. (1967). *Tidal flow in entrances; water-level fluctuations of basins in communication with seas.*

Kjerfve, & Knoppers, B. A. (1991). Tidal choking in a coastal lagoon. In *Tidal hydrodynamics* (pp. 169–181). John Wiley.

Kuang, C. P., Chen, S. Y., Zhang, Y., Gu, J., Deng, L., Pan, Y., & Huang, J. (2012). A two-dimensional morphological model based on next generation circulation solver II:

Application to Caofeidian, Bohai Bay, China. *Coastal Engineering, 59*(1), 14−27. https://doi.org/10.1016/j.coastaleng.2011.06.006

Li, L., Wang, X. H., Sidhu, H., & Williams, D. (2011). Modelling of three dimensional tidal dynamics in Darwin Harbour, Australia. *ANZIAM Journal, 52*, C103−C123.

Li, L., Wang, X. H., Williams, D., Sidhu, H., & Song, D. (2012). Numerical study of the effects of mangrove areas and tidal flats on tides: A case study of Darwin Harbour, Australia. *Journal of Geophysical Research: Oceans, 117*(6). https://doi.org/10.1029/2011JC007494

Liang, H., Kuang, C., Olabarrieta, M., Song, H., Ma, Y., Dong, Z., Han, X., Zuo, L., & Liu, Y. (2018). Morphodynamic responses of Caofeidian channel-shoal system to sequential large-scale land reclamation. *Continental Shelf Research, 165*, 12−25. https://doi.org/10.1016/j.csr.2018.06.004

Liu, C. (2008). *The cohesive sediment transport in the Yangshan Port and the study on the profile shaping process*. Master's thesis.

Lu, Y., Ji, R., & Zuo, L. (2009). Morphodynamic responses to the deep water harbor development in the Caofeidian sea area, China's Bohai Bay. *Coastal Engineering, 56*(8), 831−843. https://doi.org/10.1016/j.coastaleng.2009.02.005

MacMahan, J., van de Kreeke, J., Reniers, A., Elgar, S., Raubenheimer, B., Thornton, E., Weltmer, M., Rynne, P., & Brown, J. (2014). Fortnightly tides and subtidal motions in a choked inlet. *Estuarine, Coastal and Shelf Science, 150*, 325−331. https://doi.org/10.1016/j.ecss.2014.03.025

Moody, J. A. (1988). *Small-scale inlets as tidal filters*. Springer Science and Business Media LLC. https://doi.org/10.1007/978-1-4757-4057-8_8

Pawlowicz, R., Beardsley, B., & Lentz, S. (2002). Classical tidal harmonic analysis including error estimates in MATLAB using T_TIDE. *Computers & Geosciences, 28*(8), 929−937. https://doi.org/10.1016/S0098-3004(02)00013-4

Pugh, D., & Woodworth, P. (2014). *Sea-level science: Understanding tides, surges, Tsunamis and mean sea-level changes*. Cambridge University Press. https://doi.org/10.1017/CBO9781139235778, 9781139235778.

Rydberg, L., & Wickbom, L. (1996). Tidal choking and bed friction in Negombo Lagoon, Sri Lanka. *Estuaries, 19*(3), 540−547. https://doi.org/10.2307/1352516

Song, D., Wang, X. H., Kiss, A. E., & Bao, X. (2011). The contribution to tidal asymmetry by different combinations of tidal constituents. *Journal of Geophysical Research: Oceans, 116*(12). https://doi.org/10.1029/2011JC007270

Speer, P. E., Aubrey, D. G., & Friedrichs, C. T. (1991). Nonlinear hydrodynamics of shallow tidal inlet/bay systems. In B. B. Parker (Ed.), *Tidal hydrodynamics* (pp. 321−339). New York: John Wiley.

Stigebrandt, A. (1980). Some aspects of tidal interaction with fjord constrictions. *Estuarine and Coastal Marine Science, 11*(2), 151−166. https://doi.org/10.1016/s0302-3524(80)80038-7

Stigebrandt, A. (1999). Resistance to Barotropic tidal flow in straits by Baroclinic wave drag. *Journal of Physical Oceanography, 29*(2), 191−197. https://doi.org/10.1175/1520-0485(1999)029<0191:RTBTFI>2.0.CO;2

Suh, S. W., Lee, H. Y., & Kim, H. J. (2014). Spatio-temporal variability of tidal asymmetry due to multiple coastal constructions along the west coast of Korea. *Estuarine, Coastal and Shelf Science, 151*, 336−346. https://doi.org/10.1016/j.ecss.2014.09.007

Tianjin Academy of Water Transport Engineering of the Ministry of Transport. (2006). *Report on Numerical simulation of tidal current, sediment transport and seabed deformation in the west harbor area of Yangshan Deep-water Harbor*. Shanghai International Shipping Center, Tianjin Academy of Water Transport Engineering of the Ministry of Transport (In Chinese).

van Maren, D. S., & Winterwerp, J. C. (2013). The role of flow asymmetry and mud properties on tidal flat sedimentation. *Continental Shelf Research, 60*, S71–S84. https://doi.org/10.1016/j.csr.2012.07.010

Winterwerp, J. C., Wang, Z. B., Van Braeckel, A., Van Holland, G., & Kösters, F. (2013). Man-induced regime shifts in small estuaries - II: A comparison of rivers. *Ocean Dynamics, 63*(11–12), 1293–1306. https://doi.org/10.1007/s10236-013-0663-8

Ying, X., Ding, P., Wang, Z. B., & Van Maren, D. S. (2012). Morphological impact of the construction of an offshore Yangshan deepwater harbor in the Port of Shanghai, China. *Journal of Coastal Research, 278*, 163–173. https://doi.org/10.2112/JCOASTRES-D-11-00046.1

Ying, X., Ding, P., Wang, Z. B., & Van Maren, D. S. (2011). Morphological impact of the construction of an offshore Yangshan deepwater harbor in the Port of Shanghai, China. *Journal of Coastal Research, 28*, 163–173.

Yu, Z., & Zhang, Z. (2006). *Analysis on the scouring and silting characteristics of water and sediment movement and the influence of passage closure works in the Yangshan Deepwater Harbor, Shanghai International Shipping Center.*

Yu, Z., Li, S., Zhang, Z., Xu, H., Zhu, Q., & Zhuang, H. (2013). *Response of morphodynamics on the project of Yangshan harbor, Shanghai international shipping center* (p. 163). Science Press.

Zhou, Z., Ge, J., Wang, Z. B., van Maren, D. S., Ma, J., & Ding, P. (2019). Study of lateral flow in a stratified tidal channel-shoal system: The importance of intratidal salinity variation. *Journal of Geophysical Research: Oceans, 124*(9), 6702–6719. https://doi.org/10.1029/2019JC015307

Recent progresses in the studies of boundary upwelling

11

Fanglou Liao
Ocean Institute, Northwestern Polytechnical University, Taicang, China

1. Introduction

Upwelling is a dynamic process characterized by the upward movement of cold, nutrient-rich water toward the ocean surface (Kämpf & Chapman, 2016). It is widely recognized that upwelling plays a crucial role in cooling the upper-ocean waters and significantly contributes to primary productivity. Many renowned fishing grounds around the world are closely associated with upwelling systems (Carr & Kearns, 2003; Pauly & Christensen, 1995). Moreover, upwelling holds great importance in global three-dimensional circulations, which influence the spatial and temporal distribution of tracers such as heat and salinity (Liang et al., 2015; Tamsitt et al., 2017). The sinking of dense water masses in the Southern Ocean and at high latitudes of the North Atlantic Ocean, subsequently spreading throughout the abyssal oceans, is a well-established phenomenon (Dickson et al., 1988; Jacobs, 2004; Solodoch et al., 2022). These water masses eventually resurface through upwelling (Tamsitt et al., 2017), thus constituting a vital component of the global meridional overturning circulation. In contrast to upwelling, downwelling refers to the downward movement of seawater and occurs in extensive regions of the global ocean, including the five subtropical gyres. However, historically, downwelling has received far less attention compared to upwelling. Given the growing concerns regarding greenhouse gas accumulations in the atmosphere, it is expected that downwelling will garner increasing attention due to its vital role in regulating greenhouse gas concentrations. Overall, further research and understanding of both upwelling and downwelling processes are crucial for comprehending the complex dynamics of our oceans in the face of global environmental challenges.

The primary driving force behind upwelling is wind stress, particularly alongshore wind stress as elucidated by the dynamical framework established by Ekman (1905). In coastal areas, for instance, the alongshore wind stress plays a pivotal role in inducing upwelling. Additionally, upwelling can also occur through a process known as Ekman suction by the wind stress curl. Cyclonic mesoscale eddies are also associated with upwelling, as they lead to horizontal divergence of seawater. Furthermore, various other processes including flow-topography interaction and Kelvin waves contribute to upwelling in specific regions (Ray et al., 2022; Roughan & Middleton, 2004). Eastern boundaries, equatorial oceans, and the Southern Ocean stand out as prominent locations housing well-known upwelling systems. Notably, the Southern Ocean is the largest and vertically coherent upwelling system. Enhancing our

Current Trends in Estuarine and Coastal Dynamics. https://doi.org/10.1016/B978-0-443-21728-9.00011-9

understanding of these diverse upwelling mechanisms and their regional manifesta-
tions is essential for comprehending the intricacies of oceanic circulation patterns
and promoting effective management of marine ecosystems.

Extensive research was conducted to understand the existence, mechanisms, variabilities, and impacts of upwelling, given its crucial role in coastal physical-
biogeochemical processes, ocean circulation, and climate. Particularly along the
eastern oceanic boundaries, where equatorward alongshore wind stress drives surface
water away from the coast following Ekman dynamics, significant advancements have
been made. In such regions, the movement of surface water away from the coast leads
to the upward ascent of subsurface water toward the ocean surface, ensuring mass conservation. These upwelling systems along the eastern boundaries are commonly
referred to as Eastern Boundary Upwelling Systems (EBUSs), as depicted in
Fig. 11.1. The EBUSs encompass the California Current System (CalCS), Canary Current System (CanCS), Humboldt Current System (HumCS), and Benguela Current
System (BenCS). Despite occupying a small area in the global ocean, EBUSs make
a disproportionately large contribution to global ocean productivity, supporting over
20% of the global fish catch (Bograd et al., 2023; Pauly & Christensen, 1995). This
highlights their ecological significance and underscores the importance of studying
and understanding these unique upwelling systems to manage marine resources and
maintain ecosystem health.

The upwelling phenomenon in western boundary currents (WBCs) is relatively
less recognized, partially due to the complexity of ocean dynamics in these systems.
Unlike in EBUSs, upwelling in WBCs is not typically driven by wind, which makes
it challenging to detect using proxies like satellite-based wind. However, recent
studies have demonstrated the significant presence of upwelling in WBCs, as evidenced by findings from Castelao and Barth (2006), Roughan and Middleton
(2004), and Liao, Liang, et al. (2022). These studies underscore a greater focus on
understanding the complex upwelling processes in WBCs. A more in-depth comprehension of how upwelling manifests in WBCs could provide insights into the impact
of these processes on the global ocean circulation, biogeochemical cycling, and
ecosystem productivity.

This chapter provides an overview of the recent developments in upwelling studies
during the past decade (2013 onward). We focus on boundary upwelling systems,
including EBUSs, WBCs, and some marginal seas. Our primary objective is to
examine the variability and future trends of upwelling, which are critical concerns
for the oceanography and climate communities in the context of global warming.
The chapter is structured as follows. Section 2 presents the advancements made in
the study of boundary upwelling, including new methods, mechanisms, and the discovery of new upwelling systems. This section also covers research on the ecological
impacts of upwelling systems. Section 3 offers an extensive discussion of the temporal
variations in boundary upwelling systems, ranging from seasonal scales to long-term
trends. In Section 4, our attention shifts to an in-depth analysis of projections concerning upwelling intensification in the 21st century. Finally, Section 5 concludes with a
summary of the major research achievements made in boundary upwelling studies
over the last decade.

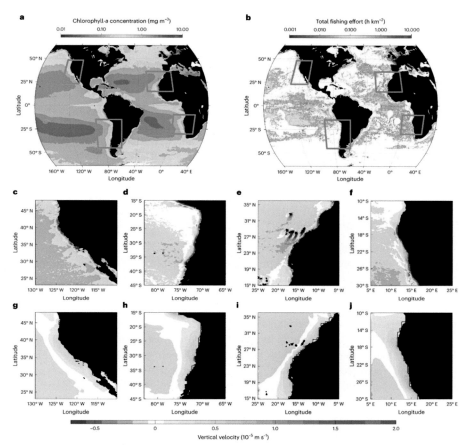

Figure 11.1 Simulated upwelling in the EBUSs in climate models. (a, b) Geographic distribution of time-mean ocean chlorophyll-a concentration during 2002—22 (a) and fishing efforts during 2012—21 (b). (c—f) The 1920—2005 annual-mean vertical velocity at 50 m in the CalCS (c), HCS (d), CanCS I (e) and BCS (f) simulated in the CESM-H ensemble mean. (g—j) The 1920—2005 annual-mean vertical velocity at 50 m in the CalCS (g), HCS (h), CanCS (i), and BCS (j) simulated in the CMIP6 CGCM ensemble mean. Regions shallower than 50 m in c—j are masked in black according to the bathymetry of CESM-H.
Sourced from Jing, Z., Wang, S., Wu, L., Wang, H., Zhou, S., Sun, B., Chen, Z., Ma, X., Gan, B., & Yang, H. (2023). Geostrophic flows control future changes of oceanic eastern boundary upwelling. Nature Climate Change, 13, 148—154. https://doi.org/10.1038/s41558-022-01588-y.

2. Upwelling around the eastern and western boundaries of the global ocean

Boundary upwelling plays a crucial role in global upwelling systems. While along-shore wind stress is the primary driving force behind coastal upwelling in many regions, the specific dynamical mechanisms of coastal upwelling can vary depending on the region and season. Previous studies have commonly utilized proxies such as wind stress or SST to estimate the presence of upwelling. Although these methods generally yield satisfactory results, they may have limitations in certain scenarios. For instance, these methods often overlook the influence of factors like eddies, coastal-trapped waves, bottom friction, and the landward intrusion of swift boundary currents. To address these limitations, a more direct approach to studying upwelling involves analyzing oceanic vertical velocity from numerical models, which theoretically encompasses all relevant dynamical processes. However, small-scale processes within the sub-grid remain subject to physical parameterizations. In this section, we highlight several noteworthy studies conducted in the past decade that have contributed to the discovery of new upwelling systems, advanced our understanding of dynamical mechanisms, or introduced novel methodologies.

2.1   Upwelling along eastern boundaries

The spatial distribution of global wind stress reveals a prominent pattern of consistent equatorward alongshore wind stress prevailing over four distinct boundary current systems. These regions, known as EBUSs, have garnered significant attention due to their crucial role in climate dynamics, ecosystem functioning, and socioeconomic development. The robust upwelling occurring in these regions is of paramount importance, impacting various facets of our planet's intricately interconnected systems.

Measuring oceanic vertical velocity, which is approximately three orders of magnitude weaker than horizontal velocity, remains a challenging task as existing oceanography measurement facilities are unable to directly measure the vertical motion at this slow scale, except for rapid upward sub-mesoscale flow. Consequently, most prior studies have resorted to indirect proxies such as alongshore wind stress and SST to identify upwelling processes. These proxy-based methods may have certain limitations, such as the possibility that a drop in SST may not necessarily result from upwelling. To increase confidence in upwelling detection, Abrahams et al. (2021) developed a novel method by combining both wind stress and SST measurements. Specifically, they introduced an upwelling index based on wind stress, wherein positive values correspond to upwelling. The presence of upwelling was then confirmed when evidence of a positive upwelling index coincided with a concurrent drop in SST. This approach provides improved accuracy in detecting upwelling and can enhance our understanding of the underlying dynamics and impacts of these systems.

In the past decade, there has been a growing concern regarding the reliability of using proxies such as alongshore wind stress to calculate upwelling. This concern stems from the recognition that multiple processes can induce offshore water transport, which can either enhance or restrict the Ekman transport driven by the alongshore wind stress. Consequently, efforts have been made to consider additional factors when studying boundary upwelling, particularly in EBUSs. Studies in this field include Jacox et al. (2018) and Jing et al. (2023), which highlight the significant role played by the geostrophic current in modulating wind-driven upwelling in EBUSs. Further details on these studies will be provided later. Recognizing that various dynamical processes can influence upwelling, some researchers have started employing oceanic vertical velocity diagnosed from ocean models to investigate the spatiotemporal characteristics and climate and ecological impacts associated with boundary upwelling. Notable examples of such studies include Xiu et al. (2018) and Liao, Gao, et al. (2022), which will be discussed in more details in subsequent sections.

In the last decade, numerous studies have delved into the question of whether upwelling in the four EBUSs will intensify in the face of future warming. Noteworthy contributions to this topic include works by Bakun et al. (2015), Wang et al. (2015), Xiu et al. (2018), Sousa et al. (2020), and Jing et al. (2023). Some of these studies have observed an increase in alongshore wind stress within the EBUSs, particularly at higher latitudes within each region. However, a growing body of evidence suggests that wind intensification does not necessarily equate to stronger upwelling in these EBUSs. Further elaboration on these studies will be provided in Section 4.

2.2 Upwelling along western boundaries

The western boundaries of the global ocean are renowned for their fast horizontal currents, which include the Kuroshio, Gulf Stream, Agulhas, East Australian, and Brazil Currents. In contrast to the eastern boundaries, there is a low occurrence of upwelling-favorable wind stress along these western boundaries. The potential for upwelling along these boundaries is likely to result from more complex dynamics, such as flow—topography interaction or eddies. Previous research has found that the zonal movement of the East Australian Current toward the shelf can generate upwelling (Roughan & Middleton, 2004). A similar mechanism has also been observed in the upwelling of subsurface Kuroshio water in the China Seas.

Like other upwelling systems, upwelling in the WBCs can also be important in marine ecosystem productivity. For instance, Cresswell et al. (2017) investigated the dynamics of the East Australian Current (EAC) along the east coast of Australia, focusing on its impact on upwelling and downwelling during the summer season. The researchers utilized satellite data and oceanographic measurements to analyze temperature, salinity, and current patterns in the region. They discovered that the EAC generates a robust offshore flow, resulting in the upwelling of colder, nutrient-rich waters from deeper layers of the ocean. These upwellings play a vital role in supporting phytoplankton growth and enhancing marine ecosystem productivity. However, the study also identified instances of downwelling in specific areas, where the EAC induces a sinking motion of surface waters. This process can deplete nutrients and ultimately lead to a decrease in productivity within those regions.

A comprehensive investigation into the WBC upwelling along the southeast coast of Australia was conducted by Huang and Wang (2019). To explore the spatiotemporal variabilities of upwelling in this region, the researchers developed a scale-independent and semiautomatic image processing technique that utilized 14 years of monthly Moderate Resolution Imaging Spectroradiometer (MODIS) SST data. The study identified two major coastal upwelling systems along the Australian southeast coast, as depicted in Fig. 11.2. The first system is located along the New South Wales (NSW) coastline, while the second system is situated along the western Victoria and adjacent South Australia (WVIC/SA) coasts.

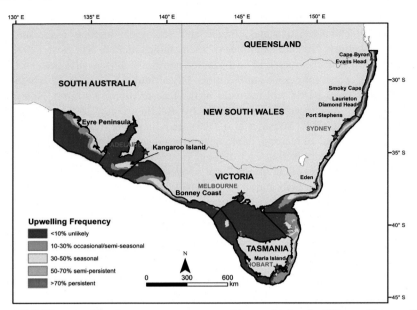

Figure 11.2 The upwelling frequency map, generated by combining the 126 monthly upwelling maps.
Sourced from Huang, Z. & Wang, X. H. (2019). Mapping the spatial and temporal variability of the upwelling systems of the Australian south-eastern coast using 14-year of MODIS data. Remote Sensing of Environment, 227, 90−109. https://doi.org/10.1016/j.rse.2019.04.002.

The NSW upwelling exhibits greater persistence, occurring from austral spring to autumn, whereas the WVIC/SA upwelling predominantly appears in the austral summer. In terms of strength, the NSW upwelling surpasses the WVIC/SA upwelling based on factors such as influence area, SST anomaly, chlorophyll-a concentrations, and upwelling speed. The authors attributed the driving forces behind these upwelling systems to the East Australian Current (EAC) and its eddies for the NSW upwelling, particularly its central and northern sections. Alternatively, wind stress was identified as the primary driver for the WVIC/SA upwelling. Additionally, the study suggested that ENSO (El Niño Southern Oscillation) may exert varying impacts on these two upwelling systems, with stronger and weaker upwelling occurring during the positive and negative phases of ENSO, respectively.

The role of wind stress in inducing upwelling in WBCs has been previously recognized. Aguiar et al. (2014) conducted a comprehensive study on the upwelling phenomenon along the Brazil coast, employing a high-resolution regional ocean model in conjunction with satellite-derived SST and reanalysis wind data. The study revealed that wind stress plays a critical role in driving the identified upwelling along this region. Specifically, Ekman transport was found to be the dominant mechanism responsible for upwelling between $17°S$ and $21°S$, while Ekman pumping took precedence between $21°S$ and $23°S$. Additionally, the study highlighted the importance of flow—topography interaction and mesoscale dynamics in the generation processes of upwelling along the Brazil coast. These factors contribute significantly to the complexity and dynamics of the upwelling system in this region.

Despite the above-mentioned studies conducted on upwelling in WBCs, there has been a noticeable disparity in attention between upwelling in WBCs and EBUSs. This discrepancy can be attributed to several key factors. Firstly, upwelling in EBUSs is predominantly driven by alongshore wind stress, which can be easily detected from satellites. Conversely, the dynamics of upwelling in WBCs are considerably more intricate and challenging to comprehend. Additionally, the absence of direct measurements for oceanic vertical velocity poses another obstacle. Till recently, none of the existing instruments were capable of directly measuring oceanic vertical velocity. However, with the emergence of advanced ocean models and reanalysis products, it is now feasible to routinely obtain oceanic vertical velocity data. This new development opens up opportunities for detailed investigations into the spatiotemporal characteristics of upwelling across the global ocean.

For instance, a study conducted by Liang et al. (2017) delved into the global ocean's vertical motions using the oceanic vertical velocity data obtained from Estimating the Circulation and Climate of the Ocean (ECCO), a dynamically consistent ocean state estimate developed by Forget et al. (2015). This study provided comprehensive insights into the spatial distribution of global oceanic vertical velocity. The researchers discovered that Eulerian-mean and eddy-induced vertical motions exhibit opposing directions in numerous regions across the global ocean. However, it was determined that Eulerian-mean vertical motion serves as the principal component of the residual vertical motion. Building upon this work, Liao, Liang, et al. (2022) conducted a subsequent study investigating upwelling phenomena associated with the major WBCs using the oceanic vertical velocities from numerical ocean models or reanalysis. Their findings revealed the presence of a robust subsurface upwelling system within all five major WBCs, namely the Kuroshio, Gulf Stream, Agulhas Current, East Australian Current, and Brazil Current (see Fig. 11.3). Notably, these subsurface WBC upwellings exhibit significant intensity below approximately 200 m and can reach much greater depths compared to EBUSs (as exemplified by the Peruvian upwelling, Fig. 11.4). Furthermore, the study demonstrated that these subsurface upwelling systems within WBCs are considerably stronger than the Peruvian upwelling (Fig. 11.4). Importantly, it was observed that these subsurface upwellings play a pivotal role in the global vertical transport of heat and salinity.

Liao, Liang, et al. (2022) taken the Gulf Stream as an example to tentatively explain the dynamical mechanisms responsible for these subsurface upwelling systems, shown in Fig. 11.5. A volume budget analysis was conducted between 55 and 2084m. The large horizontal divergences/convergences in different layers require vertical transport of seawater to conserve mass. The dynamical mechanism underpinning these horizontal divergences/convergences can be deduced from the density structure along the same cross-sections. A pronounced poleward increase in density was observed, indicating that the total density change along section BC exceeded that along section AB. The greater density change corresponds to elevated vertical shear of horizontal velocity along section BC. As the mass flow through each section should remain

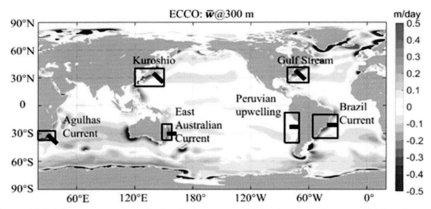

Figure 11.3 Time-averaged vertical velocity $\overline{w}$ near 300 m between January 1992 and December 2009. This figure is based on ECCO data. The black boxes in (a) show the domains of the five western boundary and one eastern boundary systems investigated in this study. Adapted from Liao, F., Liang, X., Li, Y., & Spall, M. (2022). Hidden upwelling systems associated with major western boundary currents. Journal of Geophysical Research: Oceans, 127, e2021JC017649.

constant (mass continuity), water upwelling is the sole means of maintaining mass conservation. The authors contended that these WBC upwellings share dynamic similarities with boundary downwelling in subpolar regions (Spall, 2010) and are insensitive to numerical models, subgrid-scale mixing, or bottom topography.

2.3 Upwelling in the marginal seas

Coastal upwelling in the marginal seas constitutes a pivotal component of the global upwelling system, playing a critical role in regulating local climate and ecosystems. The China Seas, encompassing the Bohai Sea, Yellow Sea, East China Sea, and South China Sea, serve as a prominent example of marginal seas, spanning a total area of approximately 4.7 million km^2. Within the coastal regions of the China Seas, upwelling is widespread and influenced by various factors, including the summer monsoon, flow-topography interaction, river discharge, and Kuroshio encroachment. Hu and Wang (2016) conducted a comprehensive literature review on the research advancements concerning upwelling in the China Seas. Through their review, they identified and delineated a total of 12 upwelling regions (refer to Fig. 2 in their study), elucidating the dynamic mechanisms driving each upwelling system. Furthermore, the study delved into the spatiotemporal characteristics of these upwelling systems and explored the associated biogeochemical responses.

Previous studies have predominantly associated upwelling in marginal seas with alongshore wind stress, wind stress curl, or eddies. However, it is crucial to recognize that flow-topography interaction can also serve as a significant mechanism driving upwelling in these regions (Gan et al., 2009, 2015). For instance, Gan et al. (2015)

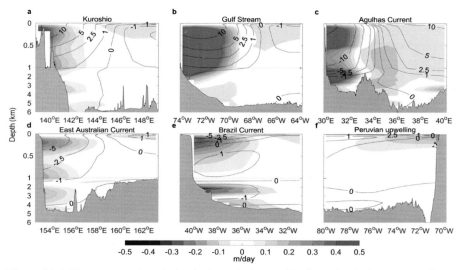

Figure 11.4 Time-averaged vertical velocity $\overline{w}$ (color) and horizontal velocity (contour lines, unit: cm/second) in selected cross sections from estimating the circulation and climate of the ocean. The cross sections are marked with thick black lines in Fig. 11.3. (a) Kuroshio. (b) Gulf Stream. (c) Agulhas current. (d) East Australian current. (e) Brazil current. (f) Peruvian upwelling. The contour lines show the horizontal velocity (cm/second) perpendicular to the cross sections, indicative of the strength of adjacent western boundary currents. Note that the depth axis is stretched for better visualization.

Sourced from Liao, F., Liang, X., Li, Y., & Spall, M. (2022). Hidden upwelling systems associated with major western boundary currents. Journal of Geophysical Research: Oceans, 127, e2021JC017649.

investigated on the formation, maintenance, and relaxation of summer upwelling in the Northeastern South China Sea (NSCS) utilizing a regional ocean model forced by realistic conditions. The authors observed that this summer upwelling primarily stemmed from an intensified westward pressure gradient force along the isobaths, coupled with bottom frictional effects that induced cross-isobath currents over the expansive NSCS shelf. Both mechanisms emerged as a result of the interaction between the eastward shelf current and the topography of the shelf. Notably, this flow-topography-induced upwelling demonstrated resilience even during periods of downwelling-favorable winds. This can be attributed to the widened shelf, which significantly impeded the retreat of the eastward shelf current. Furthermore, the authors discovered that downstream outflow pumping served as a secondary mechanism, facilitating the persistence of upwelling in the NSCS region even under downwelling-favorable wind conditions.

In addition to the aforementioned mechanisms, remote forcing from Kelvin waves can also significantly contribute to driving or modulating upwelling in the marginal seas. Li et al. (2020) utilized data from the World Ocean Atlas 2018 (WOA18) and multisource satellite datasets to identify a springtime upwelling system along Manaung Island, Myanmar. Their findings revealed that this upwelling system was primarily

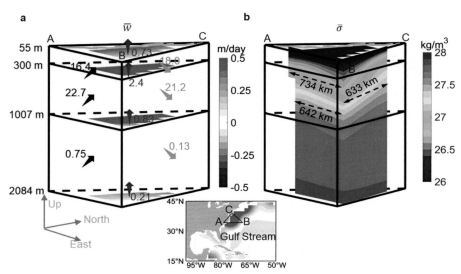

Figure 11.5 Time-averaged vertical velocity $\overline{w}$, potential density anomaly $\overline{\sigma}$, and volume flux in a triangle-shape domain in the Gulf Stream region. (a) Time-averaged vertical velocity at four depths (colors) and lateral and vertical volume fluxes. The black and blue arrows represent the lateral volume fluxes in Sv (10^6 m^3/second), and the purple arrows show the vertical volume fluxes in Sv. (b) Time-averaged potential density anomaly along the southern section and northern section of the triangle-shaped domain between 55 and 2000 m (shown in the inset). The gray curve in the 55 m section along AC represents the coastline. The results are based on ECCO product on the native grids.

Sourced from Liao, F., Liang, X., Li, Y., & Spall, M. (2022). Hidden upwelling systems associated with major western boundary currents *Journal of Geophysical Research: Oceans, 127,* e2021JC017649.

driven by remote equatorial forcing. Specifically, the equatorial easterly wind stress during January—March generated an upwelling Kelvin wave that propagated along the eastern boundary of the Bay of Bengal, further uplifting the thermocline and halo- cline in the coastal areas around the Manaung Island. Additionally, the local wind stress surrounding the island also favored the onset of upwelling, thereby intensifying the effects of remote forcing. Similarly, Ray et al. (2022) noted the crucial role played by the equatorial upwelling Kelvin wave in modulating coastal upwelling in the West- ern Bay of Bengal.

2.4 *Ecological and climate impacts of boundary upwelling*

Upwelling serves as an ecologically important phenomenon in the EBUSs. Cheresh et al. (2023) utilized a high-resolution physical-biogeochemical model and determined that coastal upwelling played a crucial role in dictating the aragonite saturation state along the central California coast. An empirical orthogonal function (EOF) analysis revealed that the intensity of coastal upwelling and the amount of dissolved inorganic carbon (DIC) present in the upwelling source water were the two principal factors influencing the interannual variability of undersaturation events. The severity of undersaturation events alone could be explained by up to 43% of the interannual variability in upwelling intensity. Moreover, upwelling has been found to exacerbate ocean acidification, which can have devastating consequences for marine ecosystems. Based on in-situ measurements, Schulz et al. (2019) observed that upwelling of water masses from depths of 200−250m off the Central East Australian shelf could significantly lower seawater pH in the Cape Byron Marine Park, Australia.

Gómez et al. (2023) made notable discoveries regarding the profound influence of upwelling on the chemical properties of a coral system. Through analysis of collected in-situ data, the authors observed that during the upwelling seasons, there was a remarkable 42% increase in $CaCO_3$ accretion. Accompanying this increase, there were also elevated levels of total alkalinity and dissolved inorganic carbon. However, aragonite saturation exhibited a decrease during these periods of upwelling. These variations provide compelling evidence for the crucial role played by upwelling in shaping and impacting the coral community.

Jiang and Wang (2018) conducted a study on the impacts of upwelling on ecosystems along the coasts of the South China Sea. Their findings demonstrated that the ecosystems located in the northeast shelf and east of the Leizhou Peninsula exhibited significant responses to the summer coastal upwelling, as evidenced by the occurrence of nearshore cold, high salinity, and nutrient-rich water. While coastal upwelling greatly enhances coastal productivity, the maximum levels of phytoplankton and chlorophyll-a tend to lag the maximum nutrient concentration by approximately 10 days. Using multiple satellite datasets, Chen, Shi, and Zhao (2021) discovered a strong correlation between wind-induced upwelling, via Ekman transport or Ekman pumping, and summer phytoplankton blooms off the Vietnamese coast. These observations highlight the essential role of upwelling in shaping regional ecosystems.

Along the Zhejiang coast in the East China Sea (ECS), Chen, Shiah, Gong, and Chen (2021) examined summertime water samples collected from 33 stations. They observed that phytoplankton blooms and hypoxia co-occurred beneath the surface water layer. The coastal upwelling appeared to elevate and expand the presence of hypoxic and low-oxygen water in the subsurface, posing a threat to marine organisms. Moreover, the upward transport of nutrients through upwelling intensified the proliferation of phytoplankton blooms, leading to increased consumption of organic matter derived from these blooms. Consequently, this exacerbated the occurrence of hypoxia.

Upwelling has also been found to facilitate the spread of river plumes through the jet produced by the upwelling front. For example, in the South China, Sea, Chen et al. (2017) investigated the dynamic mechanism behind the far-reaching transport of Pearl River plume water, which was observed extending as far as the Taiwan Bank from satellites. This belt of turbid water was carried along the shelf front, associated with the upwelling along the Guangdong coast. Through analysis of numerical simulation results, the authors discovered that the upwelling-induced horizontal density gradient generated a strong horizontal geostrophic jet, which was the dominant factor responsible for transporting the highly turbid water from the Pearl River to the Taiwan Bank over long distances. The far-reaching transport of nutrient-rich plume water plays a crucial role in promoting biophysical activities.

Upwelling has also been found to produce important climate impacts. Through the process of lifting deep cold water to the surface, upwelling is anticipated to act as a buffer against rising ocean temperatures and mitigate the severity of marine heatwaves (MHWs), defined as periods of anomalously high ocean temperatures compared to the region's climatology (Hobday et al., 2016). MHWs can persist for several days or even months, and the resulting extreme ocean temperatures pose a significant threat to marine ecosystems. Extensive research has been conducted to examine the influence of upwelling on the characteristics of MHWs, aiming to better understand their impacts.

Bograd et al. (2023) conducted a study investigating the relationship between coastal upwelling and the characteristics of MHWs along central California. Their findings highlighted the critical role of upwelling in modulating regional MHWs at shorter timescales compared to basin-wide climate modes like PDO and ENSO. Overall, upwelling was found to reduce the frequency of MHW occurrences. Additionally, the study revealed that upwelling anomalies could both initiate and sustain MHWs. Specifically, a negative upwelling anomaly was observed to initiate an MHW. This was attributed to weakened upwelling, which increased stratification and led to an interior return flow, thereby reducing the effectiveness of upwelling in providing a cooling effect in nearshore areas. These research findings provide compelling evidence supporting the essential role of coastal upwelling in regulating the characteristics of MHWs.

In a study conducted by Varela et al. (2021), high-resolution satellite SST data were utilized to analyze and compare the occurrence of MHWs in both coastal and offshore regions of the four major EBUSs. The results consistently indicated that the trends in MHW occurrence were comparatively weaker in coastal upwelling regions when compared to offshore regions. This observation suggests that upwelling plays a significant role in mitigating the intensity of MHWs to some extent. Similarly, Yao and Wang (2021) conducted a study that revealed a negative wind stress curl as a leading factor in weakening or eliminating upwelling in the midwestern South China Sea. These variations in upwelling were found to be associated with the development of severe summer MHWs on a basin-wide scale. These studies provide compelling evidence demonstrating the influence of upwelling on the occurrence and severity of MHWs, with coastal upwelling acting as a protective mechanism against extreme ocean temperatures in specific regions.

A recent research by Wang et al. (2023) has shown that upwelling regions can be significantly affected by MHWs, even more so than the surrounding oceans. Using a global eddy-resolving coupled model, the authors projected the characteristics of MHWs in EBUSs under a high-emission scenario. Their findings indicated that two EBUSs, namely, HumCS and BenCS in the southern hemisphere, are expected to be vulnerable MHW hotspots due to a weakening of the eastern boundary current caused by ocean warming. These results demonstrate that even upwelling regions, which were previously perceived as relatively resistant to MHWs, may face severe consequences from elevated ocean temperatures. This study highlights the significance of continued research in understanding the complex relationships between oceanic currents, upwelling, and MHWs, particularly in the context of future climate change scenarios.

3. Variations of boundary upwelling in the past

Due to its significant impacts on regional climate and ecosystems, changes in boundary upwelling have garnered considerable attention. Depending on the driving mechanisms, upwelling varies at various temporal scales, ranging from seasonal to interannual, decadal, or even secular trends. In this review, we will highlight some studies conducted over the past decade that have focused on examining the variations of boundary upwelling. Starting from shorter scales (seasonal and interannual) to longer scales (10 years or more), our goal is to provide an overview of the most significant findings on upwelling variability.

3.1 Seasonal and interannual variations in boundary upwelling

Several factors, including the Madden-Julian Oscillation (MJO), equatorial Kelvin waves, and ENSO, can induce atmospheric oscillations, leading to seasonal and interannual variations in boundary upwelling. Rosales Quintana et al. (2021) recently discovered significant interannual variability related to upwelling in the Peruvian system. Using a Lagrangian framework and outputs from an eddy-resolving global ocean model, the authors investigated the sources of upwelling water off the coast of northern Peru. Their analysis of released virtual particles revealed that a substantial proportion of upwelling water originates from the Equatorial Undercurrent (EUC), which exhibits significant interannual variability. Specifically, more of the Peruvian upwelling water is sourced from the EUC during its warm phases, and more locally when the EUC is at its cold phase.

Previous and recent studies have primarily focused on examining boundary upwelling using metrics derived from wind stress and SST. While these methods are generally reliable, it is important to consider potential contamination of upwelling signals by various processes, including SST changes associated with ENSO. A recent study by Liao, Gao, et al. (2022) has provided novel insights into both the seasonality and long-term trends of global ocean upwelling/downwelling, including boundary upwelling. The authors employed oceanic vertical velocity data from multiple ocean models with diverse horizontal resolutions, forcings, and numerical schemes. To the best of our knowledge, this is the first study of its kind. Notably, Liao, Gao, et al. (2022) observed significant seasonal variations in the four EBUSs, as depicted in Fig. 11.6. Regarding the long-term trend spanning from 1998 to 2017, none of the EBUSs exhibited a statistically significant increase in upwelling intensity. However, they did identify robust interannual variations associated with climate modes like ENSO in these EBUSs, as illustrated in Fig. 11.7. Importantly, these variation patterns remained consistent across all examined models, indicating their insensitivity to model configuration.

Myriad efforts have also been focused on understanding variations in upwelling within marginal seas, such as the Vietnamese upwelling system in the South China Sea, which has been a hot topic of research. One notable study by Da et al. (2019) employed a high-resolution regional model to investigate the interannual variability

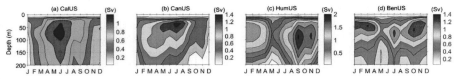

Figure 11.6 Month-depth distributions of the upward volume flux in the EBUS (Fig. 11.1c). CalUS stands for the California upwelling system (111°W−130°W, 25°N−47°N); CanUS for the Canary upwelling system (8°W−22°W, 16°N−43°N); HumUS for the Humboldt upwelling system (68°W−80°W, 17°S−41°S) and the BenUS for the Benguela upwelling system (4°E−19°E, 17°S−35°S). Only grids where upwelling appears were used to calculate vertical volume flux.
Sourced from Liao, F., Gao, G., Zhan, P., & Wang, Y. (2022). Seasonality and trend of the global upper-ocean vertical velocity over 1998−2017. Progress in Oceanography, 204, 102804. https://doi.org/10.1016/j.pocean.2022.102804.

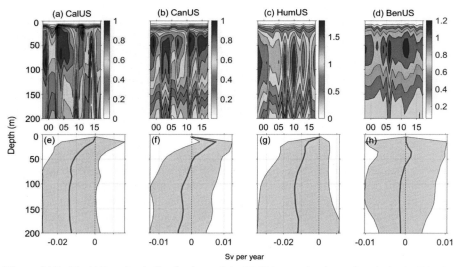

Figure 11.7 (Top) Year-depth distributions of the UVF (upward volume flux) in in the EBUSs; (bottom) vertical profiles of UVF trend from 1998 to 2017 (red lines; the gray shadow shows the 95% confidence interval). Only grids where upwelling appears were used to calculate vertical volume flux. The ticks in the horizontal axis stands for the year from 1998 to 2017, with 95 indicating 1995 and 00 meaning 2000, etc.
Sourced from Liao, F., Gao, G., Zhan, P., & Wang, Y. (2022). Seasonality and trend of the global upper-ocean vertical velocity over 1998−2017. Progress in Oceanography, 204, 102804. https://doi.org/10.1016/j.pocean.2022.102804.

of the summer South Vietnam Upwelling (SVU), with upwelling represented by an SST-derived index. The study revealed that there are two distinct SVU regimes, one located in the coastal ocean and the other offshore. Furthermore, their results demonstrated that summer wind and the eastward jet play major roles in controlling coastal

SVU variability, while wind stress curl is the dominant factor influencing offshore SVU variability. Intrinsic ocean variability can also impact SVU variability through background eddy circulation fluctuations. Notably, ENSO can affect summer winds, resulting in significant influences on SVU.

Ngo and Hsin (2021) conducted an analysis of a long-term satellite SST dataset covering the period from 1982 to 2019. Their focus was on investigating the interannual variations in summer upwelling along the southern coast of Vietnam. The authors identified three distinct regions of upwelling: the Southern Coastal Upwelling (SCU) located south of 12.5°N, the Northern Coastal Upwelling (NCU) situated north of 12.5°N, and the Offshore Upwelling (OU) positioned east of 110°E. To quantify the interannual variabilities, the authors developed an adaptive SST-based upwelling index. Their findings revealed that the variations in these coastal upwelling systems were primarily influenced by the alongshore wind stress, wind stress curl dipole, and the associated currents. These factors played key roles in driving the observed interannual fluctuations. By utilizing the satellite SST data and introducing an adaptive upwelling index, Ngo and Hsin (2021) provides valuable insights into the interannual variations in summer upwelling off the southern coast of Vietnam. Understanding the dynamics and drivers of these coastal upwelling systems contributes to our knowledge of regional oceanic processes and has implications for various marine ecosystems in the area.

Recent studies have also investigated seasonal and interannual variations in other upwelling systems within the South China Sea region. For instance, Zhu et al. (2023) explored interannual variations in the coastal upwelling around Hainan Island using satellite SST data from 2003 to 2021. Their analysis revealed significant interannual fluctuations in this coastal upwelling system. In another study, Sun et al. (2016) analyzed the seasonal and interannual variability of upwelling along the southern East China Sea coast (ECSC). They used observed wind data to calculate the upwelling index based on Ekman transport dynamics and mass continuity equations. Their findings suggested that upwelling is crucial for local productivity, as evidenced by the high correlation between the upwelling index and observed chlorophyll concentration. Additionally, they noted that large-scale temporal variations in ECSC upwelling are associated with natural climate modes such as ENSO and Pacific Decadal Oscillation (PDO). Through these investigations, we gain a better understanding of the dynamics of upwelling systems in the South China Sea and their relationship to regional oceanic and climatic variability. Such information can aid in managing marine ecosystems and forecasting responses to changes in both natural and anthropogenic factors.

3.2 Long-term variations in boundary upwelling

Over the past decade, there has been a significant focus on researching the long-term intensification or weakening of upwelling systems, which has the potential to have remarkable impacts on regional hydrodynamics and ecosystems. One prominent hypothesis that has motivated much of this research is Bakun's hypothesis, proposed by Bakun (1990). Bakun's hypothesis suggests that the accumulation of greenhouse gases can lead to stronger alongshore wind stress, subsequently intensifying EBUS based on Ekman's dynamics. This hypothesis has garnered widespread attention and interest in understanding the long-term trends of EBUS under past, present, and future climate change scenarios. The study of long-term upwelling trends has become a hot topic due to its implications for marine ecosystems and the need to assess the potential impacts of climate change. Investigating the dynamics of upwelling systems over periods longer than 10 years provides valuable insights into the response of coastal regions to environmental shifts. By studying the intensification or weakening of EBUSs, researchers aim to understand the complex relationship between greenhouse gas accumulation, wind stress, and the resulting changes in upwelling patterns.

To test the canonic Bakun's hypothesis, Sydeman et al. (2014) conducted a meta-analysis of the research on the topic of coastal wind intensification along the major EBUSs. This analysis revealed that most prior studies on upwelling-favorable wind demonstrated its intensification over time scales ranging up to 60 years in all EBUSs except the Iberian one, where wind was weakening. The authors concluded that this comprehensive analysis supported Bakun's hypothesis. Furthermore, this study noted that the upwelling intensification was more pronounced at higher latitudes. To illustrate a typical example of wind intensification, Varela et al. (2015) examined changes in coastal upwelling between 1982 and 2010 across worldwide coasts from the perspective of alongshore upwelling-favorable wind. Their findings showed that alongshore upwelling-favorable wind significantly intensified during this period, particularly along EBUS and the western Australian and southern Caribbean coasts.

Two accompanying studies led by Polonsky and Serebrennikov (2021a, 2021b) employed satellite SST and wind datasets to examine the long-term changes in upwelling intensity along the EBUSs in the Atlantic and Pacific Oceans, respectively. The findings revealed a consistent intensification of upwelling in these EBUSs since the 1980s. However, it is important to note that the rate of intensification was not consistently linear. Specifically, the Atlantic EBUSs, namely, the Canary and Benguela upwelling systems, exhibited a cessation of upwelling intensification since the late 1990s. Similarly, the Chile upwelling system in the Pacific Ocean showed a lack of intensification since approximately 2008. This study suggests that in addition to long-term impacts from global warming, natural interdecadal variations may play a significant role in regulating coastal upwelling in major EBUSs. These natural variability patterns could potentially influence the overall trend of upwelling intensification observed over the past few decades.

In the study conducted by Bonino et al. (2019), an eddy-permitting global ocean model was employed to investigate the low-frequency variability within the four EBUSs from 1958 to 2015. The findings revealed distinct differences in both the variability patterns and underlying dynamical mechanisms among the four EBUSs. Comparing the Pacific Ocean's California and Humboldt upwelling systems, the only consistent variability observed was attributed to the El Niño-Southern Oscillation (ENSO). In contrast, the variability in the two Atlantic EBUSs exhibited a predominant association with long-term trends, and these trends were found to be independent of each other. This research highlights the unique characteristics of each EBUS and underscores the need to recognize the specific drivers of variability within different systems. The presence of coherent ENSO-related variability in the Pacific EBUSs suggests a strong influence of large-scale climate phenomena.

In line with numerous previous studies, Aguirre et al. (2018) demonstrated that the upwelling-favorable winds have intensified at the higher latitudes of the Humboldt upwelling system in recent decades. Additionally, there has been a notable increase in chlorophyll-a levels over the past two decades, likely resulting from the intensified upwelling.

Some studies, however, did not observe an intensification in the EBUS upwelling. A typical study is conducted by Barton et al. (2013), who examined historical trends in the Canary upwelling system by using multiple wind and SST datasets spanning primarily from the 1960s to the 2000s. These datasets did not present consistent evidence supporting the intensification of upwelling along the coasts of Northwest Africa or Iberia. Consequently, this particular study does not lend support to Bakun's hypothesis.

Research efforts have also been directed toward studying coastal upwelling in marginal seas. Xu et al. (2021) investigated the Vietnam coastal upwelling and observed a rapid intensification since the 1950s, as evidenced by sedimentary changes in a marine core collected from the Vietnamese coastline. The authors attributed this intensified upwelling to global warming, which they suggest increased summer winds and in turn amplified the upwelling process.

In a study conducted by Zhang (2021), the focus was placed on the long-term changes in summer upwelling within the Taiwan Strait over the course of four decades, from 1982 to 2019. The analysis involved examining SST and wind data. Despite experiencing nonlinear fluctuations, the study found a remarkable decrease in upwelling intensity in the Taiwan Strait, estimated at approximately 35%, after the year 2000 when compared to the period prior to 2000. This significant weakening of upwelling was primarily attributed to the reduction in Ekman transport, which is responsible for driving the upward movement of colder and nutrient-rich water toward the surface. The author observed a robust negative correlation between the occurrence of ENSO Modoki events (central Pacific ENSO) and the intensity of summer upwelling in the Taiwan Strait.

For the coastal upwelling around the South China Sea, Liu et al. (2023) investigated the long-term trends of coastal upwelling around the South China Sea region using satellite and reanalysis SST (SST) data spanning from 1982 to 2020. The authors discovered a significant intensification in all three typical upwelling systems situated along the eastern Guangdong, eastern Hainan, and eastern Vietnam coasts, as depicted in Fig. 11.8. An analysis of wind data revealed that the increase in wind stress was dominant in driving the upwelling intensification of the first two regions. Meanwhile, both wind stress curl and alongshore wind stress were identified as contributing factors behind the increased upwelling effect observed along the eastern Vietnam coast.

4. Will boundary upwelling intensify in the 21st century

The escalating issue of global warming has raised significant concerns across various fields, including rising sea levels, extreme MHWs, and severe threats to ecosystems. Within the oceanography and climate science communities, the projection of

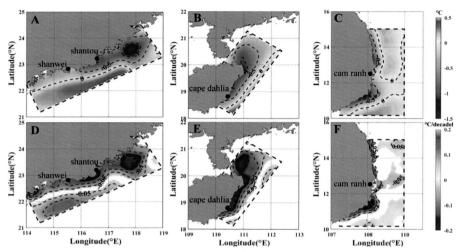

Figure 11.8 Spatial distribution of time-averaged TPI (topographic position index) and trends calculated from OSTIA data in June and July during 1982−2020. Climatologic TPI of UEG (a), of UEH (b), and of UEV (c); TPI trend of UEG (d), of UEH (e), and of UEV (f). Only the area that passes the 95% significance test is shown in (d, e). UEG means eastern of Guangdong; UEH means eastern coast of Hainan; UEV means southeastern coast of Vietnam.
Sourced from Liu, S., Zuo, J., Shu, Y., Ji, Q., Cai, Y., & Yao, J. (2023). The intensified trend of coastal upwelling in the South China Sea during 1982-2020. Frontiers in Marine Science, 10, 1084189. https://doi.org/10.3389/fmars.2023.1084189.

upwelling in EBUSs under future warming has become an extensively debated topic over the past few decades. Numerous studies have been conducted to provide insights into this question. Traditionally, the intensification of alongshore wind stress has served as a widely utilized metric to indicate upwelling intensification since wind stress is generally the primary driving force behind upwelling in EBUSs. However, wind stress alone offers an incomplete understanding of the upwelling mechanism. Consequently, some studies have employed alternative methods, such as utilizing oceanic vertical velocity directly obtained from climate models, to calculate upwelling. These different data sources and methodologies have yielded mixed results regarding the future projection of upwelling in EBUSs. In this discussion, we will introduce several notable studies that have explored the potential intensification of boundary upwelling in the future. Additionally, we will delve into research advancements concerning the climate and ecological impacts associated with variations in future upwelling patterns.

4.1 Projections of boundary upwelling in the 21st century

The intensification of boundary upwelling, particularly along EBUSs, has been a topic of extensive debate since Bakun's seminal work in 1990. Understanding the potential intensification of boundary upwelling under various scenarios has garnered significant attention in the past decade. Some studies have indicated that the upwelling-favorable wind will indeed increase due to global warming, lending support to Bakun's hypothesis of enhanced boundary upwelling. However, recent research has shed light on the role of nonwind factors that may influence the response of coastal upwelling to climate change. In this concise literature review, we provide an overview of studies conducted since 2013, highlighting the recent progress made in this dynamic field. Although not exhaustive, this review aims to discuss key findings and advancements in understanding the complex interplay between climate change and boundary upwelling.

Bakun's hypothesis has received support from numerous studies. For instance, García-Reyes et al. (2015) conducted an analysis of various studies focused on EBUSs. They argued that the alongshore wind stress at the high latitudes of each EBUS has intensified and is projected to continue intensifying in the future due to global warming. These changes are believed to be correlated with the meridional movement of atmospheric high-pressure cells. However, it is important to note that an increase in upwelling intensity does not necessarily translate to enhanced productivity. The warming of the ocean leads to an increase in vertical stratification, which can partially impede the upward transport of nutrient-rich deep water. As a result, the effects of upwelling intensification may be counteracted to some extent. It is crucial to consider both the intensification of upwelling and the potential limitations imposed by ocean warming-induced stratification in order to accurately assess the overall impact on productivity. This highlights the complexity involved in understanding the response of marine ecosystems to changes in upwelling patterns under global warming.

Miranda et al. (2013) provide further evidence in support of Bakun's hypothesis. In their study, they examined the upwelling system near the Iberian coast using two 30-year simulations based on a regional ocean model with a horizontal resolution of 1/12 degrees. The simulations were compared between a historical period (1961−90) and a future period (1971−2100). Their findings showed that the future scenario is expected to experience significantly stronger upwelling, particularly along the northern coast, compared to the historical scenario. This provides additional support for the potential intensification of coastal upwelling due to global warming.

Wang et al. (2015) represents another typical study that reinforces Bakun's hypothesis. The researchers conducted an investigation on the projections of four EBUSs under the RCP8.5 scenarios. They utilized daily wind stress data from 22 climate models participating in the Coupled Model Intercomparison Project Phase 5 (CMIP5). This wind stress data were then used to calculate the offshore Ekman transport, serving as an indicator of upwelling. The study's findings indicated that the EBUSs off the Canary, Benguela, and Humboldt coasts are projected to exhibit earlier onset, prolonged duration, and increased intensity at higher latitudes, while the lower latitudes may experience less noticeable changes. This spatially uneven trend could potentially reduce the characteristic latitudinal variations observed in each EBUS (refer to Fig. 2 in their study). Regarding the upwelling system off the California coast, the authors suggested that the lack of a significant trend might be attributed to atmospheric feedback in response to climate change in the future. Overall, the study by Wang et al. (2015) demonstrates a strong connection between the intensification of upwelling and the escalating greenhouse warming, thus providing substantial support for Bakun's classical hypothesis.

Sousa et al. (2017) further contributed to the evidence supporting the future intensification of upwelling along the EBUSs in the 21st century, with the western Iberian Peninsula as an example. Their study utilized a regional climate model that incorporated the RCP8.5 emission scenario. Through their analysis, the researchers calculated upwelling indices and observed a significant increase in upwelling, particularly at high latitudes in the region. This change was primarily attributed to the displacement of the Azores High. It is projected that the Azores High will undergo intensification at a rate of 0.03 hPa per decade and drift northeastward at a rate of 10 km per decade throughout the 21st century.

A recent study by Xiu et al. (2018) employed a high-resolution coupled physical—biological model with a horizontal resolution of 7 km to assess the response of California's coastal upwelling ecosystems to future global warming under the RCP8.5 scenario. The authors defined two distinct time periods, 1st period spanning 1990—2009 and 2nd period spanning 2030—49. Their findings showed that stronger alongshore winds will promote coastal upwelling (<150 km from the coast) at a depth of 100 m by 20% in central California's upwelling regions during 2nd period, compared to 1st period. Further offshore (150—800 km), their results indicated that the wind-curl-induced Ekman pumping dominated the changes in upwelling at 100 m depth. As the open ocean warms, the isotherms will move downward, resulting in an increase in nutrient concentration in the upward water masses. However, despite these intensified upwelling events and more nutritious water masses, there will not be a linear increase in the plankton community. The plankton community responds to increased nutrient inputs in a complex and nonlinear manner. Fig. 11.9 outlines the key findings of this study.

However, concerns have been raised regarding the intensification of upwelling under future global warming. Shortly after the publication of Wang et al.'s (2015) study, Di Lorenzo (2015) expressed concerns about the findings and conclusions presented in Wang et al. (2015). Di Lorenzo (2015) pointed out that the conclusions of Wang et al. heavily relied on wind-based upwelling indices, neglecting other physical processes that contribute to upwelling but are not wind-driven. This limitation in Wang et al. (2015) study raises questions about the discovered upwelling intensification in the future.

In contrast to the findings of Wang et al. (2015), Rykaczewski et al. (2015) conducted a study that tested Bakun's hypothesis using an ensemble of ocean-atmosphere models. Their research revealed that an increase in the temperature difference between land and sea does not necessarily lead to an intensification of upwelling-favorable winds. Specifically, their study projected that winds in the summer season would strengthen at higher latitudes within climatological upwelling regions while weakening at lower latitudes. This observed pattern is attributed to the poleward displacement of major atmospheric high-pressure cells, which aligns with the findings of García-Reyes et al. (2015). The wind intensification detected at higher latitudes corresponds with the majority of other studies, including Wang et al. (2015).

Several other studies also challenge the idea that wind stress in EBUSs will intensify under global warming. For instance, Belmadani et al. (2014) utilized dynamical downscaling to examine past, present, and future trends in the Peru—Chile upwelling

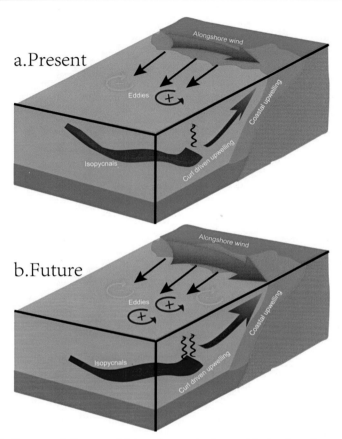

Figure 11.9 Conceptual diagram of future changes in deep nutrients, alongshore winds, wind stress curl, coastal upwelling, and eddy activity. In the future, differences in isopycnal deepening between the gyre and coastal regions mean that more nutrients are transported to the California current system. Enhanced local alongshore winds promote coastal upwelling, and changes in mesoscale eddies and wind stress curl promote upward nutrient flux in the offshore region.

Sourced from Xiu, P., Chai, F., Curchitser, E. N., & Castruccio, F. S. (2018). Future changes in coastal upwelling ecosystems with global warming: The case of the California Current System. Scientific Reports, 8, 2866. https://doi.org/10.1038/s41598-018-21247-7.

system (PCUS). They utilized an atmospheric model on a stretched grid to simulate changes in winds under different climate scenarios. The results showed a strengthening of winds off central Chile south of 35°S (at 30°S-35°S) in austral summer (winter) and a weakening elsewhere. These findings were contrary to Bakun's (1990) hypothesis, which proposed that wind weakening (strengthening) off Peru and northern Chile (off central Chile) would be associated with cross-shore gradients. Belmadani et al. (2014) found that alongshore pressure gradients, rather than cross-shore gradients, were responsible for the observed wind weakening or strengthening. Moreover, the authors demonstrated that a larger land-sea temperature gradient did not necessarily promote alongshore wind stress. The study suggests that the poleward displacement and intensification of the South Pacific Anticyclone is the primary mechanism behind the wind stress off Chile, while increased precipitation over the tropics and associated convective anomalies may contribute to the weakening of wind stress off Peru. Overall, this research does not support Bakun's hypothesis.

The future trajectory of EBUS upwelling is expected to exhibit an intricate pattern. Sousa et al. (2017) examined wind and SST data from both regional and global climate models to investigate the potential climate response of the Canary Upwelling Ecosystem. Their findings revealed that this upwelling system would undergo seasonal variations in its long-term trend as a result of future warming. Specifically, the study indicated that the upwelling intensity would strengthen during the period from October to April, while weakening from May to August. This seasonal weakening of the upwelling is projected to contribute to higher rates of warming from May to August in the future.

The majority of the previous studies examining coastal wind projections in EBUSs have utilized relatively coarse-resolution models. In light of this limitation, Chamorro et al. (2021) developed regional atmospheric models with three nested levels, with the innermost model featuring a horizontal resolution of 7 km specifically for the Peruvian upwelling region. Their study revealed that both climate change and the alongshore SST gradient may impact the coastal wind in this upwelling region. Overall, the results showed that under the RCP8.5 scenario, there would be a moderate weakening of summer coastal wind off Peru, but an increase in coastal wind off central Peru. This reduction in summer wind intensity could result in weakened upwelling and a less productive summer season along the Peruvian coast.

While Wang et al. (2015) presented compelling findings, it is worth noting that the study primarily relied on the derived Ekman transport from the wind stress of global climate models, which tend to have relatively coarse horizontal resolutions. As such, other dynamic processes may compensate for or increase the offshore water transport via the Ekman transport, leading to complex variations in future EBUSs. Jacox et al. (2018) attempted to address this issue by incorporating the geostrophic current in their investigation of upwelling off the west coast of the USA. Their findings indicate that the geostrophic current may play a crucial role in regulating the intensity and variation of upwelling (see their Fig. 1). This suggests that future studies should consider a wider range of dynamic processes and physical factors when examining upwelling in EBUSs, in order to better understand and predict changes in these important ecosystems.

Similarly, Ding et al. (2021) conducted a study that identified the wind-driven Ekman transport and the geostrophic flow as the two primary processes influencing future changes in coastal upwelling along the California coast. Furthermore, they highlighted the potential for the geostrophic transport to play an even more significant role than the wind-driven Ekman transport along certain sections of the California coast. Building upon this research, a more recent study by Jing et al. (2023) reevaluated the projections of four EBUSs under a high-emission scenario using an eddying global climate model. Their findings presented a distinct perspective, demonstrating that the long-term upwelling trend in these EBUSs is predominantly governed by the geostrophic flow. Consequently, this resulted in weakened or unchanged upwelling conditions, despite the presence of intensified upwelling-favorable winds, as depicted in Fig. 11.10. This study serves as a crucial complement to Wang et al. (2015), offering valuable insights into the verification of Bakun's hypothesis.

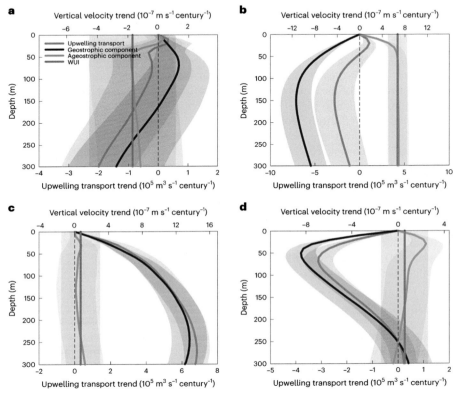

Figure 11.10 Effects of geostrophic flows on the long-term changes of upwelling transport in the EBUSs under greenhouse warming. (a–d) The slope of the linear trend of annual-mean upwelling transport (red) during 2006–2100 and its decomposition into components caused by horizontal mass-flux divergence of geostrophic (purple) and ageostrophic (blue) flows for one member of CESM-H in the CalCS (a), HCS (b), CanCS (c) and BCS (d). The gray line denotes the slope of the linear trend of WUI (total Ekman transport into/out of the integration region). The shading denotes the 90% confidence interval for the slope. The axes of the vertical velocity trend measure the values of the upwelling transport trend divided by the integration area. The sample size (the number of time records used to derive the confidence interval) is 95. CalCS means California current system; HCS means Humboldt current system; CanCS means Canary current system; BCS means Benguela current system. CESM-H means community Earth system model with the high-resolution.
Sourced from Jing, Z., Wang, S., Wu, L., Wang, H., Zhou, S., Sun, B., Chen, Z., Ma, X., Gan, B., & Yang, H.: Geostrophic flows control future changes of oceanic eastern boundary upwelling. Nature Climate Change, 13, 148–154. https://doi.org/10.1038/s41558-022-01588-y.

Furthermore, other studies have also highlighted a curious phenomenon where alongshore wind stress intensifies, but coastal upwelling weakens. Sousa et al. (2020) conducted a comparative analysis of upwelling along the northwest Iberian Peninsula during historical (1976–2005) and future (2070–99) periods, with the latter simulated using the forcing under the RCP8.5 scenario. The study projected that while the alongshore upwelling-favorable wind will indeed intensify in the future period, the upwelling itself would become weaker due to increased stratification caused by ocean warming. This paradoxical phenomenon underscores the complex interactions and feedback loops between physical processes in coastal regions.

A comparable conclusion was reached by Oyarzún and Brierley (2019) when examining the projection of the Humboldt coastal upwelling system under the RCP8.5 scenario. In a manner akin to Liao, Gao, et al. (2022), this study utilized oceanic vertical velocity data derived from global climate models participating in CMIP5. This approach offers a direct representation of upwelling dynamics. The findings indicated that although the alongshore wind stress was projected to increase, the upwelling itself did not necessarily exhibit the same trend. The underlying reason for this discrepancy lies in the heightened vertical stratification resulting from ocean warming, which serves as a limiting factor for upwelling intensification.

Prior studies have extensively investigated the potential future changes in coastal upwelling across marginal seas, and several findings have emerged supporting Bakun's hypothesis. A notable illustration is the research conducted by deCastro et al. (2016). Their study focused on examining the projections of upwelling along the Somali coast under two distinct scenarios: RCP4.5 and RCP8.5. Through an analysis of an ensemble of global and regional climate models, the authors revealed a projected intensification of coastal upwelling at a rate even higher than that observed along major EBUSs. Intriguingly, they discovered that this intensification rate increased as latitude increased. The authors attributed this phenomenon to global warming, which resulted in larger land-sea thermal and pressure gradients, consequently augmenting alongshore winds and promoting coastal upwelling. This finding closely aligned with Bakun's hypothesis. Furthermore, similar observations of intensified coastal upwelling have been made in other studies. For instance, Praveen et al. (2016) compared numerical simulations of the upwelling system along the Oman coast between two distinct periods: 1981–2000 and 2080–99. The former represented historical forcing, while the latter employed the RCP8.5 scenario forcing. Their results distinctly demonstrated an intensification of the upwelling system as a direct response to the poleward shift in the monsoon low-level jet. This intensified upwelling is anticipated to enhance marine productivity accordingly. These collective findings provide valuable insights into the potential impacts of climate change on coastal upwelling dynamics and underscore the importance of understanding these changes for managing and preserving marine ecosystems and resources.

4.2   Climate consequences of future changes in the boundary upwelling

Upwelling involves the ascent of deep, cold water toward the ocean surface, resulting in a significant reduction in upper-ocean temperature. This process can have impactful consequences on the ecological environment. Traditionally, it has been believed that upwelling regions serve as critical sanctuaries for marine life in the face of global warming. Through upwelling's cooling effect on the ocean surface, it creates an ideal habitat for phytoplankton, which forms the basis of the marine food chain. The resultant increase in primary productivity augments the abundance and diversity of marine species, making upwelling areas important sites for commercial fishing.

Bakun et al. (2015) examined the potential implications of climate change on ecosystems in EBUSs, which are particularly sensitive to changes in coastal upwelling. The study revealed that ecosystem productivity in these regions is threatened by climate change, as upwelling intensities and offshore advection rates are expected to increase during spring and summer months. While this could offset warming habitat effects, it may also lead to more frequent hypoxic events and lower densities of food particles suitable for fish larvae. Moreover, intensified upwelling will result in increased ocean acidity levels, thus adversely affecting organisms with carbonate structures. Moreover, even without changes in upwelling, large-scale climate episodes such as ENSO (El Niño–Southern Oscillation) could affect near-surface stratification, turbulent diffusion rates, source water origins, and thermocline depths, potentially causing significant ecological impacts. Major impacts on pelagic fish resources seem unlikely unless coupled with overfishing, but compositional shifts toward subtropical communities are probable. Additionally, marine mammals and seabirds associated with limited nesting or resting grounds could face difficulties in accessing prey resources or adaptively respond by migrating to more favorable biogeographical regions. In conclusion, Bakun's study highlights the impact of climate change on EBUSs and emphasizes the need for conservation strategies to preserve marine ecosystems and resources.

Varela et al. (2022) examined the potential impacts of future variations in the Canary upwelling system on SST under a business-as-usual emission scenario. Their findings revealed a notable shift in the meridional position of the Canary upwelling system. The study indicated that this change in the upwelling system would have contrasting effects on the coastal region. In the middle and northern parts of the Canary Current System, the upwelling system's buffering effect on surface warming would be enhanced. This means that these areas would experience a more effective mitigation of surface warming due to the intensified upwelling. Conversely, the southern part of the coastal region would face a reduced buffering effect of upwelling due to a decrease in upwelling intensity. As a result, this region is expected to be more vulnerable to the impacts of surface warming.

The coupled ocean-biogeochemical model developed in Pozo Buil et al. (2021) projected the future trends of upwelling and chlorophyll levels in the California upwelling region. The researchers discovered a robust correlation between the changes in upwelling and chlorophyll concentrations. Specifically, the study revealed that in the northern part of the upwelling region, both upwelling and chlorophyll concentrations are projected to increase. This suggests a positive relationship, where intensified upwelling leads to higher chlorophyll levels, indicating enhanced productivity in that area. On the other hand, in the southern part of the upwelling system, a weakening of upwelling is anticipated. As a result, the study predicts a decrease in chlorophyll levels in this region. This negative correlation indicates a reduced productivity associated with the weakened upwelling.

5. Conclusions

In this chapter, we introduced and discussed the progresses that have been made over the last decade on the boundary upwelling. The principal findings of this literature review are listed as follows.

(1). There has been a growing recognition of the limitations posed by relying on a single metric derived solely from wind stress or SST in assessing upwelling processes. In response, researchers have proposed novel upwelling metrics that consider multiple factors. For example, an upwelling index has been introduced that combines wind stress and SST data, providing a more comprehensive understanding of upwelling conditions. Furthermore, advancements in technology and data availability have enabled the direct calculation of upwelling using oceanic vertical velocity. This approach utilizes three-dimensional velocity data obtained from ocean models or reanalysis, allowing for a potentially more accurate and robust estimation of upwelling intensity. By incorporating additional variables and utilizing multiple data sources, these new upwelling metrics provide a more holistic assessment of upwelling processes and their associated impacts on marine ecosystems.

(2). A recent research study has shed light on the presence of robust upwelling systems in the subsurface waters of major western boundary currents (WBCs). While previous studies have provided insights through isolated cases, the persistent nature of subsurface upwelling has historically been underestimated. This underappreciation can be attributed to the complexity of the underlying dynamics and limited data availability. The findings of this new study reveal that the subsurface upwelling occurring in WBCs is not only stronger but also deeper than the upwelling systems observed in EBUSs. The driving force behind this phenomenon is believed to be the meridional density gradient along the WBCs. Importantly, these discoveries highlight the limitations associated with relying solely on wind stress or SST to identify and analyze upwelling processes. The traditional metrics derived from wind stress or SST fail to capture the full extent and complexity of subsurface upwelling in WBCs. By uncovering the significant role of subsurface upwelling in WBCs and emphasizing its distinct characteristics, this study emphasizes the need for a more comprehensive approach when studying and evaluating upwelling systems. Incorporating the meridional density gradient and considering subsurface dynamics are crucial for a comprehensive understanding of the upwelling processes occurring in WBCs.

(3). Boundary upwelling is widely acknowledged for its ecological significance. It facilitates the vertical transport of nutrients, which in turn promotes primary productivity in marine ecosystems. However, it is important to recognize that upwelling can also have negative consequences, such as the occurrence of hypoxic events. In light of the ongoing warming trends, upwelling regions have been considered potential refuges for marine organisms in certain studies. This perspective stems from the belief that these regions may provide relatively cooler and more favorable conditions amid rising temperatures. However, recent findings indicate that EBUSs in the southern hemisphere might actually become hotspots for MHWs. The projected increase in MHW events in EBUSs is attributed to the weakening of horizontal boundary currents, primarily driven by global warming. As these currents weaken, the capacity to transport cooler waters to the surface diminishes, making EBUSs more susceptible to prolonged periods of elevated temperatures. This emerging understanding challenges the previous notion of upwelling regions as solely beneficial for marine life under a warming climate. Instead, it highlights the potential risks associated with MHWs in EBUSs, emphasizing the need to consider the complex interactions between upwelling dynamics and climate change. Moving forward, further research is required to comprehensively assess the ecological consequences of these changing conditions in upwelling regions. Such studies will support the development of effective management strategies to mitigate the ecological impacts and enhance the resilience of marine ecosystems in the face of global warming.

(4). In the past decade, considerable research has focused on predicting the future intensification of upwelling systems. Specifically, the impact of higher CO_2 emissions scenarios on major EBUSs has garnered significant attention. It is generally anticipated that under these scenarios, alongshore wind stress in EBUSs will increase, particularly at higher latitudes. However, it is worth noting that the California Current System (CCS) may exhibit different behavior compared to other EBUSs. Some studies have indicated that the wind stress in the CCS may not follow the same upward trend as observed in other regions. Furthermore, recent research incorporating higher-resolution data, particularly direct oceanic vertical velocity data, has shed new light on this matter. These recent studies suggest that factors beyond wind stress alone might significantly influence the climate response of EBUS upwelling. Of particular importance is the modulation by the geostrophic current, which can potentially lead to negligible changes or even weakening of EBUS upwelling despite enhanced alongshore wind stress.

Boundary upwelling has emerged as a topic of great interest within the oceanography and climate communities over the past decade, with significant progresses being made in this field. Nonetheless, there remains much to be explored in order to gain a more comprehensive understanding of boundary upwelling dynamics. To this end, we outline some areas of potential research that could prove crucial in the future.

(1). Compared to EBUSs, the upwelling in western boundary currents (WBCs) has received relatively less research attention. Although some studies have elucidated the dynamical mechanisms of subsurface upwelling in the Gulf Stream using a volume flux budget analysis, these investigations often overlooked the impact of other crucial factors. Therefore, further research is needed to develop a more comprehensive understanding of subsurface upwelling in WBCs. One of the factors that require detailed investigation is the role of mesoscale eddies in subsurface upwelling formation, maintenance, and relaxation. Mesoscale eddies greatly influence oceanic circulation and therefore have the potential to significantly impact the upwelling processes in WBCs. Investigating the interactions between mesoscale eddies and upwelling could provide valuable insights into the dynamics of these regions. Additionally, it is important to consider the impact of other physical factors on subsurface upwelling in WBCs, such as variations in water temperature, salinity, and density.

These factors can significantly influence the vertical advection of nutrients and have a profound effect on the ecology of upwelling regions.

(2). While much attention has been given to future projections of upwelling along eastern boundary systems, there has been relatively limited research into the future of subsurface upwelling in western boundary currents (WBCs) under global warming. As the ocean continues to warm, it is critical to study how these changes will affect subsurface upwelling, which plays a vital role in ocean ecology. Another important question that requires exploration is whether there is a correlation between the observed slowdown of the meridional overturning circulation and a weakening of subsurface upwelling in WBCs. This could provide valuable insights into the complex interactions between oceanic circulation and upwelling processes. Understanding the future of subsurface upwelling in WBCs is crucial as it has significant implications for regional fisheries, ecosystems, and climate.

(3). The use of oceanic vertical velocity as a diagnostic tool for studying boundary upwelling has gained increasing popularity due to its superior performance compared to proxy-based methods. However, current research primarily relies on global climate models, which may not capture fine-scale details accurately. Therefore, there is a pressing need to employ regional coupled models with higher resolutions to uncover more intricate aspects of boundary upwelling dynamics. By utilizing regional coupled models, we can enhance our understanding of the complex processes underlying boundary upwelling on a smaller scale. These models provide more localized information, allowing us to explore the specific characteristics and mechanisms of upwelling in different regions more accurately. Fine-resolution models can better capture the influence of coastal features, such as coastline geometry and bathymetry, which have significant implications for upwelling dynamics. Furthermore, regional coupled models enable us to investigate the interactions between local atmospheric conditions and oceanic dynamics, providing a more comprehensive understanding of the factors driving boundary upwelling. This integrated approach will offer valuable insights into the complex feedbacks and processes that govern upwelling phenomena.

(4). Considerable efforts have been dedicated to investigating the impacts of boundary upwelling on MHWs. However, further research is needed to comprehensively study all major upwelling systems, including both the subsurface upwelling in western boundary currents (WBCs) and the upwelling occurring along the marginal seas. To gain a more complete understanding of the relationship between boundary upwelling and MHWs, it is crucial to explore the specific characteristics and mechanisms underlying each upwelling system. This includes assessing how subsurface upwelling in WBCs influences the occurrence and intensity of MHWs, as well as examining the role of upwelling in the marginal seas. In-depth investigations into these systems will provide valuable insights into the complex interactions between upwelling processes and the development of MHWs. By considering the unique oceanic and atmospheric conditions associated with each region, we can gain a more holistic understanding of the impacts of boundary upwelling on MHWs. Furthermore, expanding research to encompass the marginal seas is important as these areas often exhibit distinct physical and ecological characteristics. Investigating the influence of upwelling in these regions on the occurrence and duration of MHWs will enhance our knowledge of the broader impacts of boundary upwelling on marine ecosystems.

References

Abrahams, A., Schlegel, R. W., & Smit, A. J. (2021). Variation and change of upwelling dynamics detected in the world's eastern boundary upwelling systems. *Frontiers in Marine Science, 8*. https://doi.org/10.3389/fmars.2021.626411

Aguiar, A. L., Cirano, M., Pereira, J., & Marta-Almeida, M. (2014). Upwelling processes along a western boundary current in the Abrolhos–Campos region of Brazil. *Continental Shelf Research, 85*, 42–59. https://doi.org/10.1016/j.csr.2014.04.013

Aguirre, C., García-Loyola, S., Testa, G., Silva, D., & Farías, L. (2018). Insight into anthropogenic forcing on coastal upwelling off south-central Chile. *Elementa: Science of the Anthropocene, 6*, 59. https://doi.org/10.1525/elementa.314

Bakun, A., Black, B. A., Bograd, S. J., García-Reyes, M., Miller, A. J., Rykaczewski, R. R., & Sydeman, W. J. (2015). Anticipated effects of climate change on coastal upwelling ecosystems. *Current Climate Change Reports, 1*, 85–93. https://doi.org/10.1007/s40641-015-0008-4

Bakun, A. (1990). Global climate change and intensification of coastal ocean upwelling. *Science, 247*, 198–201. https://doi.org/10.1126/science.247.4939.198

Barton, E. D., Field, D. B., & Roy, C. (2013). Canary current upwelling: More or less? *Progress in Oceanography, 116*, 167–178. https://doi.org/10.1016/j.pocean.2013.07.007

Belmadani, A., Echevin, V., Codron, F., Takahashi, K., & Junquas, C. (2014). What dynamics drive future wind scenarios for coastal upwelling off Peru and Chile? *Climate Dynamics, 43*, 1893–1914. https://doi.org/10.1007/s00382-013-2015-2

Bograd, S. J., Jacox, M. G., Hazen, E. L., Lovecchio, E., Montes, I., Pozo Buil, M., Shannon, L. J., Sydeman, W. J., & Rykaczewski, R. R. (2023). Climate change impacts on eastern boundary upwelling systems. *Annual Review of Marine Science, 15*, 303–328. https://doi.org/10.1146/annurev-marine-032122-021945

Bonino, G., Di Lorenzo, E., Masina, S., & Iovino, D. (2019). Interannual to decadal variability within and across the major eastern boundary upwelling systems. *Scientific Reports, 9*, Article 19949. https://doi.org/10.1038/s41598-019-56514-8

Carr, M.-E., & Kearns, E. J. (2003). Production regimes in four eastern boundary current systems. *Deep Sea Research Part II: Topical Studies in Oceanography, 50*, 3199–3221. https://doi.org/10.1016/j.dsr2.2003.07.015

Castelao, R. M., & Barth, J. A. (2006). Upwelling around Cabo Frio, Brazil: The importance of wind stress curl. *Geophysical Research Letters, 33*. https://doi.org/10.1029/2005GL025182

Chamorro, A., Echevin, V., Dutheil, C., Tam, J., Gutiérrez, D., & Colas, F. (2021). Projection of upwelling-favorable winds in the Peruvian upwelling system under the RCP8.5 scenario using a high-resolution regional model. *Climate Dynamics, 57*, 1–16. https://doi.org/10.1007/s00382-021-05689-w

Chen, C.-C., Shiah, F.-K., Gong, G.-C., & Chen, T.-Y. (2021). Impact of upwelling on phytoplankton blooms and hypoxia along the Chinese coast in the East China Sea. *Marine Pollution Bulletin, 167*, Article 112288. https://doi.org/10.1016/j.marpolbul.2021.112288

Chen, Y., Shi, H., & Zhao, H. (2021). Summer phytoplankton blooms induced by upwelling in the western south China sea. *Frontiers in Marine Science, 8*, Article 740130.

Chen, Z., Pan, J., Jiang, Y., & Lin, H. (2017). Far-reaching transport of Pearl River plume water by upwelling jet in the northeastern South China Sea. *Journal of Marine Systems, 173*, 60–69. https://doi.org/10.1016/j.jmarsys.2017.04.008

Cheresh, J., Kroeker, K. J., & Fiechter, J. (2023). Upwelling intensity and source water properties drive high interannual variability of corrosive events in the California Current. *Scientific Reports, 13*, Article 13013. https://doi.org/10.1038/s41598-023-39691-5

Cresswell, G. R., Peterson, J. L., & Pender, L. F. (2017). The East Australian Current, upwellings and downwellings off eastern-most Australia in summer, Mar. *Freshwater Research, 68*, 1208−1223.

Da, N. D., Herrmann, M., Morrow, R., Niño, F., Huan, N. M., & Trinh, N. Q. (2019). Contributions of wind, ocean intrinsic variability, and ENSO to the interannual variability of the South Vietnam upwelling: A modeling study. *Journal of Geophysical Research: Oceans, 124*, 6545−6574. https://doi.org/10.1029/2018JC014647

deCastro, M., Sousa, M. C., Santos, F., Dias, J. M., & Gómez-Gesteira, M. (2016). How will Somali coastal upwelling evolve under future warming scenarios? *Scientific Reports, 6*, Article 30137. https://doi.org/10.1038/srep30137

Di Lorenzo, E. (2015). The future of coastal ocean upwelling. *Nature, 518*, 310−311. https://doi.org/10.1038/518310a

Dickson, R. R., Meincke, J., Malmberg, S.-A., & Lee, A. J. (1988). The "great salinity anomaly" in the Northern North Atlantic 1968−1982. *Progress in Oceanography, 20*, 103−151. https://doi.org/10.1016/0079-6611(88)90049-3

Ding, H., Alexander, M. A., & Jacox, M. G. (2021). Role of geostrophic currents in future changes of coastal upwelling in the California current system. *Geophysical Research Letters, 48*, e2020GL090768. https://doi.org/10.1029/2020GL090768

Ekman, V. W. (1905). *On the influence of the earth's rotation on ocean-currents.*

Forget, G., Campin, J.-M., Heimbach, P., Hill, C. N., Ponte, R. M., & Wunsch, C. (2015). ECCO version 4: An integrated framework for non-linear inverse modeling and global ocean state estimation. *Geoscientific Model Development, 8*, 3071−3104. https://doi.org/10.5194/gmd-8-3071-2015

Gómez, C. E., Acosta-Chaparro, A., Bernal, C. A., Gómez-López, D. I., Navas-Camacho, R., & Alonso, D. (2023). Seasonal upwelling conditions modulate the calcification response of a tropical scleractinian coral. *Oceans, 4*, 170−184. https://doi.org/10.3390/oceans4020012

Gan, J., Li, L., Wang, D., & Guo, X. (2009). Interaction of a river plume with coastal upwelling in the northeastern South China Sea. *Continental Shelf Research, 29*, 728−740.

Gan, J., Wang, J., Liang, L., Li, L., & Guo, X. (2015). A modeling study of the formation, maintenance, and relaxation of upwelling circulation on the Northeastern South China Sea shelf. *Deep Sea Research Part II: Topical Studies in Oceanography, 117*, 41−52. https://doi.org/10.1016/j.dsr2.2013.12.009

García-Reyes, M., Sydeman, W. J., Schoeman, D. S., Rykaczewski, R. R., Black, B. A., Smit, A. J., & Bograd, S. J. (2015). Under pressure: Climate change, upwelling, and eastern boundary upwelling ecosystems. *Frontiers in Marine Science, 2*. https://doi.org/10.3389/fmars.2015.00109

Hobday, A. J., Alexander, L. V., Perkins, S. E., Smale, D. A., Straub, S. C., Oliver, E. C. J., Benthuysen, J. A., Burrows, M. T., Donat, M. G., Feng, M., Holbrook, N. J., Moore, P. J., Scannell, H. A., Sen Gupta, A., & Wernberg, T. (2016). A hierarchical approach to defining marine heatwaves. *Progress in Oceanography, 141*, 227−238. https://doi.org/10.1016/j.pocean.2015.12.014

Hu, J., & Wang, X. H. (2016). Progress on upwelling studies in the China seas. *Reviews of Geophysics, 54*, 653−673. https://doi.org/10.1002/2015RG000505

Huang, Z., & Wang, X. H. (2019). Mapping the spatial and temporal variability of the upwelling systems of the Australian south-eastern coast using 14-year of MODIS data. *Remote Sensing of Environment, 227*, 90−109. https://doi.org/10.1016/j.rse.2019.04.002

Jacobs, S. (2004). Bottom water production and its link with the thermohaline circulation. *Antarctic Science, 16*, 427—437. https://doi.org/10.1017/S095410200400224X

Jacox, M. G., Edwards, C. A., Hazen, E. L., & Bograd, S. J. (2018). Coastal upwelling revisited: Ekman, Bakun, and improved upwelling indices for the U.S. West coast. *Journal of Geophysical Research: Oceans, 123*, 7332—7350. https://doi.org/10.1029/2018JC014187

Jiang, R., & Wang, Y.-S. (2018). Modeling the ecosystem response to summer coastal upwelling in the northern South China Sea. *Oceanologia, 60*, 32—51. https://doi.org/10.1016/j.oceano.2017.05.004

Jing, Z., Wang, S., Wu, L., Wang, H., Zhou, S., Sun, B., Chen, Z., Ma, X., Gan, B., & Yang, H. (2023). Geostrophic flows control future changes of oceanic eastern boundary upwelling. *Nature Climate Change, 13*, 148—154. https://doi.org/10.1038/s41558-022-01588-y

Kämpf, J., & Chapman, P. (2016). *Upwelling systems of the world*. Springer.

Li, Y., Qiu, Y., Hu, J., Aung, C., Lin, X., & Dong, Y. (2020). Springtime upwelling and its formation mechanism in coastal waters of Manaung island, Myanmar. *Remote Sensing, 12*. https://doi.org/10.3390/rs12223777

Liang, X., Spall, M., & Wunsch, C. (2017). Global Ocean vertical velocity from a dynamically consistent ocean state estimate. *Journal of Geophysical Research: Oceans, 122*, 8208—8224. https://doi.org/10.1002/2017JC012985

Liang, X., Wunsch, C., Heimbach, P., & Forget, G. (2015). Vertical redistribution of oceanic heat content. *Journal of Climate, 28*, 3821—3833. https://doi.org/10.1175/JCLI-D-14-00550.1

Liao, F., Gao, G., Zhan, P., & Wang, Y. (2022). Seasonality and trend of the global upper-ocean vertical velocity over 1998—2017. *Progress in Oceanography, 204*, Article 102804. https://doi.org/10.1016/j.pocean.2022.102804

Liao, F., Liang, X., Li, Y., & Spall, M. (2022). Hidden upwelling systems associated with major western boundary currents. *Journal of Geophysical Research: Oceans, 127*, e2021JC017649.

Liu, S., Zuo, J., Shu, Y., Ji, Q., Cai, Y., & Yao, J. (2023). The intensified trend of coastal upwelling in the South China Sea during 1982-2020. *Frontiers in Marine Science, 10*, Article 1084189. https://doi.org/10.3389/fmars.2023.1084189

Miranda, P. M. A., Alves, J. M. R., & Serra, N. (2013). Climate change and upwelling: Response of iberian upwelling to atmospheric forcing in a regional climate scenario. *Climate Dynamics, 40*, 2813—2824. https://doi.org/10.1007/s00382-012-1442-9

Ngo, M.-H., & Hsin, Y.-C. (2021). Impacts of wind and current on the interannual variation of the summertime upwelling off southern Vietnam in the South China sea. *Journal of Geophysical Research: Oceans, 126*, e2020JC016892. https://doi.org/10.1029/2020JC016892

Oyarzún, D., & Brierley, C. M. (2019). The future of coastal upwelling in the Humboldt current from model projections. *Climate Dynamics, 52*, 599—615. https://doi.org/10.1007/s00382-018-4158-7

Pauly, D., & Christensen, V. (1995). Primary production required to sustain global fisheries. *Nature, 374*, 255—257. https://doi.org/10.1038/374255a0

Polonsky, A. B., & Serebrennikov, A. N. (2021a). Long-term tendencies of intensity of eastern boundary upwelling systems assessed from different satellite data. Part 1: Atlantic upwellings. *Izvestiya - Atmospheric and Oceanic Physics, 57*, 1658—1669. https://doi.org/10.1134/S0001433821120161

Polonsky, A. B., & Serebrennikov, A. N. (2021b). Long-term tendencies of intensity of eastern-boundary upwelling systems assessed from different satellite data. Part 2: Pacific

upwellings. *Izvestiya - Atmospheric and Oceanic Physics, 57*, 1670–1679. https://doi.org/
10.1134/S0001433821120173

Pozo Buil, M., Jacox, M. G., Fiechter, J., Alexander, M. A., Bograd, S. J., Curchitser, E. N.,
Edwards, C. A., Rykaczewski, R. R., & Stock, C. A. (2021). A dynamically downscaled
ensemble of future projections for the California current system. *Frontiers in Marine
Science, 8.* https://doi.org/10.3389/fmars.2021.612874

Praveen, V., Ajayamohan, R. S., Valsala, V., & Sandeep, S. (2016). Intensification of upwelling
along Oman coast in a warming scenario. *Geophysical Research Letters, 43*, 7581–7589.
https://doi.org/10.1002/2016GL069638

Ray, S., Swain, D., Ali, M. M., & Bourassa, M. A. (2022). Coastal upwelling in the western Bay
of bengal: Role of local and remote windstress. *Remote Sensing, 14.* https://doi.org/
10.3390/rs14194703

Rosales Quintana, G. M., Marsh, R., & Icochea Salas, L. A. (2021). Interannual variability in
contributions of the Equatorial Undercurrent (EUC) to Peruvian upwelling source water.
Ocean Science, 17, 1385–1402. https://doi.org/10.5194/os-17-1385-2021

Roughan, M., & Middleton, J. H. (2004). On the East Australian current: Variability,
encroachment, and upwelling. *Journal of Geophysical Research: Oceans, 109.* https://
doi.org/10.1029/2003JC001833

Rykaczewski, R. R., Dunne, J. P., Sydeman, W. J., García-Reyes, M., Black, B. A., &
Bograd, S. J. (2015). Poleward displacement of coastal upwelling-favorable winds in the
ocean's eastern boundary currents through the 21st century. *Geophysical Research Letters,
42*, 6424–6431. https://doi.org/10.1002/2015GL064694

Schulz, K. G., Hartley, S., & Eyre, B. (2019). Upwelling amplifies ocean acidification on the
East Australian shelf: Implications for marine ecosystems. *Frontiers in Marine Science, 6.*
https://doi.org/10.3389/fmars.2019.00636

Solodoch, A., Stewart, A. L., Hogg, A. M. C., Morrison, A. K., Kiss, A. E., Thompson, A. F.,
Purkey, S. G., & Cimoli, L. (2022). How does Antarctic bottom water cross the Southern
Ocean? *Geophysical Research Letters, 49*, e2021GL097211. https://doi.org/10.1029/
2021GL097211

Sousa, M. C., deCastro, M., Alvarez, I., Gomez-Gesteira, M., & Dias, J. M. (2017). Why coastal
upwelling is expected to increase along the western Iberian Peninsula over the next century? *The Science of the Total Environment, 592*, 243–251. https://doi.org/10.1016/
j.scitotenv.2017.03.046

Sousa, M. C., Ribeiro, A., Des, M., Gomez-Gesteira, M., deCastro, M., & Dias, J. M. (2020).
NW Iberian Peninsula coastal upwelling future weakening: Competition between wind
intensification and surface heating. *The Science of the Total Environment, 703*, Article
134808. https://doi.org/10.1016/j.scitotenv.2019.134808

Spall, M. A. (2010). Dynamics of downwelling in an eddy-resolving convective basin. *Journal
of Physical Oceanography, 40*, 2341–2347. https://doi.org/10.1175/2010JPO4465.1

Sun, Y., Dong, C., He, Y., Yu, K., Renault, L., & Ji, J. (2016). Seasonal and interannual
variability in the wind-driven upwelling along the southern East China Sea coast. *Ieee
Journal of Selected Topics in Applied Earth Observations and Remote Sensing, 9*,
5151–5158. https://doi.org/10.1109/JSTARS.2016.2544438

Sydeman, W. J., García-Reyes, M., Schoeman, D. S., Rykaczewski, R. R., Thompson, S. A.,
Black, B. A., & Bograd, S. J. (2014). Climate change and wind intensification in coastal
upwelling ecosystems. *Science, 345*, 77–80. https://doi.org/10.1126/science.1251635

Tamsitt, V., Drake, H. F., Morrison, A. K., Talley, L. D., Dufour, C. O., Gray, A. R.,
Griffies, S. M., Mazloff, M. R., Sarmiento, J. L., Wang, J., & Weijer, W. (2017). Spiraling

pathways of global deep waters to the surface of the Southern Ocean. *Nature Communications, 8*, 172. https://doi.org/10.1038/s41467-017-00197-0

Varela, R., Álvarez, I., Santos, F., deCastro, M., & Gómez-Gesteira, M. (2015). Has upwelling strengthened along worldwide coasts over 1982-2010? *Scientific Reports, 5*, Article 10016. https://doi.org/10.1038/srep10016

Varela, R., Rodríguez-Díaz, L., de Castro, M., & Gómez-Gesteira, M. (2022). Influence of Canary upwelling system on coastal SST warming along the 21st century using CMIP6 GCMs. *Global and Planetary Change, 208*, Article 103692. https://doi.org/10.1016/j.gloplacha.2021.103692

Varela, R., Rodríguez-Díaz, L., de Castro, M., & Gómez-Gesteira, M. (2021). Influence of Eastern Upwelling systems on marine heatwaves occurrence. *Global and Planetary Change, 196*, Article 103379. https://doi.org/10.1016/j.gloplacha.2020.103379

Wang, D., Gouhier, T. C., Menge, B. A., & Ganguly, A. R. (2015). Intensification and spatial homogenization of coastal upwelling under climate change. *Nature, 518*, 390—394. https://doi.org/10.1038/nature14235

Wang, S., Jing, Z., Wu, L., Sun, S., Peng, Q., Wang, H., Zhang, Y., & Shi, J. (2023). Southern hemisphere eastern boundary upwelling systems emerging as future marine heatwave hotspots under greenhouse warming. *Nature Communications, 14*, 28. https://doi.org/10.1038/s41467-022-35666-8

Xiu, P., Chai, F., Curchitser, E. N., & Castruccio, F. S. (2018). Future changes in coastal upwelling ecosystems with global warming: The case of the California Current System. *Scientific Reports, 8*, 2866. https://doi.org/10.1038/s41598-018-21247-7

Xu, L., Ji, C., Kong, D., & Guo, M. (2021). Abrupt change in Vietnam coastal upwelling as a response to global warming. *Journal of Quaternary Science, 36*, 488—495. https://doi.org/10.1002/jqs.3292

Yao, Y., & Wang, C. (2021). Variations in summer marine heatwaves in the South China sea. *Journal of Geophysical Research: Oceans, 126*, e2021JC017792. https://doi.org/10.1029/2021JC017792

Zhang, C. (2021). Responses of summer upwelling to recent climate changes in the Taiwan Strait. *Remote Sensing, 13*. https://doi.org/10.3390/rs13071386

Zhu, J., Zhou, Q., Zhou, Q., Geng, X., Shi, J., Guo, X., Yu, Y., Yang, Z., & Fan, R. (2023). Interannual variation of coastal upwelling around Hainan Island. *Frontiers in Marine Science, 10*. https://doi.org/10.3389/fmars.2023.1054669

Influences of extreme rainfall events on the nutrient and chlorophyll-a dynamics in coastal regions

12

Guandong Gao [1,2,3], Rushui Xiao [4] and Yunhuan Li [5]
[1]Key Laboratory of Ocean Observation and Forecasting, Key Laboratory of Ocean Circulation and Waves, Institute of Oceanology, Chinese Academy of Sciences, Qingdao, China; [2]CAS Key Laboratory of Ocean Circulation and Waves, Institute of Oceanology Chinese Academy of Sciences, Qingdao, China; [3]Laoshan Laboratory, Qingdao, China; [4]National Marine Environment Monitoring Center, Dalian, China; [5]Dalian University of Technology, Panjin, China

1. Introduction

Extreme precipitation events are defined in meteorology as an extreme weather hazard with a maximum daily rainfall greater than 50 mm (Cloern, 1987; Rapport & Friend, 1979, pp. 11−510). During extreme precipitation events, particulate matter, nutrients, and pollutants are exported from the entire watershed to estuaries, bays, lagoons, and coastal oceans (Chen, Lin, et al., 2012; Chen, Wu, et al., 2012; Lewitus et al., 2008; Paerl & Peierls, 2008; Rabalais et al., 2009). Even with a single extreme precipitation event, large amounts of nutrients may exacerbate annual nutrient loads (Mckee et al., 2000; Vidon et al., 2010). The importance of investigating the effects of extreme precipitation events on coastal ecosystems lies in enhancing our comprehension of the ramifications of climate change. For marine areas in proximity to urban centers, augmented precipitation elevates the likelihood of flash flooding, rapid and recurrent surges in river flows (Malta et al., 2017; Panton et al., 2020), and the discharge of highway runoff. The highway runoff means that water discharged from roads, often ladened with pollutants such as road salts, grease, nitrogen, and phosphorus, directly into the ocean (Pereira et al., 2007). These factors can have severe repercussions for marine ecosystems as levels of urbanization rises (Manhique et al., 2015; Rözer et al., 2021). Extreme precipitation events in the vicinity of intensively cultivated agricultural land can result in elevated nutrient and sediment discharge, with consequential implications for water quality in estuarine environments (Herbeck et al., 2011; Mitchell et al., 1997). Therefore, extreme precipitation can exert a considerable ecological influence on coastal regions in proximity to both urban and rural areas, with potential ramifications for the local economy, particularly in sectors such as fisheries and tourism.

Current Trends in Estuarine and Coastal Dynamics. https://doi.org/10.1016/B978-0-443-21728-9.00012-0

Phytoplankton biomass is predicted to rise in response to increased precipitation at numerous sites around the world ( ) because they are strongly related (Smetacek & Cloern, 2008; Thompson et al., 2015; Zhai et al., 2021). For example, there is a significant (above 60%) positive relationship between precipitation and phosphate concentration and between long-term mean salinity and chlorophyll-a concentration (Thompson et al., 2015). The most significant sources of nutrients during typhoon occurrences are flooding and precipitation (Zhang et al., 2009; Zhao et al., 2013; Zheng & Tang, 2007). Changes in extreme precipitation are closely related to future climate change. Over the past few decades, global warming has slowed typhoon advection in the western North Pacific by 21%, which may further increase precipitation frequency and intensity (Kossin, 2018; Peierls et al., 2003; Solomon et al., 2007; Statham, 2012). The global land mean precipitation frequency, intensity, and/or quantity are predicted to continue increasing with future warming (Masson-Delmotte et al., 2019). This is especially obvious for winter precipitation in certain regions of northern Europe and the east coast of North America, summer precipitation on the east coast of Asia, and year-round precipitation in the southeastern region of South America (IPCC, 2014). In addition to climate warming, extreme weather events will also affect future extreme precipitation events. According to studies, the likelihood of extreme precipitation during future La Niña events is growing (Packett, 2017) (Fig. 12.1).

Global variations in precipitation are predicted to exacerbate eutrophication in North America, India, China, and Southeast Asia (Sinha et al., 2017). This eutrophication can cause more frequent or severe algal blooms (Chen et al., 2018; Gallegos et al., 1992; Pinckney et al., 1997), as well as hypoxic in estuarine and coastal waters (Diaz & Rosenberg, 2008; Paerl, 2006; Rabalais et al., 2010; Statham, 2012). Thus, investigating the effects of extreme precipitation events on coastal ecosystems and generating future projections can facilitate the formulation of more scientifically informed ecological conservation policies and measures to mitigate the impact of extreme precipitation events on ecosystems and ensure the sustainable development of human society. We will synthesize the effects of extreme precipitation on nearshore environments from both physical and biochemical perspectives.

2.   Impacts of extreme precipitation events

2.1   Physical impacts of extreme precipitation events

Precipitation influences a number of environmental states that could impact on diatoms such as residence time (flushing), turbidity (irradiance), salinity, and stratification (Cloern, 1987). As precipitation injects freshwater, salinity decreases (Sahoo et al., 2017). For large estuaries heavily influenced by oceanic circulation, such as the lower Yangtze delta, typhoon-induced precipitation causes a sudden increase in freshwater discharge from rivers and a general decrease in surface salinity and density. However, strong onshore winds induce saltwater intrusion in offshore areas, only to increase water salinity and density in the mesopelagic and benthic layers, thereby intensifying stratification (Wang, 2015).

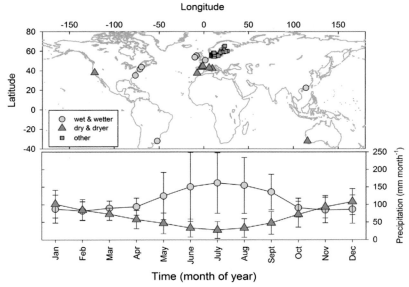

Figure 12.1 Sites with long-term time series of phytoplankton that reside in areas identified (IPCC, 2014) as currently or predicted to be: (1) Decreasing in precipitation (triangles). (2) Increasing in precipitation (circles), other additional sites in the Baltic region squares were included in some analyses. Lower panels NOAA's Cath System Research laboratory (ESRL) monthly average rainfall and calculated standard deviations for the dryer sites and wetter sites. From Thompson, P. A., O'Brien, T. D., Paerl, H. W., Peierls, B. L., Harrison, P. J. & Robb, M. (2015). Precipitation as a driver of phytoplankton ecology in coastal waters: A climatic perspective. *Estuarine, Coastal and Shelf Science, 162*, 119−129. https://doi.org/10.1016/j.ecss. 2015.04.004. Copyright of this figure was obtained via CCC (Copyright Clearance Center https://www.copyright.com/)

The extreme precipitation also influences the rate and direction of flow of the seawater due to the weather processes that often occur with heavy rain, such as strong winds. The wind mixing on physical structure of the water column also plays an important role for the ecosystem response to the extreme precipitation events ( ). Rainfall and runoff enhance the flushing volume of water and decrease its residence time. Mendoza-Salgado et al. (2005) found that summer rainfall disturbances in coastal lagoons in arid regions can momentarily alter water volume and nutrient exchange with the adjacent ocean (Fig. 12.2).

Extreme precipitation events may exacerbate littoral erosion (Anon, 1987; Milliman & Meade, 1983). Deep soil moisture increases during the most intense rainfall events, and when the soil moisture is >20%, the loss of soil and nutrient increases due to increased storm runoff that exceeds the soil filtration capacity (Ramos & Martínez-Casasnovas, 2009). Concentrations of particulate matter (PN, PP, suspended sediment) and the turbidity of seawater increase (Lee et al., 2016; Mitchell et al., 1997; Sahoo et al., 2017).

GRAPHICAL ABSTRACT

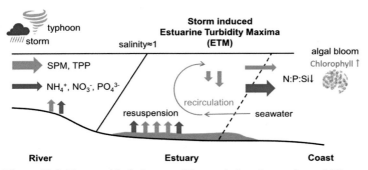

Figure 12.2 The graphical abstract of "storm induced estuarine turbidity maxima and controls on nutrient fluxes across river-estuary-coast continuum".
From Chen, N., Krom, M. D., Wu, Y., Yu, D. & Hong, H. (2018). Storm induced estuarine turbidity maxima and controls on nutrient fluxes across river-estuary-coast continuum. *Science of the Total Environment, 628–629*, 1108–1120. https://doi.org/10.1016/j.scitotenv.2018.02. 060. Copyright of this figure was obtained via CCC (Copyright Clearance Center https://www. copyright.com/)

2.2   Biological impacts of extreme precipitation events

2.2.1   Direct effects on biochemical concentrations

Due to intense rainfall events, large quantities of freshwater, particulate and dissolved organic matter (DOM), nutrients, metals, herbicides, and other chemical pollutants can be imported into aquatic systems (Paerl et al., 2001; Pereira et al., 2007; Southwick et al., 2009; Vizzo et al., 2021). There are two main types of transport processes: wet atmospheric deposition and riverine runoff. Rainfall can effectively transport airborne material directly into surface seawater and leach material from the soil into surface runoff. These materials can then be transported to the sea via rivers or dispersed directly into the sea.

Sea surface nutrient concentrations tend to increase after typhoons (Herbeck et al., 2011; Li et al., 2013; Zhao et al., 2009), and flooding and precipitation are regarded as the most important sources of nutrients during typhoon events (Zhang et al., 2009; Zhao et al., 2013; Zheng & Tang, 2007). The nutrient concentrations have been observed to be up to 10 times higher than usual (Morrison et al., 2001; Naidu & Morrison, 1994; Pratap et al., 2020). In large estuaries with stronger vertical stratification, however, nutrient concentrations in the bottom water decrease due to the outflow of surface water to the sea via runoff, resulting in the invasion of the bottom by low nutrient water from the sea, particularly dissolved inorganic nitrogen and silicate (Kalnejais et al., 2010; Tengberg et al., 2003). The ratio of nutrients will also change. The nutrient uptake by plankton community may not immediately increase due to low molar nitrogen (N) to phosphorus (P) ratios (i.e., N/P ratios) and/or low nutrient availability (e.g., DIN or silicate), which worsens the level of nutrient enrichment and allows it to persist for longer periods (Meng et al., 2014).

Nitrogen (N) is an essential element for the growth and primary productivity of aquatic ecosystems (Damashek & Francis, 2018; Rabalais, 2002). The nitrogen carried in by precipitation can directly influence the dynamics of offshore nutrients. For instance, across the Discovery Bay reef area in Jamaica, widespread heavy rainfall events have greatly increased NO_3 concentrations, with elevated NO_3 concentrations sometimes lasting for months (Greenaway & Gordon-Smith, 2006). Precipitation accounts for roughly 30% of the total input of dissolved inorganic nitrogen to the southern Great Barrier Reef river basin (Packett, 2017). The isotopic composition of storm water discharge suggests that most precipitation-derived nitrate is derived from terrestrial nitrogen and sewage rather than direct atmospheric deposition (Balint et al., 2021). There is also evidence that atmospheric nitrogen deposition in coastal systems may be underestimated (Joyce et al., 2020; Li et al., 2016). In Narragansett Bay, Rio de Janeiro, Brazil, the concentrations of inorganic nitrogen in rain water were approximately 240% higher than in riverine discharge (Balint et al., 2021). Southeast Asia keeps experiencing an increase in atmospheric nitrogen pollution (Kanakidou et al., 2016; Packett, 2017). Denitrification is a microbially facilitated process, whereby nitrate (NO_3^-) is reduced and ultimately becomes molecular nitrogen such as N_2 (Zumft, 1997). Notably, in estuarine environments, such as Louisiana estuary (Seo et al., 2008), the Douro River estuary, Portugal (Magalhães et al., 2005), and the Danish estuary (Jørgensen & Sørensen, 1985), extreme precipitation occurs and large amounts of nutrients enter through freshwater discharge. Denitrification rates appear to be the highest under freshwater conditions with salinity close to 0 PSU and decrease with increasing salinity.

Phosphorus (P) as an energy currency is fundamental to life, serving an integral role in aspects of cellular metabolism ranging from energy storage, to cellular structure, to the very genetic material that encodes all life on the planet (Acker et al., 2022; Dyhrman, 2016, pp. 155−183). For example, agricultural drainage systems accessing river tributaries during storm events contain high concentrations of total phosphorus in southern Baltic coastal waters (Zimmer et al., 2016). Heavy precipitation and freshwater discharge may stimulate the release of benthic P fluxes to coastal ecosystems (Chen et al., 2015). Extreme precipitation has been identified as a significant driver of N and P accumulation (Ballard et al., 2019; Gentry et al., 2007; Nausch et al., 2017; Sinha & Michalak, 2016). Some studies have found an immediate decrease in N/P ratios (up to 15-fold) following storm events in some marine areas due to higher riverine P inputs than N, suggesting that climate change may temporarily exacerbate nitrogen limitation in some estuarine and coastal ecosystems (Chen et al., 2015; Correll et al., 1999).

Oceanic silicate, or silicic acid (H_4SiO_4), is another essential nutrient. In contrast to phosphate and nitrate, which are required by nearly all marine plankton, silicate is only required by very specific biota, including diatoms, radiolaria, silicoflagellates, and siliceous sponges. These organisms require silicates to construct their siliceous cell walls (DeMaster, 1981; Jr, 2005; Tréguer & Pondaven, 2000) (Boggs). Intense precipitation caused by Typhoon Phailin in 2013 led to increased river flows that carried large amounts of silicate-rich silt into the Chilika Lagoon in the eastern tropics of peninsular India. This influx of nutrients fueled preferential diatom growth (Sahoo et al., 2017).

However, inorganic silicate concentrations would decrease significantly after precipitation due to algal depletion following algal blooms caused by extreme precipitation (Meng et al., 2014). Extreme precipitation has also been shown to oversupply Si in rivers, preventing P from binding to Fe and thereby increasing P release (Mok et al., 2019).

In addition to major inorganic nutrients, extreme precipitation can also have an impact on enhanced concentrations of soluble organic matter of terrestrial origin. (Dupouy et al., 2020; Koliyavu et al., 2021; Martias et al., 2018; Mok et al., 2019) studied the effects of three consecutive typhoon-induced heavy rainfall events and resultant freshwater runoff on the partitioning of organic carbon (C_{org}) oxidation and nutrient dynamics. The results suggest that the intensified storm events and resultant riverine runoff induce a shift of C_{org} oxidation pathways in the sediments, which ultimately alters C$-$N$-$P$-$S$-$Fe dynamics and may deepen N-limiting conditions in coastal ecosystems of the Yellow Sea. In the semi-annular Neuse Estuary (NRE), North Carolina, USA, extreme precipitation has been found to be linked to elevated dissolved organic carbon (DOC) fluxes (Apple et al., 2011; Paerl et al., 2018).

Heavy precipitation also increases the transport of trace metals (e.g., Fe, Al, and Mn), which play a crucial role in the proliferation of phytoplankton in the ocean (Martino et al., 2014; Okin et al., 2011). Iron can stimulate phytoplankton nutrient assimilation in regions with high nutrient concentrations and low chlorophyll-a (Chl-a) concentrations (Behrenfeld et al., 1996; Martin & Fitzwater, 1988). During and after intense rainfall, the Ganghwa intertidal wetland (one of the largest intertidal zones in the Yellow Sea, located on the west coast of Korea) exhibited a significant decrease in iron reduction (i.e., FeR) (Mok et al., 2019). The primary determinant of iron reduction in coastal sediments with enough organic carbon availability is considered to be the presence of iron oxide, as suggested by previous studies (Hyun et al., 2009; Kostka et al., 2002; Meiggs & Taillefert, 2011). Hence, it is probable that the significance of FeR as a pathway for Corg oxidation increased during the occurrence of the rainfall event. This can be attributed to the rejuvenation of the pool of Fe oxide, which was stimulated by alterations in redox conditions induced by rainfall (Mark (Jensen et al., 2003), as well as the delivery of Fe oxide to surface sediments through intense rainfall and extensive runoff (Poulton & Raiswell, 2002).

Additionally, extreme precipitation can influence the concentration and distribution of sulfates. In Ganghwa intertidal wetland (Mok et al., 2019) and in some tropical regions (Alongi, 1995; Alongi et al., 1999, 2004), precipitation and freshwater infiltration caused significant reductions in pore water sulfate concentrations. During and after the heavy rainfall, sulfate (i.e., SR) was substantially reduced due to abrupt fluctuations in salinity (approximately 3.7-fold reduction) and redox oscillations. The oxidation of precipitation and freshwater runoff is caused by a combination of sediment splash by raindrops, infiltration of the rainwater, and tidal inundation by low salinity waters (Mok et al., 2019). In the sediments of coastal lagoons in the Gulf of Mexico, rainfall and riverine inputs reduce salinity and sulfate content and diminish the abundance of sulfate-reducing bacteria (Torres-Alvarado et al., 2016). In the intertidal wetlands of Jacksonville, Florida, USA, freshwater pulses led to a decrease in sulfate and consequently sulfate reduction rates in sediments (Chambers et al., 2013).

2.2.2 Effects on biodiversity and ecosystem function

The response of phytoplankton to precipitation varies according to the seasons and geographical locations. Phytoplankton in summer exhibits a greater response to increased precipitation compared to winter. By comparing long-term time series data for Chl-a, diatoms, methanogens, chloroplasts, chrysophytes, and chloroplasts from 106 sites around the world, the results show that increased summer precipitation causes pulsed changes in nutrients, pollution and other materials associated with "hot-spots" such as estuaries and nearshore bays (Fig. 12.3). This not only leads to eutrophy in coastal ecosystems and oligotrophic oceans but also to increased Chl-a concentrations (Chang et al., 1996; Fujii & Yamanaka, 2008; Hubertz & Cahoon, 1999; Shi & Wang, 2007; Subrahmanyam et al., 2002; Zhang et al., 2009), and the changes in phytoplankton community structure (Fogel et al., 1999; Lin et al., 2003; Zhao et al., 2009), as well as increases in total phytoplankton abundance (Chen et al., 2018; Gallegos et al., 1992; Meng et al., 2014; Morse et al., 2011; Paerl et al., 2001; Peierls et al., 2003; Pinckney et al., 1997; Tester et al., 2003).

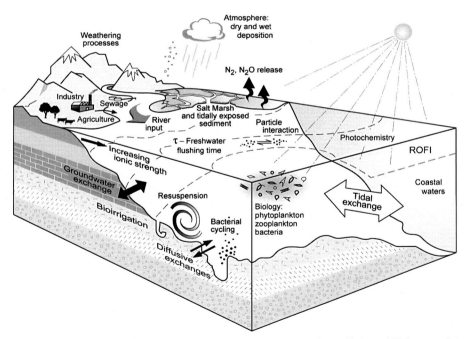

Figure 12.3 Processes and exchanges influencing the macronutrients Si, P, and N in estuarine systems. *ROFI*, region of freshwater influence.
From Statham, P. J. (2012). Nutrients in estuaries - An overview and the potential impacts of climate change. *Science of the Total Environment*, *434*, 213−227. https://doi.org/10.1016/j. scitotenv.2011.09.088. Copyright of this figure was obtained via CCC (Copyright Clearance Center https://www.copyright.com/)

The phytoplankton responses to precipitation depend upon the seasons and regions. In summer, phytoplankton responds more positively to increased precipitation than winter. By comparing long-term time series data for Chl-a, diatoms, dinoflagellates, chlorophytes, chrysophytes, and euglenophytes from 106 sites worldwide, the results show that increased summer precipitation may increase Chl-a and favor chlorophytes, whereas increased winter water may decrease chlorophyll *a*, diatoms, and chryso-phytes (Thompson et al., 2015). For the majority of estuarine and coastal regions, spring precipitation is low, or there is little rain or snow throughout the winter. Diatoms are extremely sensitive to increases in spring precipitation, possibly due to their dependence on a small window of opportunity. The decreased turbidity and increased irradiance in the spring also favor the early emergence of spring diatoms (Middelkoop et al., 2001; Wang, 2005). Within regions that are wet and getting wetter, diatom abundances were reduced in wet springs, while Chl-a increases and dinoflagellate abundances are reduced in wet autumns. In dry and drying ecosystems, the abundances of chlorophytes decreased during dry springs and summers (Thompson et al., 2015). In general, rainfall is associated with elevated phytoplankton production; however, in two subtropical South African estuaries, contradictory results were observed. In the Kariega Estuary, there was no association between freshwater flow and Chl-a. In contrast, in the Great Fish Estuary, where Chl-a concentration was associated with freshwater flow, phytoplankton biomass in the mid-estuary increased after rainfall (Grange & Allanson, 1995; Grange et al., 2000).

Rainfall can not only affect 476 biomass but also alter the composition and structure of phytoplankton communities. For example, at the coastal Neuse River Estuary (NRE) in North Carolina, USA, riverine discharge from Hurricane Florence resulted in a more than 100-fold reduction in Picophytoplankton (PicoP) biomass over a period of more than 2 weeks (Johnson & Sieburth, 1979; Stockner & Antia, 1986; Waterbury et al., 1979). PicoP are phytoplankton $<2-3$ μm in diameter that are present in nearly all aquatic systems. Changes in the composition of PicoP communities may affect ecosystem health and function as well as estuarine carbon cycling. In the NRE basin, increased precipitation has also substantially altered the composition of PicoP communities, with cyanobacteria decreasing and ephemeral "blooms" of picoeukaryotic phytoplankton (Paerl, Hall, et al., 2020; Paerl, Venezia, et al., 2020).

A similar decline in cyanobacterial biomass was observed in the Rodrigo de Freitas Lagoon (RFL), a shallow marine system in southeastern Brazil that is surrounded by urban areas, and increased the contribution of both cryptophytes and diatoms (Neves & Santos, 2022). Heavy rainfalls have already been related to reductions in cyanobacteria biomass, consequently decreasing its relative contribution and dominance in phyto-plankton, with the emergence of diatoms, cryptophytes, and chlorophytes (Pedusaar et al., 2010). Therefore, the shifts in the phytoplankton contribution evidenced here after the very heavy rainfall events seem to be beneficial to planktonic community by the increase in the relative contribution of more palatable microalgae (diatoms and cryp-tophytes) to mesozooplankton consumers (calanoid copepods (Souza et al., 2011). The decrease of dinoflagellates was found in particular increased precipitation in the Swan River estuary, a shallow microtidal water body in south-western Western Australia with long residence times (>365 days) and long periods of stratification during summer (Chan & Hamilton, 2001). In particular, increased precipitation during November was very disruptive to dinoflagellates possibly because it disrupted their preferred habitat of a stable, stratified water column (Chan & Hamilton, 2001; Jephson et al., 2011, 2012; Margalef, 1978). Dapeng Lagoon (Taiwan), the largest and hyper-eutrophic lagoon along the southwestern coast of Taiwan Island, was discovered to have a higher proportion of diatoms, especially *C. curvisetus*, following precipitation (Meng et al., 2014). Prior to the occurrence of extreme precipitation, dominance by dinoflagellates has been observed; the growth of diatoms is limited by low silicate levels (Tew et al., 2010). When extreme precipitation occurs, phytoplankton is domi-nated by cyanobacteria 2 days after precipitation due to nutrient depletion, the removal of silicates, and the cessation of diatom growth (Meng et al., 2014).

While the majority of studies confirm that heavy precipitation increases phyto-plankton biomass and enriches the composition of phytoplankton in offshore systems, a few studies have shown that extremely heavy rainfall has a negative impact on phyto-plankton in low nutrient marine environments (Lin, 2012) and that hypoxia is the pri-mary reason for the reduction (Neves & Santos, 2022), especially the shallow marine systems surrounded by cities (Dos Santos & Teixeira, 2019). Specifically, the sudden onset of heavy rainfall during extended dry periods can lead to a decrease in surface salinity and an increase in seawater stratification. This can result in longer and more extensive low dissolved oxygen in the deeper waters of estuaries and coastal seas (Diaz & Rosenberg, 2008). Large areas of hypoxia are extremely prone to fish mortal-ity (Kragh et al., 2020). For example, there is a strong correlation between summer pre-cipitation (June–August) and the total number of summer hypoxia days in Narragansett Bay, New England, in the northeastern United States (Oviatt et al., 2017). In addition to hypoxia, factors such as physical flushing, sinking, grazing or viral infection, and parasitism can cause significant loss of phytoplankton abundance following the cessation of precipitation (Banse, 1994; Crumpton & Wetzel, 1982; Meng et al., 2014), particularly a sudden decrease in the abundance of *N. scintillans* (Baek et al., 2009; Huang & Qi, 1997).

Escherichia coli density (*E. coli*) was also detected in aquatic systems after mega-storm events (Neves & Santos, 2022). Fecal bacteria indicators (FBIs) originated from domestic sewage (e.g., total coliforms, *E. coli*, *Enterococcus*) are proxies of water contamination by potential pathogenic microorganisms (Heaney et al., 2012). Increased fecal microorganisms, such as *E. coli*, are associated with diarrhea. FBI occurrence and/or density are used for predictive water quality modeling and imple-mentation of coastal area management strategies related to pollution and sanitary con-ditions for both recreational and seafood harvesting purposes (Ribeiro & Kjerfve, 2002). Significant increases in FBI concentrations have been observed in eutrophic aquatic systems such as estuaries (Coulliette & Noble, 2008), rivers (García-Aljaro et al., 2017), and lagoons (Neves & Santos, 2022), where these pathogenic microor-ganisms can cause outbreaks of diarrheal disease if they contaminate shellfish and drinking water (Carlton et al., 2014; Chen, Lin, et al., 2012; Chen, Wu, et al., 2012; Drayna et al., 2010).

There is little research on the effects of extreme precipitation events on zooplankton. For example, within 1 day of an extreme precipitation event, a considerable increase in zooplankton was observed at Dee Why Lagoon on the northern beaches of Sydney, Australia, with a twofold increase in the adult stage of the copepod *Oithona* sp., followed a week later by nauplii and adult *Acartia bispinosa* (Rissik et al., 2009). In the RFL (Brazil), zooplankton may change their diet in response to increased food availability in the water column. The mesopelagic zooplankton consumers (copepods) increased in the relative contribution of more palatable microalgae (diatoms and cryptophytes) (Souza et al., 2011). In addition, it may favor energy transfer through the classical aquatic food chain (phytoplankton-zooplankton-fish), which are positive to ecosystem functioning (Sarmento, 2012).

3. Case studies of extreme precipitation events on coastal ecosystems

Coastal ecosystems are susceptible to storm-driven pulses of freshwater, particle, and nutrient inputs and associated environmental changes (Bruesewitz et al., 2013; Paerl, 2006; Paerl et al., 2001). Estuaries, bays, and lagoons are considered marine biodiversity hotspots that are severely impacted by extreme precipitation and are sensitive to global climate change. Heavy precipitation in these areas is often accompanied by flooding, cyclones, mudslides, and other hazards. These areas often host additional functions such as farming, tourism, and nature reserves. Therefore, extreme precipitation frequently has repercussions and economic losses for the local population and economy. In this section, we summarize the impact of extreme precipitation on the coastal environment in three typical affected areas: bays, estuaries, and lagoons.

3.1 Impact of extreme precipitation events on bays

Bays are generally defined as bodies of water encompassed on three sides by land. They are characterized by semi-enclosures, weaker water exchange, and shallower water depths, all of which exacerbate the local effects of extreme precipitation and lengthen recovery periods. For example, extreme precipitation is frequently accompanied by increased nutrient input from rivers, adding to its burden. Jiaozhou Bay (JZB) on the southern coast of the Shandong Peninsula, China, is a typical semi-enclosed bay surrounded on three sides by urban Qingdao, Jiaozhou, and Huangdao districts and connected to the Yellow Sea on one side.

3.1.1 Jiaozhou Bay

Jiaozhou Bay is located on the south coast of the Shandong Peninsula, with an area of 302.9 km^2 and an average depth of 7 m. Heavy rainfall events may increase the risk of eutrophication in the JZB (Xing et al., 2017). Observations show that the fluxes of DIN, SiO$_3$, and PO$_4$ through rivers and wet deposition were 2.7, 19.7, and 25.6 times higher than before the rainfall, respectively, and the average concentrations of DIN, SiO$_3$, and PO$_4$ in the surface seawater of JZB reached 1.5, 2.4, and 2.3 times

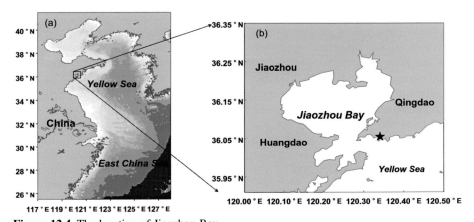

Figure 12.4 The location of Jiaozhou Bay.
From Xing, J., Song, J., Yuan, H., Li, X., Li, N., Duan, L., Kang, X. & Wang, Q. (2017). Fluxes, seasonal patterns and sources of various nutrient species (nitrogen, phosphorus and silicon) in atmospheric wet deposition and their ecological effects on Jiaozhou Bay, North China. *Science of the Total Environment*, *576*, 617–627. https://doi.org/10.1016/j.scitotenv.2016.10.134. Copyright of this figure was obtained via CCC (Copyright Clearance Center https://www. copyright.com/)

higher than before the rainfall, respectively. The ratio of nutrients also changed significantly, with N/Si/P changing from 24:8:1 to 19:10:1, which was closer to the Redfield value (Zhang et al., 2016). Heavy precipitation caused a significant decrease in zooplankton abundance and a significant change in its distribution pattern in JZB (Gu et al., 2016). Coupled physical-biological modeling revealed that the input of nitrogen and phosphorus from river runoff was 1.80 and 2.14 times greater than atmospheric deposition, respectively. The riverine inputs sustain 2.14 times more productivity than atmospheric deposition (Han et al., 2023) (Fig. 12.4).

3.1.2 Laucala Bay

Laucala Bay (LBW) is located between the western side of the Suva peninsula and the eastern side of the Rewa River delta in Suva, Fiji. Similar to JZB, LBW in the Fiji Islands is surrounding with rapid economic development regions in developing country. The climate conditions are mainly influenced by the South Pacific Convergence Zone (SPCZ) with a warm and wet season from November to April and a dry and cold season from May to October. Rainfall varies greatly from region to region and is mainly influenced by the topography of the islands and the south-easterly trade winds (Singh & Aung, 2008). Episodic rainfall events accelerate nutrient availability and variability (Koliyavu et al., 2021; Morrison et al., 2001; Naidu & Morrison, 1994; Pratap et al., 2020).

Heavy rainfall of about 74.4 mm was observed on December 1, 2017, while the monthly average rainfall for the same month was 13.65 mm. Snapshots of the optical fingerprints of DOM source and selected nutrient, biogeochemical, and physical variables were characterized for the month of December 2017 at 10 stations in LBW. The results indicate that large freshwater flows from the Rewa River and from small rivers along the west coast during precipitation strongly influence the distribution of suspended particulate matter (SPM) (Fichez et al., 2006).

Four fluorescent intensities were also identified, type M, linked to the degradation of organic matter; fulvic-like, protein-like, and humic-like. The results show that the overall dynamics and nature of the biogeochemistry and bio-optical characteristics of the LBW are heavily dictated by the frequency and intensities of the rainfall. Noteworthy findings include the formation of the southwestern plume caused by heavy rainfall, colored dissolved organic matter (CDOM) fluorescence, and absorption parameters (as well as other biogeochemical parameters), which is characterized by a large inflow of river discharge into the LBW and a moderate outflow the narrow Nasesese passage into Suva Harbor waters. The Suva Barrier Reef system limits the dispersal of particles into the western Pacific Ocean waters, thus supplementing the accumulation of particles in the western part of the bay. Meanwhile, the observed high Chl-a concentrations coincided with the accumulation of nutrients in the bay, which can originate from terrigenous (riverine discharges) and anthropogenic inputs. The observed association between Chl-a and dissolved oxygen (DO) implies a potential relationship between primary production and decomposition within the marine ecosystem. The presence of negative gradients in salinity and temperature along the coast can be attributed to the dilution of freshwater resulting from riverine discharges, which, in turn, is proportional to rainfall intensities. The presence of tidal input and seafloor resuspension demonstrated the occurrence of mixing, which exhibited a negative correlation with the variables under investigation. Additionally, wind played a role in shaping the spatial distribution patterns of the optical characteristics associated with the CDOM. DOM was being utilized as both energy and nutrients sources in the LBW as observed from the responses of DOC or CDOM to the nutrients variables NOx, PO_4, and Si $(OH)_4$. Furthermore, it was observed that dissolved organic nitrogen (DON) played a pivotal role as the primary supplier of organic nutrients in LBW with its responses to the tested nutrient variables. The high nutrient loads (i.e., in the form of DON, etc.), have the potential to contribute to the substantial biomass of phytoplankton (specifically Chl-a) within the LBW (Koliyavu et al., 2021).

3.1.3 Cook's Bay

Cook's Bay (also known as Paopao Bay) is a 3-km long bay on the north coast of the island of Moorea, Tahiti. Between January 13 and January 22, 2017, the South Pacific Island of Moorea, French Polynesia experienced an extreme rainfall event. About 57 cm of rain was delivered over a 10-day storm. The concentrations of $NO_3 + NO_2$ exhibited a range of $1-43$ μM, surpassing the average value recorded at the Moorea Long-Term Ecological Research (LTER) since 2005 by over 100-fold (Alldredge, 2019). Similarly, the amounts of phosphate observed during the storm event exhibited a range of $0.25-2.8$, which is 27 times greater than the average value recorded (Alldredge, 2019). These pulsed inputs were of short duration and were either flushed out of the bay or taken up biotically after 9 days. Observations also found that pulsed sediments affected water clarity during the 6 days following the storm, with the greatest impacts closest to the estuary. Pulsed nutrient and sediment inputs may serve as a potential mechanism underlying the occurrence of macroalgae blooms on coral reefs, despite typically low nutrient and sediment availability. The pulsed nutrient and sediment supplies can also influence the outcome of competition among macroalgal species on coral reef ecosystems (Fong & Fong, 2018) and may drive a shift in communities toward species with rapid uptake and growth (Haan et al., 2016). In temperate communities, pulsed nutrient subsidies frequently drive macroalgal blooms (Cohen & Fong, 2004; Kennison et al., 2011); if the nutrient subsidy continues to escalate, rainfall events have the potential to exacerbate the occurrence of macroalgal blooms on coral reefs. Hence, it is imperative to thoroughly document and effectively manage the enduring effects of nutrient and sediment influxes resulting from intense rainfall events on benthic ecosystems inside coral reefs. Elevated tissue N in *Padina boryana* and δN15 indicate that sewage was the source of nutrients, suggesting that nutrients were transferred to the producer community.

Rainfall events can impact top-down control of short algal turf communities, potentially changing benthic communities. (Hayes et al., 2021) compared the results of three herbivore exclosure experiments from previously published experiments without rainfall (Fong & Fong, 2018), with light rainfall (Fong et al., 2020), and with an extreme rainfall on the same fringing reef along Cook's Bay. The results show that storms may cause short-term reductions in herbivory leading to long, sediment laden algal turfs, thereby reducing services provided. Herbivory is a critical ecological force supporting coral reef resilience through strong top-down control on algal communities like turf (Bellwood et al., 2004; Hughes et al., 2007; McClanahan et al., 2009). Turf, however, can be divided into short productive algal turf (SPATs) and long sediment-laden algal turf (LSATs) (Goatley et al., 2016). Short algal turfs are highly productive benthic communities that provide substantial nutrient support (Adey & Goertemiller, 1987). Generally, macroalgae is considered to be a degraded, undesirable state on coral reefs (Hughes et al., 2007). Storms may drive transitions from SPATs to either LSATs or macroalgae. One mechanism by which heavy precipitation is disrupting the strong top-down control of SPATs by herbivores and potentially contributing to the transition to LSATs is the increased flux of terrestrial sediments to the nearshore environment, often through visible sediment plumes (Fong et al., 2020). For example, increased sedimentation can potentially deter grazing behavior by making turf unpalatable. The accumulation of fine particles within algae can discourage some herbivorous and detritivorous fishes (Clausing et al., 2014; Goatley & Bellwood, 2012; Gordon et al., 2016; Rusuwa et al., 2006; Tebbett et al., 2017). Alternatively, herbivores may have left areas affected by the plume. Several studies have indicated that herbivores may engage in migratory behavior to avoid storm disruptions (Kaufman, 1983; Walsh, 1983) and enhanced sedimentation (Chew et al., 2013). An alternative hypothesis for the disruption of herbivory by rain could be attributed to variations in temperature that occur concurrently with precipitation and runoff events. If rain causes the water temperature to drop, herbivory by poikilothermic fishes might decrease as a result (Smith, 2008).

In conclusion, precipitation may suppress the top-down control that is essential for the maintenance of the ecosystem services provided by algal turf. A similar shift to the proliferation of tall algal turf has been investigated on Australian reefs (Goatley et al., 2016): Severe Tropical Cyclone Yasi increased sediment loading and decreased herbivory, leading to longer algal turf that lasted for a few years after the storm. Therefore, storm events may result in a shift of coral reefs toward algal dominance, thereby reducing food chain support (Clausing et al., 2014; Tebbett & Bellwood, 2020).

3.1.4 Other typical bays

In Tokyo Bay in Japan, heavy precipitation and high surface temperatures in the late spring and summer gave rise to a highly stratified water-column and stimulated a series of phytoplankton blooms (Bouman et al., 2010). In the Atsumi Bay estuary, Japan, heavy rains caused by three typhoons, Chaba, Ma-On, and Roke, increased the discharge of freshwater and surface nutrient concentrations, resulting in subsequent phytoplankton blooms (Ernawaty et al., 2014). In 2006, the Northwest Bay (NY, USA) experienced unusually heavy rains that caused river discharge and phosphorus loads more than three times greater than in 2007. The additional phosphorus load had a positive effect on Chl-a concentrations in 2006 (Swinton & Boylen, 2014).

3.2 Impact of extreme precipitation events on estuaries

An estuary is an area where a freshwater river or stream meets the ocean. A bay is a body of water partially surrounded by land. Bays frequently have river inflows as well. To differentiate them, estuaries in this section refer to open areas that rivers flow directly into the open ocean. Although the open ocean can help to better clean up the pollution caused by precipitation, it is still a significant area subject to extreme precipitation. For example, the Pearl River Estuary (PRE) in the South China Sea (SCS) (Qiu et al., 2019; Zhao et al., 2009), the Jiulong River estuary in the Taiwan Strait (Chen et al., 2018), the intertidal zone of the Han River estuary in the Yellow Sea (Mok et al., 2019), southern Baltic Sea coastal water bodies (CWB) (Berthold et al., 2018), and NRE (Bales, 2003; Paerl et al., 2018, 2019; Paerl, Hall, et al., 2020; Paerl, Venezia, et al., 2020).

3.2.1 The Pearl River Estuary

The Pearl River Estuary is located in the central area of the Guangdong-Hong Kong-Marco Greater Bay Area of China, which is one of the most urbanized and industrialized regions in the world, with an area of 453,700 km^2 and a population of over 200 million (Watkins et al., 2021). Higher total suspended solids (TSS) concentration observed near western inlets was mainly caused by the large amounts of rainfall discharge during a Category 3/4 typhoon Vicente (Ye et al., 2014). Previous research found that typhoons can trigger phytoplankton blooms in and near the PRE, as well as the offshore regions of the SCS: Remote sensing and in situ datasets were used to study two phytoplankton blooms triggered by a moderate Category two typhoon Nuri in the

SCS near or around the PRE and to clarify the important role of extreme precipitation (Zhao et al., 2009); Intense precipitation is one of the significant factors influencing the short-term dynamics (the growth, reproduction, and expansion) of phytoplankton (especially *Synechococcus*) in the PRE during two successive typhoons: Nangka and Soudelor. Enhanced freshwater input to the estuary or easterly winds will cause a high abundance of phytoplankton in conjunction with the brackish water—enhanced extension from the estuary into the nearshore areas when Chl-a peaks in the estuary (Qiu et al., 2019); On August 23, 2017, Hato was a strong typhoon that formed in the western Pacific and rapidly intensified into a Category 3 intense typhoon. An integrated modeling system was reproduced the satellite-observed variability of sea surface salinity and Chl-a before, during, and after the passage of Hato. The high surface Chl-a concentration extended over the 30-m isobath near the western bank 1 week after Hato, and 70-m isobath 2 weeks after Hato. The increased discharge induced by heavy rainfall from the storm halved the residence time and doubled the decay rate of the water mass within Lingding Bay (a). However, the low phytoplankton biomass within Lingding Bay after Hato was primarily because the residence time was greatly shortened by the typhoon-induced heavy rainfall (Feng et al., 2022) (Fig. 12.5).

Another study was conducted to examine the abundance and composition of microplastics in sediments and surface waters along the Pearl River (Wu et al., 2023). The findings revealed that persistent rainfall led to an increase in both the abundance and diversity of microplastics in surface waters. Conversely, an opposite pattern was exhibited in sediments, with a decrease in microplastic abundance detected. The presence of significant hydrodynamic disturbances can lead to a temporary shift in sediments, causing them to transition from acting as sinks for microplastics to potentially becoming sources. Furthermore, the results obtained from high throughput sequencing analysis revealed notable dissimilarities in both the microbial community composition and functionality of the plastisphere before and after the persistent rain. According to a runoff experiment, increasing the volume of rainfall from 20 to 60 mm enhanced the transport of plastics, whereas repeated 20 mm rainfall every 3 days had no significant effect on the transport of plastics. Transport of plastics and soil particles by rainfall-induced surface discharge may involve the same processes (Han et al., 2022). The concentration and transport of microplastics in the environment can also be affected by other meteorological events, such as storm (Bäuerlein et al., 2022), typhoon (Chen et al., 2021), and flood (Gündoğdu et al., 2018) (Fig. 12.6).

3.2.2   The Jiulong River Estuary

The (JRE) is a typical subtropical macro-tide estuary in southeast China, with a subtropical climate and a drainage area of 14,740 km^2. Three tributaries (North River, West River, and South River) discharge water into Xiamen Bay through the estuary Fig. 12.7.

Three typical storm events which occurred in May 2013 (storm A), June 2013 (extreme storm B), and July 2013 (extreme storm C, caused by Typhoon Soulik) were analyzed through continuous sampling of dissolved and particulate P (Chen

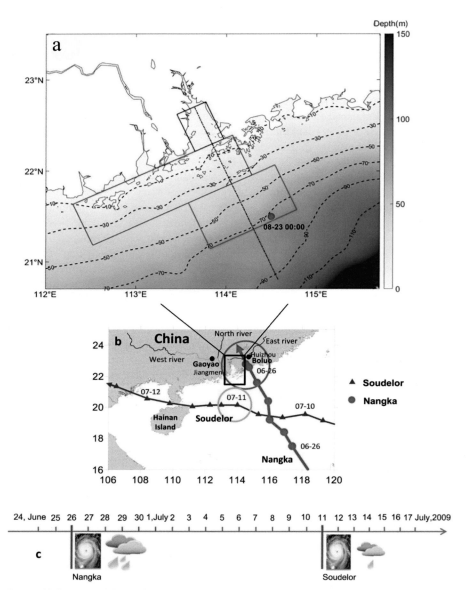

Figure 12.5 (a) The blue, red, and magenta boxes are Lingding Bay, nearshore (0−30 m), and offshore (30−70 m) regions (Feng et al., 2022). (b) Map showing location of study area in southern China and storm tracks for typhoons Nangka (25−27 June 2009) and Soudelor (10−12 July 2009) in the south China sea (SCS): (c) Schematic diagram of the timing of typhoons Nangka and Soudelor. Vertical lines indicate date when the typhoons were closest to the PRE. *Blue* line indicates a transection (13 stations) from the Humen outlet to near the Wanshan Islands. From Qiu, D., Zhong, Y., Chen, Y., Tan, Y., Song, X. & Huang, L. (2019). Short-term phytoplankton dynamics during typhoon season in and near the Pearl River Estuary, South China Sea. *Journal of Geophysical Research*: *Biogeosciences*, *124*(2), 274−292. https://doi.org/10.1029/2018jg004672. Copyright of this figure was obtained via CCC (Copyright Clearance Center https://www.copyright.com/) print it in color

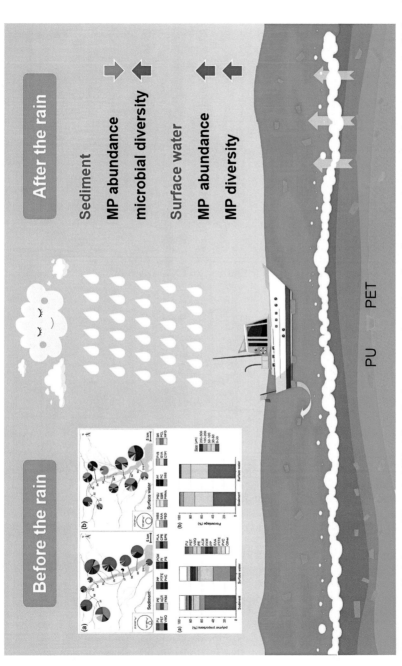

Figure 12.6 The graphical abstract (Wu et al., 2023). Microplastics concentration and type in sediment samples (a) and surface water samples (b) from the Pearl River in the size range of 10–500 µm. The *red* dots represent sampled locations, major streams are indicated with *blue* lines, microplastics concentrations correspond to the area of the circles. Percentage of microplastics determined by polymer type (c) and size fraction (d) in sediment and surface water samples.

From Wu, J., Ye, Q., Sun, L., Liu, J., Huang, M., Wang, T., Wu, P. & Zhu, N. (2023). Impact of persistent rain on microplastics distribution and plastisphere community: A field study in the Pearl River, China. *Science of the Total Environment, 879.* https://doi.org/10.1016/j.scitotenv.2023.163066. Copyright of this figure was obtained *via* CCC (Copyright Clearance Center https://www.copyright.com/)

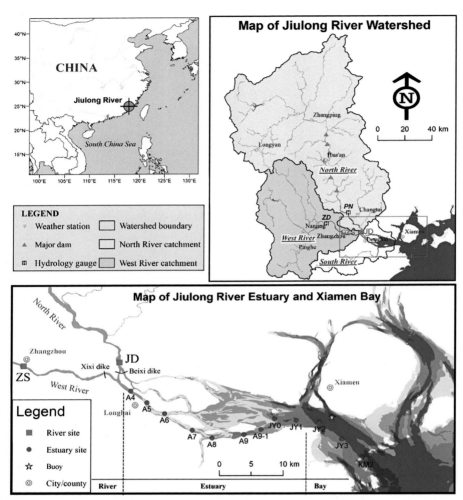

Figure 12.7 Map of study area showing sampling sites (*red circle*) along the Jiulong river-estuary gradient. Jiangdong (JD) at the lower north river and Zhongshan (ZS) at the lower west river were selected as the fixed sampling sites for continuous measurements during storm periods. The upper Estuary was defined as the part of the normal estuary, which was occupied by freshwater as a result of the storm.
From Chen, N., Krom, M. D., Wu, Y., Yu, D. & Hong, H. (2018). Storm induced estuarine turbidity maxima and controls on nutrient fluxes across river-estuary-coast continuum. *Science of the Total Environment, 628–629*, 1108–1120. https://doi.org/10.1016/j.scitotenv.2018.02. 060. Copyright of this figure was obtained via CCC (Copyright Clearance Center https://www. copyright.com/) print it in color

et al., 2015). Results show that storm-induced extreme hydrological condition increase P fluctuations. The P concentrations and total flux in storm runoff are greatest in human perturbed watersheds due to higher proportions and absolute amounts of overland flows from croplands and forests, the discharge of human and animal wastes and P loss from fertilized land, which carry more eroded soil particles from surface soils to the stream (Correll et al., 1999). Total particulate P (TPP) export was greatest toward the early to high stages of the storm events. During high flow periods, P (both dissolved and particulate) is quickly transported downstream by storm flow, and the retention capacity of reservoirs is low. The amount of TPP per gram total suspended matter (TSM) during high-flow period is also lower compared with later falling limb of hydrograph, largely due to the fact that coarse particles contain relative less P compared with fine particles. Moreover, particulate P became more easily entrained during extreme storms because the fast-flowing water did not allow much opportunity for particle bound P to sediment out or for particle-bound P to be converted into dissolved forms under the energetic conditions (Fraser et al., 1999).

In summary, the response of P dynamics in the downriver during storms represents a combination of the relative contribution of transport pathways (surface runoff, subsurface runoff) and location of the sources (within the channel and across the catchment). The key process is associated with rainfall pattern (size, timing, and spatial distribution), and antecedent soil moisture (saturated or unsaturated) determining P dynamics with respect to enrichment versus dilution. The storm events had different effects on the two dissolved P components: dissolved organic phosphorus (DOP) and dissolved reactive phosphorus (DRP). Generally, the greatest variation in stream water P concentrations often occur over storm events, mainly a dilution of DRP & DOP and an increase in TPP as previous findings (Evans & Johnes, 2004; Haygarth et al., 2005). The dilution of DRP might be offset by a rapid supplement from external sources. DOP can be transported via runoff and become enriched in river water during the rising limb of the hydrograph, before subsequent dilution by excess rainwater in the falling limb. Moreover, mean concentration of storm A was the highest of the three storms even though rainfall was not as great as in storm B and C. That is because the first significant rain (Storm A) after a long dry period is likely to have mobilized accumulated waste and usually such events generate high concentrations of dissolved matter in runoff (Ribarova et al., 2008). The "first-flush" effect is considered to be more extreme in human perturbed watersheds than in natural ones (Yoon & Stein, 2008). However, DRP and DOP exhibited a similar clockwise trajectory around the high flow period, which implied that within-channel mobilization of P can be diluted by increasing storm runoff.

Three storm events, which occurred on July 13–14, 2013 (Storm C, caused by Typhoon Soulik), May 21–23, 2014 (Storm D, the first major storm of the year) and July 23–24, 2014 (Storm E) were also studied (Chen et al., 2018). During major storms, the freshwater-saltwater front in the JRE shifts approximately 10 km downstream. This storm-induced runoff increases nutrient fluxes into the JRE, altering nutrient availability and N:P:Si ratios. A sharp increase in DIN loads and changes in N composition (relatively more ammonium) in the extreme storm also occurred in June 2010 (Chen & Hong, 2012). As a result, there is an increased potential for eutrophication, enhancement of the P-limitation, and increased Chl-a concentrations (Chen et al., 2018).

3.2.3   The Han River estuary

During the July in 2006, three consecutive typhoons (Ewiniar, Bilis, and Kaemi) landed on the intertidal zone of the Han River estuary and released heavy rains (>100 mm/d; total 603 mm in a month) in the SeoulIncheon-Ganghwa area on the west coast of Korea. The penetration of rain and freshwater induced a distinct decrease of salinity and sulfate concentrations in the pore water. Low concentrations and the vertically uniform distributions of NH4+ and Fe^{2+} in the pore water during the heavy precipitation on July 19 and 25 directly indicated oxidative effects by the rainfall and freshwater runoff, which were likely caused by a combination of sediment splash by raindrops, infiltration of the rainwater, and tidal inundation by low salinity waters. As the typhoon-induced heavy rainfall continued, a significant shift in anaerobic organic carbon (C_{org}) oxidation pathways was identified, from sulfate reduction dominated conditions to iron reduction and denitrification prevailed condition (Mok et al., 2019).

3.2.4   The southern Baltic Sea

The CWB of the southern Baltic Sea in western Germany consist of shallow fjords and bays. In years with greater precipitation, phytoplankton growth was greater, and DIP concentrations were lower than in years with average precipitation. (Berthold et al., 2018) during years with above-average precipitation the water bodies were affected seasonally. Because bioavailable nitrogen was also delivered into coastal systems during extreme weather events, the DIN:DIP ratio increased (reaching up to 60:1 during storms) (Rees et al., 2009). Therefore, N is the cause of eutrophication in the majority of coastal marine ecosystems. The increased demand for DIP during growth in conjunction with simultaneous DIN loading may account for the lower DIP levels. Even though nitrogen is most likely the leading cause of eutrophication, optimal management of postrain eutrophication calls for a reduction of both nitrogen and phosphorus (Howarth & Marino, 2006).

The Gulf of Gdańsk is a bay in the southeast of the Baltic Sea. The values of total iron in 30 rainwater samples collected between December 2002 and November 2003 showed that the yearly flux of Fe was 11.22 mgm/2 year. It suggested that wet deposition could be a source of iron for phytoplankton in the southern Baltic. Iron in rain originated from both land and saltwater. During winds from the east and southeast, the swash zone saturated with marine aerosols was a substantial source of iron. After long and intense rainfall, the subsequent sample exhibited very low iron concentrations despite the high volume of rain collected. This suggests that earlier precipitation was more efficient at removing aerosols from the atmosphere (Dunajska et al., 2006).

Mercury (Hg) is a highly hazardous contaminant that poses a substantial threat to global wildlife and human health, as well as the global economy (Mergler et al., 2007; Swain et al., 2007). Water samples collected from four rivers in the Southern Baltic region (the Reda, the Zagórska Struga, the Putnica, and the Gizdepka) reveal a lower retention rate of mercury in the river catchment following intense rainfall, indicating mercury elution from the catchment (Saniewska et al., 2018). The downpours, which occurred in the summer of 2016, had insignificant effect on Hg concentration but played a key role in the transportation of Hg to the rivers, which resulted in a surface run-off that was rich in Hg in SPM (Babiarz et al., 2012; Lyons et al., 2006; Saniewska et al., 2014), similar to other downpour events in the Southern Baltic region (Dominika Saniewska et al., 2014; Jedruch et al., 2017).

3.2.5   The Neuse River Estuary

The NRE, United States, is the second largest estuary in the southern United States, which has experienced 38 tropical storms and hurricanes since 1996 and has been increasingly impacted by extreme weather events leading to record precipitation and flooding events (Bales, 2003; Paerl et al., 2018, 2019). Increased precipitation is a key regulator of microalgal abundance and community composition on time scales of 1 week to 1 month and 1 year (Paerl, Hall, et al., 2020; Paerl, Venezia, et al., 2020). In September 2018, Hurricane Florence caused an extreme surge in river flow, with salinity in the mid-estuary dropping from ∼ 10 to ≤2.5 16 days after landfall and remaining <7.5 for three months after landfall. CDOM concentrations increased 1.75−2 folds relative to prestorm levels. Total dissolved nitrogen (TDN) increased ∼ 3-fold, remaining >750 μg/NL (Rudolph et al., 2020; Fig. 12.8).

3.3   Impact of extreme precipitation events on lagoons

Coastal lagoons are shallow bodies of water separated from the ocean by sandbars, barrier islands, or coral reefs. They occupy 13% of the coastal areas worldwide (Kjerfve, 1994). The inner waters of lagoons are more enclosed than those of bays, with a narrower area connected to the open sea. In some cases, the inner waters are only intermittently connected to the open sea. The enclosed characteristics of lagoons make them more susceptible to internal riverine runoff and eutrophication. Nutrients can be taken up by biota for extended periods, and organic matter can be decomposed by bacteria, releasing additional bioavailable nutrients (Livingston, 2001). This can result in a vicious cycle, especially under stratified and anoxic conditions (Eyre & Ferguson, 2002). Different from the bays or estuaries, even small changes in water levels due to precipitation can significantly impact the volume and surface area of a lagoon. Its effect of dilution on the concentration of nutrients generated by precipitation is often overlooked or underemphasized. This can significantly affect calculations of concentrations, leading to an underestimation of the impact of rainfall on lagoonal systems (Rissik et al., 2009; Thomas et al., 2005).

3.3.1   The Ria Formosa lagoon

The Ria Formosa lagoon is a mesotidal lagoon (average depth 2 m) extending along the south coast of Portugal for approximately 55 km. The region is subjected to hot dry summers and moderate winters, typical of Mediterranean climate (Domingues et al., 2023). Particularly, short-term predictions (2020−2050) indicate a possible increase in the volume and intensity of winter precipitation (Stigter et al., 2014).

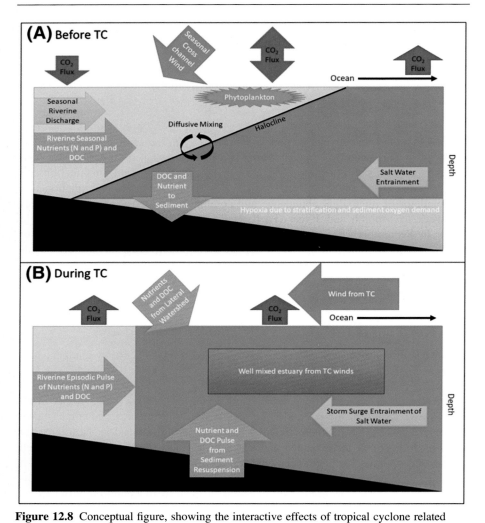

Figure 12.8 Conceptual figure, showing the interactive effects of tropical cyclone related freshwater discharge, DOC and nutrient loading and wind mixing on physical structure of the water column, phytoplankton and associated microbial activities and biogeochemical responses, including hypoxia and air–water CO_2 exchange in a lagoonal estuary like the Neuse River Estuary. Illustrated are the before, during, and after TC scenarios.
From Paerl, H. W., Hall, N. S., Hounshell, A. G., Rossignol, K. L., Barnard, M. A., Luettich, R. A., Rudolph, J. C., Osburn, C. L., Bales, J. & Harding, L. W. (2020). Recent increases of rainfall and flooding from tropical cyclones (TCs) in North Carolina (USA): Implications for organic matter and nutrient cycling in coastal watersheds. *Biogeochemistry*, *150*(2), 197–216. https://doi.org/10.1007/s10533-020-00693-4; Paerl, R. W., Venezia, R. E., Sanchez, J. J. & Paerl, H. W. (2020). Picophytoplankton dynamics in a large temperate estuary and impacts of extreme storm events. *Scientific Reports*, *10*(1), 22026. https://doi.org/10.1038/s41598-020-79157-6. Copyright of this figure was obtained via CCC (Copyright Clearance Center https://www.copyright.com/)

Increased stochastic events may have a significant impact on the annual nutrient budget and, consequently, the ecological function of the Ria Formosa lagoon (Malta et al., 2017). The coastal ocean is continuously fertilized by particulate and dissolved organic (carbon) and inorganic (N and P) nutrients. Following a period of significant rainfall in late autumn, this pattern was reversed: a net import of nutrients from the ocean was observed during this time. The high percentage of nitrates in the water flowing into the lagoon was also notably impacted by agricultural runoff. This may be due to insufficient mixing of outflowing stream freshwater with denser oceanic water, resulting in the reimport of nutrient-rich water with the next incoming tide. This theory is supported by the high-nutrient concentrations and reduced salinity of incoming ocean water during the morning flood.

It is worth noting that Ria Formosa Lagoon is the most important bivalve production region of the South Portuguese coast. However, the microbial contamination has been increasing in the north-west area of the lagoon and rain was the most influential abiotic factor (Bettencourt et al., 2013). In the fall and winter, precipitation increases the transport of contaminants via runoff and that will flow into the lagoon (Campos & Cachola, 2007). For example, the highest values of *E. coli* in bivalves were associated with periods of heaviest rainfall from 1990 to 2009 (Almeida & Soares, 2012). The runoff of waters caused by precipitation carries a variety of organic and inorganic chemicals that will flow into the coastline. It may lead the runoff from contaminated lands to the discharge of untreated or partially treated sewage from sewage treatment plants (STPs), and other intermittent discharges (Kay et al., 2008; Lee et al., 2003). Furthermore, the effectiveness of STPs may be reduced by high flow rates during periods of severe rainfall (Ackerman & Weisberg, 2003; Younger et al., 2003).

3.3.2 Songkhla lagoon

Songkhla lagoon is southeastern Thailand's largest natural lagoon. Rainfed crop production is the major source of livelihood surrounded by sugar palm hedges (Dumrongrojwatthana et al., 2020). Since the early 1980s, rainfall depth, number of rainy days, and maximum daily rainfall have increased in January, as well as the Gulf of Thailand's yearly frequency of extreme rainfall events (Limsakul et al., 2010). The extensive multidecadal variability of rainfall is affected by the El Niño—Southern Oscillation (Räsänen & Kummu, 2013) and the North Pacific Oscillation (Wang et al., 2007). Therefore, climate change-induced rising sea levels and heavy rainfall events threaten this coastal agricultural system and other low-lying rice-growing regions in South and Southeast Asia (Salamanca & Rigg, 2017, pp. 280—297; Yen et al., 2019).

According to the Intergovernmental Panel on Climate Change's definition of vulnerability to climate change (IPCC, 2007), a vulnerability framework was used to assess the sensitivity and response capacity to these trends of the great Songkhla spit agroecosystem (Dumrongrojwatthana et al., 2022). Three dimensions of vulnerability were examined: exposure to hazards, sensitivity to hazards, and adaptive capacity. The impacts of greater exposure to extreme rainfall events include primarily a reduction in rice-growing area, the destruction of the sandbar that protects the rice-growing flood plain, the elimination of coastal flora, and an increase in the vulnerability of shrimp-farming regions. A realistic and robust strategy, comprising short-term, long-term, nonstructural, and structural measures, is developed to mitigate the impact (Dumrongrojwatthana et al., 2022) (Figs. 12.9 and 12.10).

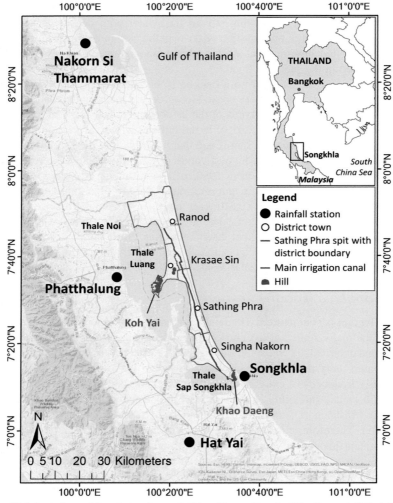

Figure 12.9 Location of the study site on Sathing Phra spit and of the four rainfall stations of Songkhla lagoon area.
Topographic base map from ESRI, ArcGIS 10.4 From Dumrongrojwatthana, P., Lacombe, G. & Trébuil, G. (2022). Increased frequency of extreme rainfall events threatens an emblematic cultural coastal agroecosystem in Southeastern Thailand. *Regional Environmental Change*, 22(2). https://doi.org/10.1007/s10113-021-01868-x. Copyright of this figure was obtained via CCC (Copyright Clearance Center https://www.copyright.com/)

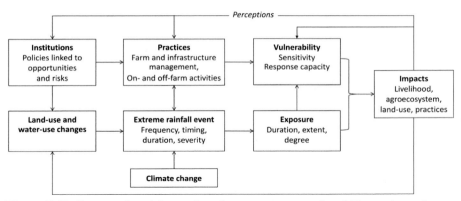

Figure 12.10 Conceptual model to analyze the agroecosystem vulnerability to change in two and potential adaptive responses.
From Dumrongrojwatthana, P., Lacombe, G. & Trébuil, G. (2022). Increased frequency of extreme rainfall events threatens an emblematic cultural coastal agroecosystem in Southeastern Thailand. *Regional Environmental Change*, 22(2). https://doi.org/10.1007/s10113-021-01868-x. Copyright of this figure was obtained via CCC (Copyright Clearance Center https://www. copyright.com/)

3.3.3 The Rodrigo de Freitas Lagoon

The RFL is a coastal shallow system of approximately 2.2 km^2 located within an urban landscape at Rio de Janeiro city, Southeastern Brazil. Very heavy and extreme rainfall events (VHREs) are defined as 99th percentile precipitation >80.7 mm/day and from 200 to 300 mm/day (Lima et al., 2021; Pristo et al., 2018). From 2018 to 2019, a total of 15 occurrences of VHREs, directly interfering on the storm-water input into RFL. Moreover, ammonia emissions increase from the lagoon to atmosphere during rainfall events (Santos et al., 2012). The majority of the VHREs were affected by the mild El Niño phenomenon (2018−19), and two consecutive VHREs were affected by the moderate La Niña phenomenon (2018−19). Rainfall, freshwater intake from rivers, and water exchange with the South Atlantic Ocean all influence the level of the lagoon (Neves et al., 2020). Even a week after VHREs, there was a noticeable rise in water levels in the lagoon. When these levels exceed 80 cm (i.e., the critical water level), flooding events on RFL can occur, having an impact on the neighboring metropolitan area.

VHREs modify lagoon water parameters, decreasing the overall quality of this shallow coastal system that supports artisanal fishery and many aquatic sport activities. Particularly, the high values of water transparency, higher contribution of *E. coli* density, low values of water temperature, dissolved oxygen, and phytoplankton abundance were characteristics of the water conditions following heavy rainfall (Neves & Santos, 2022). The increase in the entrance of diffuse sources of nutrient-enriched inputs, such as effluents from sewage treatment plants, may be the cause of the positive signal of *E. coli* contribution. Reductions in dissolved oxygen after heavy rainfalls have been proofed by a laboratory study (Dos Santos & Teixeira, 2019). In addition, VHREs caused changes in the relative contributions of phytoplanktonic community. *Cyanobacteria* made up the majority of the phytoplankton community, with unidentified *picocyanobacteria* (<1.5 μm) being the most dominant taxa (78.57%) before VHREs. The amount of *cyanobacteria* present in RFL has significantly decreased after VHREs, while the presence of both *cryptophytes* and *diatoms* has increased. Similar reductions in cyanobacteria biomass with the emergence of diatoms, *cryptophytes* and *chlorophytes*, have also been found in Ülemiste, the fourth largest lake in Estonia (Pedusaar et al., 2010). It is worth mentioning that the shifts in the phytoplankton contribution here following VHREs seem to be favorable to planktonic community by the increase in the relative contribution of more palatable microalgae (*diatoms* and *cryptophytes*) to mesozooplankton consumers (rotifers and calanoid nauplii (Souza et al., 2011).

3.3.4   Other typical lagoons

Phytoplankton growth was usually controlled by nitrogenous nutrients. However, prior to the advent of rainfall, Dapeng Lagoon (Taiwan) had relatively low phosphate concentrations and phytoplankton levels. Large quantities of dissolved inorganic phosphate nutrients have been discharged into the lagoon following extreme precipitation, causing phytoplankton blooms (Meng et al., 2014); Chilika lagoon on the east coast of India is the largest brackish water lagoon in Asia and is one of the largest tropical lagoons in the world. The lagoon, located along the eastern coast of India and connected to the Bay of Bengal (BoB), serves as a significant ecotone due to the presence of marine, brackish, and freshwater micro- and macroflora. Heavy precipitation resulted in an influx of nutrient-rich freshwater from extreme events and increased monsoon rainfall through augmented river flow and terrestrial run-off. The subsequent increase in nutrients carried by the run-off stimulated phytoplankton growth, albeit in a lag phase. The addition of SiO_4 through terrigenous run-off facilitated preferential growth of diatoms (Sahoo et al., 2017); Dee Why Lagoon, located on the northern beaches of Sydney, Australia, is an intermittently closed and open lagoon encircled by highly urbanized areas with predominantly concrete-lined drains and stormwater infrastructure. Heavy rainfall in the catchment had the potential to advect planktonic species from freshwater habitats into the lagoon, influencing the plankton composition of the receiving waters. However, this scenario to be unlikely as the catchment of Dee Why Lagoon is highly urbanized and is dominated by concrete lined drains and storm water infrastructure, with few water bodies where plankton communities could develop. Therefore, heavy precipitation primarily affects the zooplankton in the lagoon by introducing nutrients that promote the assimilative growth of phytoplankton. The lagoon has a limited plankton community, and the introduction of drifting plankton species directly from freshwater habitats into the lagoon is not feasible. Heavy precipitation primarily affects the zooplankton in the lagoon by introducing nutrients that promote the assimilative growth of phytoplankton (Rissik et al., 2009).

3.4 Impact of extreme precipitation events on other types of marine areas

In addition to the aforementioned bays, estuaries, and lagoons, there are other types of marine areas that are vulnerable to extreme precipitation events, such as the central Yellow Sea oligotrophic zone (Zou et al., 2000) and the Seto Inland Sea (Abe et al., 2013). Due to its distance from the estuary and limited upwelling, the central Yellow Sea is a typical oligotrophic region. Atmospheric wet deposition may account for 65% and 70% of the total input of nutrients to the DIN and DIP, respectively (Zhang, 1994; Zhang & Liu, 1994). The observations indicate that rainwater high in N and low in P is transported to the central Yellow Sea and plays an important role in phytoplankton growth (Zou et al., 2000). In July 2012, there was a transient increase in nutrient concentrations in the coastal zone as a result of heavy rainfall in the central Seto Inland Sea. The river outflow was increased and its low salinity and elevated nutrient concentrations were detected near the river mouths (Abe et al., 2013).

4. Conclusion

Ecosystems in coastal ecosystem following extreme precipitation do not immediately return to the preprecipitation state, but generally take several days to recover (Balint et al., 2022; Chen et al., 2019; Liu et al., 2004; Mendoza-Salgado et al., 2005; Rissik et al., 2009; Shi et al., 2011; Wang et al., 2009; Y.;Yang & Wang, 2013; Zhao et al., 2002). For example, prerainfall hydraulic balance conditions are re-established about 10 days after the storm in an arid Gulf of California coastal lagoon (Mendoza-Salgado et al., 2005). In the Dee Why Lagoon on the northern beaches of Sydney, Australia, phytoplankton biomass increased 10-fold in 25−30°C water 1 week after the initial rainfall and 2 weeks later decreased to near initial levels. General nutrients (ammonia and nitrogen oxides) increase significantly the day after the initial rainfall and then return to prestorm conditions within 5 days, whereas phytoplankton biomass reaches a maximum and recovery takes longer. Within 2 weeks of the rainfall, the zooplankton progressively returned to its original community after experiencing dramatic changes after one and 6 days (Rissik et al., 2009).

Resilience is a measure of the persistence of systems and of their ability to absorb change and disturbance and still maintain the same relationships between populations or state variables (Holling, 1973), which is also a concept linked to vulnerability—as the socio-ecological system's capacity to absorb recurrent disturbances to retain essential structures, processes, and feedbacks, and then to reorganize into a fully functioning system (Adger et al., 2005; Cutter et al., 2008). For extreme precipitation, marine ecosystem resilience incorporates two dynamic processes: the ability of ecosystems to resist and absorb disturbances from extreme precipitation, and their ability to recover to preprecipitation levels (Darling & Côté, 2018). Global studies have shown that at least 50% of DIN circulating in the atmosphere originates from fertilizer use, fossil fuel combustion, and industrial processes (Vet et al., 2014). A national study of reactive nitrogen in rainfall conducted by the Chinese Agricultural University's

Sediment Network found that DIN concentrations in rainfall nearly doubled between 1980 and 2010. Furthermore, the concentration of DIN in precipitation is rising at an average rate of 0.063 mg/L per year, with ammonium as the principal contributor (Liu et al., 2013). Consequently, enhancing marine ecological resilience is once again closely associated with energy conservation and emission reduction.

We should take steps to increase the resilience of marine ecosystems and prevent or mitigate natural disasters (such as inundation and fish kills): (1) the complete eradication of illegal sources of sewage residues originating from urban areas; (2) continuous dredging interventions to maintain the functionality of lagoons or semi-enclosed bays as connecting channels to the outer sea; (3) the establishment of long-term monitoring networks to augment our comprehension of the impacts of extreme precipitation on coastal regions characterized by rapid population growth, industrial development, highly sensitive ecosystems or intensified agricultural activity; (4) the completion of emergency management measures after extreme precipitation events. For example, the affected areas can be temporarily controlled after extreme precipitation, including reducing recreational (tourism) and resource extraction activities (shrimp farming, shellfish harvesting, etc.), and limiting sewage discharge. This emergency management measure aids in mitigating increased pollution levels and enables a more scientific evaluation of the impact of heavy precipitation on marine ecology.

Current research has focused on the short-term effects of extreme precipitation on biological and hydrological parameters (i.e., one and 2 weeks after heavy precipitation). We recommend further investigation into the long-term impacts on water quality and phytoplankton communities. Considering projected increases in precipitation intensity and the frequency of extreme events, it is imperative to fill the knowledge gap of the impacts of extreme precipitation on offshore ecosystems, including harm to aquatic biodiversity and fisheries, as well as the actual efficacy of measures implemented to facilitate ecological recovery.

AI disclosure

During the preparation of this work the author(s) used [NAME TOOL/SERVICE] in order to [REASON]. After using this tool/service, the author(s) reviewed and edited the content as needed and take(s) full responsibility for the content of the publication.

References

Abe, K., Nakagawa, N., Abo, K., & Tsujino, M. (2013). High nutrients in the coastal area after heavy rain observed in the central Seto Inland Sea in July 2012. *Journal of Oceanography, 69*(2), 269–275. https://doi.org/10.1007/s10872-013-0171-y

Acker, M., Hogle, S. L., Berube, P. M., Hackl, T., Coe, A., Stepanauskas, R., Chisholm, S. W., & Repeta, D. J. (2022). Phosphonate production by marine microbes: Exploring new sources and potential function. *Proceedings of the National Academy of Sciences, 119*(11). https://doi.org/10.1073/pnas.2113386119

Ackerman, D., & Weisberg, S. B. (2003). Relationship between rainfall and beach bacterial concentrations on Santa Monica Bay beaches. *Journal of Water and Health, 1*(2), 85−89. https://doi.org/10.2166/wh.2003.0010

Adey, W. H., & Goertemiller, T. (1987). Coral reef algal turfs: Master producers in nutrient poor seas. *Phycologia, 26*(3), 374−386. https://doi.org/10.2216/i0031-8884-26-3-374.1

Adger, W. N., Hughes, T. P., Folke, C., Carpenter, S. R., & Rockström, J. (2005). Social-ecological resilience to coastal disasters. *Science, 309*(5737), 1036−1039. https://doi.org/10.1126/science.1112122

Alldredge, A. (2019). *Mcr LTER: Coral reef: Water column: Nutrients, ongoing since 2005.* Environmental Data Initiative. https://doi.org/10.6073/PASTA/9328A024F2BF16EC C66024F07DBCC574

Almeida, C., & Soares, F. (2012). Microbiological monitoring of bivalves from the Ria Formosa Lagoon (south coast of Portugal): A 20 years of sanitary survey. *Marine Pollution Bulletin, 64*(2), 252−262. https://doi.org/10.1016/j.marpolbul.2011.11.025

Alongi, D. M. (1995). Effect of monsoonal climate on sulfate reduction in coastal sediments of the central Great Barrier Reef lagoon. *Marine Biology, 122*(3), 497−502. https://doi.org/10.1007/bf00350884

Alongi, D. M., Tirendi, F., Dixon, P., Trott, L. A., & Brunskill, G. J. (1999). Mineralization of organic matter in intertidal sediments of a tropical semi-enclosed delta. *Estuarine, Coastal and Shelf Science, 48*(4), 451−467. https://doi.org/10.1006/ecss.1998.0465

Alongi, D. M., Wattayakorn, G., Boyle, S., Tirendi, F., Payn, C., & Dixon, P. (2004). Influence of roots and climate on mineral and trace element storage and flux in tropical mangrove soils. *Biogeochemistry, 69*(1), 105−123. https://doi.org/10.1023/B:BIOG.0000031043.06245.af

Anon. (1987). Land-sea boundary flux of contaminants: Contributions from rivers. *Reports and Studies No., 32*, 32.

Apple, J. K., Strom, S. L., Palenik, B., & Brahamsha, B. (2011). Variability in protist grazing and growth on different marine Synechococcus isolates. *Applied and Environmental Microbiology, 77*(9), 3074−3084. https://doi.org/10.1128/AEM.02241-10

Babiarz, C., Hoffmann, S., Wieben, A., Hurley, J., Andren, A., Shafer, M., & Armstrong, D. (2012). Watershed and discharge influences on the phase distribution and tributary loading of total mercury and methylmercury into Lake Superior. *Environmental Pollution, 161*, 299−310. https://doi.org/10.1016/j.envpol.2011.09.026

Baek, S. H., Shimode, S., Kim, H.c., Han, M. S., & Kikuchi, T. (2009). Strong bottom-up effects on phytoplankton community caused by a rainfall during spring and summer in Sagami Bay, Japan. *Journal of Marine Systems, 75*(1−2), 253−264. https://doi.org/10.1016/j.jmarsys.2008.10.005

Bales, J. D. (2003). Effects of hurricane Floyd Inland flooding, September−October 1999, on tributaries to Pamlico Sound, North Carolina. *Estuaries, 26*(5), 1319−1328. https://doi.org/10.1007/BF02803634

Balint, S. J., Joyce, E., Pennino, S., Oczkowski, A., McKinney, R., & Hastings, M. G. (2021). Identifying sources and impacts of precipitation-derived nitrogen in Narragansett Bay, RI. *Estuaries and Coasts.* https://doi.org/10.1007/s12237-021-01029-7

Balint, S. J., Joyce, E., Pennino, S., Oczkowski, A., McKinney, R., & Hastings, M. G. (2022). Identifying sources and impacts of precipitation-derived nitrogen in Narragansett Bay, RI. *Estuaries and Coasts, 45*(5), 1287−1304. https://doi.org/10.1007/s12237-021-01029-7

Ballard, T. C., Sinha, E., & Michalak, A. M. (2019). Long-term changes in precipitation and temperature have already impacted nitrogen loading. *Environmental Science and Technology, 53*(9), 5080−5090. https://doi.org/10.1021/acs.est.8b06898

Banse, K. (1994). Grazing and zooplankton production as key controls of phytoplankton production in the open ocean. *Oceanography, 7*(1), 13−20. https://doi.org/10.5670/oceanog.1994.10

Bäuerlein, P. S., Hofman-Caris, R. C. H. M., Pieke, E. N., & ter Laak, T. L. (2022). Fate of microplastics in the drinking water production. *Water Research, 221*. https://doi.org/10.1016/j.watres.2022.118790

Behrenfeld, M. J., Bale, A. J., Kolber, Z. S., Aiken, J., & Falkowski, P. G. (1996). Confirmation of iron limitation of phytoplankton photosynthesis in the equatorial Pacific Ocean. *Nature, 383*(6600), 508−511. https://doi.org/10.1038/383508a0

Bellwood, D. R., Hughes, T. P., Folke, C., & Nyström, M. (2004). Confronting the coral reef crisis. *Nature, 429*(6994), 827−833. https://doi.org/10.1038/nature02691

Berthold, M., Karsten, U., von Weber, M., Bachor, A., & Schumann, R. (2018). Phytoplankton can bypass nutrient reductions in eutrophic coastal water bodies. *Ambio, 47*(S1), 146−158. https://doi.org/10.1007/s13280-017-0980-0

Bettencourt, F., Almeida, C., Santos, M. I., Pedroso, L., & Soares, F. (2013). Microbiological monitoring of Ruditapes decussatus from Ria Formosa lagoon (south of Portugal). *Journal of Coastal Conservation, 17*(3), 653−661. https://doi.org/10.1007/s11852-013-0264-1

Bouman, H. A., Nakane, T., Oka, K., Nakata, K., Kurita, K., Sathyendranath, S., & Platt, T. (2010). Environmental controls on phytoplankton production in coastal ecosystems: A case study from Tokyo Bay. *Estuarine, Coastal and Shelf Science, 87*(1), 63−72. https://doi.org/10.1016/j.ecss.2009.12.014

Bruesewitz, D. A., Gardner, W. S., Mooney, R. F., Pollard, L., & Buskey, E. J. (2013). Estuarine ecosystem function response to flood and drought in a shallow, semiarid estuary: Nitrogen cycling and ecosystem metabolism. *Limnology & Oceanography, 58*(6), 2293−2309. https://doi.org/10.4319/lo.2013.58.6.2293

Campos, C. J. A., & Cachola, R. A. (2007). Faecal coliforms in bivalve harvesting areas of the Alvor lagoon (Southern Portugal): Influence of seasonal variability and urban development. *Environmental Monitoring and Assessment, 133*(1−3), 31−41. https://doi.org/10.1007/s10661-006-9557-2

Carlton, E. J., Eisenberg, J. N. S., Goldstick, J., Cevallos, W., Trostle, J., & Levy, K. (2014). Heavy rainfall events and diarrhea incidence: The role of social and environmental factors. *American Journal of Epidemiology, 179*(3), 344−352. https://doi.org/10.1093/aje/kwt279

Chambers, L. G., Osborne, T. Z., & Reddy, K. R. (2013). Effect of salinity-altering pulsing events on soil organic carbon loss along an intertidal wetland gradient: A laboratory experiment. *Biogeochemistry, 115*(1), 363−383. https://doi.org/10.1007/s10533-013-9841-5

Chan, T. U., & Hamilton, D. P. (2001). Effect of freshwater flow on the succession and biomass of phytoplankton in a seasonal estuary. *Marine and Freshwater Research, 52*(6), 869−884. https://doi.org/10.1071/MF00088

Chang, J., Chung, C. C., & Gong, G. C. (1996). Influences of cyclones on chlorophyll a concentration and Synechococcus abundance in a subtropical western Pacific coastal ecosystem. *Marine Ecology Progress Series, 140*(1−3), 199−205. https://doi.org/10.3354/meps140199

Chen, L., Li, J., Tang, Y., Wang, S., Lu, X., Cheng, Z., Zhang, X., Wu, P., Chang, X., & Xia, Y. (2021). Typhoon-induced turbulence redistributed microplastics in coastal areas and reformed plastisphere community. *Water Research, 204*. https://doi.org/10.1016/j.watres.2021.117580

Chen, M. J., Lin, C. Y., Wu, Y. T., Wu, P. C., Lung, S. C., & Su, H. J. (2012). Effects of extreme precipitation to the distribution of infectious diseases in Taiwan, 1994—2008. *PLoS One, 7*(6). https://doi.org/10.1371/journal.pone.0034651

Chen, N., & Hong, H. (2012). Integrated management of nutrients from the watershed to coast in the subtropical region. *Current Opinion in Environmental Sustainability, 4*(2), 233—242. https://doi.org/10.1016/j.cosust.2012.03.007

Chen, N., Krom, M. D., Wu, Y., Yu, D., & Hong, H. (2018). Storm induced estuarine turbidity maxima and controls on nutrient fluxes across river-estuary-coast continuum. *Science of the Total Environment, 628—629,* 1108—1120. https://doi.org/10.1016/j.scitotenv.2018.02.060

Chen, N., Wu, J., & Hong, H. (2012). Effect of storm events on riverine nitrogen dynamics in a subtropical watershed, southeastern China. *Science of the Total Environment, 431,* 357—365. https://doi.org/10.1016/j.scitotenv.2012.05.072

Chen, N., Wu, Y., Chen, Z., & Hong, H. (2015). Phosphorus export during storm events from a human perturbed watershed, southeast China: Implications for coastal ecology. *Estuarine, Coastal and Shelf Science, 166,* 178—188. https://doi.org/10.1016/j.ecss.2015.03.023

Chen, Y. Y., Song, D. H., Bao, X. W., & Yan, Y. H. (2019). Impact of the cross-bay bridge on water exchange in Jiaozhou Bay, Qingdao, China. *Oceanologia et Limnologia Sinica, 50*(4), 707—718. https://doi.org/10.11693/hyhz20180900211

Chew, C. A., Hepburn, C. D., & Stephenson, W. (2013). Low-level sedimentation modifies behaviour in juvenile Haliotis iris and may affect their vulnerability to predation. *Marine Biology, 160*(5), 1213—1221. https://doi.org/10.1007/s00227-013-2173-0

Clausing, R. J., Annunziata, C., Baker, G., Lee, C., Bittick, S. J., & Fong, P. (2014). Effects of sediment depth on algal turf height are mediated by interactions with fish herbivory on a fringing reef. *Marine Ecology Progress Series, 517,* 121—129. https://doi.org/10.3354/meps11029

Cloern, J. E. (1987). Turbidity as a control on phytoplankton biomass and productivity in estuaries. *Continental Shelf Research, 7*(11—12), 367—1381. https://doi.org/10.1016/0278-4343(87)90042-2

Cohen, R. A., & Fong, P. (2004). Nitrogen uptake and assimilation in Enteromorpha intestinalis (L.) Link (Chlorophyta): Using 15N to determine preference during simultaneous pulses of nitrate and ammonium. *Journal of Experimental Marine Biology and Ecology, 309*(1), 67—77. https://doi.org/10.1016/j.jembe.2004.03.009

Correll, D. L., Jordan, T. E., & Weller, D. E. (1999). Effects of precipitation and air temperature on phosphorus fluxes from Rhode River watersheds. *Journal of Environmental Quality, 28*(1), 144—154. https://doi.org/10.2134/jeq1999.00472425002800010017x

Coulliette, A. D., & Noble, R. T. (2008). Impacts of rainfall on the water quality of the Newport River Estuary (eastern North Carolina, USA). *Journal of Water and Health, 6*(4), 473—482. https://doi.org/10.2166/wh.2008.136

Crumpton, W. G., & Wetzel, R. G. (1982). Effects of differential growth and mortality in the seasonal succession of phytoplankton populations in Lawrence Lake, Michigan. *Ecology, 63*(6), 1729—1739. https://doi.org/10.2307/1940115

Cutter, S. L., Barnes, L., Berry, M., Burton, C., Evans, E., Tate, E., & Webb, J. (2008). A place-based model for understanding community resilience to natural disasters. *Global Environmental Change, 18*(4), 598—606. https://doi.org/10.1016/j.gloenvcha.2008.07.013

Damashek, J., & Francis, C. A. (2018). Microbial nitrogen cycling in estuaries: From genes to ecosystem processes. *Estuaries and Coasts, 41*(3), 626—660. https://doi.org/10.1007/s12237-017-0306-2

Darling, E. S., & Côté, I. M. (2018). Seeking resilience in marine ecosystems. *Science, 359*(6379), 986–987. https://doi.org/10.1126/science.aas9852

DeMaster, D. J. (1981). The supply and accumulation of silica in the marine environment. *Geochimica et Cosmochimica Acta, 45*(10), 1715–1732. https://doi.org/10.1016/0016-7037(81)90006-5

den Haan, J., Huisman, J., Brocke, H. J., Goehlich, H., Latijnhouwers, K. R. W., van Heeringen, S., et al. (2016). Nitrogen and phosphorus uptake rates of different species from a coral reef community after a nutrient pulse. *Scientific Reports, 6*(1), Article 28821. https://doi.org/10.1038/srep28821

Diaz, R. J., & Rosenberg, R. (2008). Spreading dead zones and consequences for marine ecosystems. *Science, 321*(5891), 926–929. https://doi.org/10.1126/science.1156401

Domingues, R. B., Nogueira, P., & Barbosa, A. B. (2023). Co-limitation of phytoplankton by N and P in a shallow coastal lagoon (Ria Formosa): Implications for eutrophication evaluation. *Estuaries and Coasts, 46*(6), 1557–1572. https://doi.org/10.1007/s12237-023-01230-w

Dos Santos, N. de O., & Teixeira, L. (2019). Accelerated reoxygenation of water bodies using hydrogen peroxide. *International Journal of Environmental Studies, 76*(4), 558–570. https://doi.org/10.1080/00207233.2018.1494929

Drayna, P., McLellan, S. L., Simpson, P., Li, S. H., & Gorelick, M. H. (2010). Association between rainfall and pediatric emergency department visits for acute gastrointestinal illness. *Environmental Health Perspectives, 118*(10), 1439–1443. https://doi.org/10.1289/ehp.0901671

Dumrongrojwatthana, P., Lacombe, G., & Trebuil, G. (2022). Increased frequency of extreme rainfall events threatens an emblematic cultural coastal agroecosystem in Southeastern Thailand. *Regional Environmental Change, 22*(2), 36. https://doi.org/10.1007/s10113-021-01868-x

Dumrongrojwatthana, P., Wanich, K., & Trébuil, G. (2020). Driving factors and impact of land-use change in a fragile rainfed lowland rice-sugar palm cultural agroforestry system in southern Thailand. *Sustainability Science, 15*(5), 1317–1335. https://doi.org/10.1007/s11625-020-00819-5

Dunajska, D., Falkowska, L., Siudek, P., Sikorowicz, G., Lewandowska, A., Pryputniewicz, D., Magulski, R., & Kowacz, M. (2006). Iron wet deposition in the coastal zone of the Gulf of Gdańsk (Poland). *Polish Journal of Environmental Studies, 15*(1), 53–60.

Dupouy, C., Röttgers, R., Tedetti, M., Frouin, R., Lantoine, F., Rodier, M., Martias, C., & Goutx, M. (2020). Impact of contrasted weather conditions on CDOM absorption/fluorescence and biogeochemistry in the eastern lagoon of New Caledonia. *Frontiers in Earth Science, 8*. https://doi.org/10.3389/feart.2020.00054

Dyhrman, S. T. (2016). *Nutrients and their acquisition: Phosphorus physiology in microalgae.* Springer Science and Business Media LLC. https://doi.org/10.1007/978-3-319-24945-2_8

Ernawaty, R., Takanobu, I., Shinichi, A., Kuriko, Y., Yoshitaka, M., & Yoko, O. (2014). Nutrient enrichment and physical environmental effects caused by typhoons in a semi-enclosed bay. *Journal of Ecotechnology Research.* https://www.semanticscholar.org/paper/Nutrient-Enrichment-and-Physical-Environmental-by-a-Ernawaty-Takanobu/2559d741b5bd87e8b07d00e5ccc184597b47ff5c.

Evans, D. J., & Johnes, P. J. (2004). Physico-chemical controls on phosphorus cycling in two lowland streams. Part 1 — the water column. *Science of the Total Environment, 329*(1–3), 145–163. https://doi.org/10.1016/j.scitotenv.2004.02.018

Eyre, B. D., & Ferguson, A. J. P. (2002). Comparison of carbon production and decomposition, benthic nutrient fluxes and denitrification in seagrass, phytoplankton, benthic microalgae-

and macroalgae-dominated warm-temperate Australian lagoons. *Marine Ecology Progress Series, 229,* 43−59. https://doi.org/10.3354/meps229043

Feng, Y., Huang, J., Du, Y., Balaguru, K., Ma, W., Feng, Q., Wan, X., Zheng, Y., Guo, X., & Cai, S. (2022). Drivers of phytoplankton variability in and near the Pearl River Estuary, South China Sea during Typhoon Hato (2017): A numerical study. *Journal of Geophysical Research: Biogeosciences, 127*(10). https://doi.org/10.1029/2022jg006924

Fichez, R., Douillet, Chevillon, C., Torréton, J.-P., Aung, T., Chifflet, S., et al. (2006). The Suva lagoon environment: An overview of a joint IRD Camélia research unit and USP study. *At the Cross Roads: Science and Management of the Suva Lagoon, 1,* 93−105.

Fogel, M. L., Aguilar, C., Cuhel, R., Hollander, D. J., Willey, J. D., & Paerl, H. W. (1999). Biological and isotopic changes in coastal waters induced by Hurricane Gordon. *Limnology & Oceanography, 44*(6), 1359−1369. https://doi.org/10.4319/lo.1999.44.6.1359

Fong, C. R., & Fong, P. (2018). Nutrient fluctuations in marine systems: Press versus pulse nutrient subsidies affect producer competition and diversity in estuaries and coral reefs. *Estuaries and Coasts, 41*(2), 421−429. https://doi.org/10.1007/s12237-017-0291-5

Fong, C. R., Gaynus, C. J., & Carpenter, R. C. (2020). Complex interactions among stressors evolve over time to drive shifts from short turfs to macroalgae on tropical reefs. *Ecosphere, 11*(5), Article e03130. https://doi.org/10.1002/ecs2.3130

Fraser, A. I., Harrod, T. R., & Haygarth, P. M. (1999). The effect of rainfall intensity on soil erosion and particulate phosphorus transfer from arable soils. *Water Science and Technology, 39*(12), 41−45. https://doi.org/10.1016/S0273-1223(99)00316-9

Fujii, M., & Yamanaka, Y. (2008). Effects of storms on primary productivity and air-sea CO_2 exchange in the subarctic western North Pacific: A modeling study. *Biogeosciences, 5*(4), 1189−1197. https://doi.org/10.5194/bg-5-1189-2008

Gallegos, C. L., Jordan, T. E., & Correll, D. L. (1992). Event-scale response of phytoplankton to watershed inputs in a subestuary: Timing, magnitude, and location of blooms. *Limnology & Oceanography, 37*(4), 813−828. https://doi.org/10.4319/lo.1992.37.4.0813

García-Aljaro, C., Martín-Díaz, J., Viñas-Balada, E., Calero-Cáceres, W., Lucena, F., & Blanch, A. R. (2017). Mobilisation of microbial indicators, microbial source tracking markers and pathogens after rainfall events. *Water Research, 112,* 248−253. https://doi.org/10.1016/j.watres.2017.02.003

Gentry, L. E., David, M. B., Royer, T. V., Mitchell, C. A., & Starks, K. M. (2007). Phosphorus transport pathways to streams in tile-drained agricultural watersheds. *Journal of Environmental Quality, 36*(2), 408−415. https://doi.org/10.2134/jeq2006.0098

Goatley, C. H. R., & Bellwood, D. R. (2012). Sediment suppresses herbivory across a coral reef depth gradient. *Biology Letters, 8*(6), 1016−1018. https://doi.org/10.1098/rsbl.2012.0770

Goatley, C. H. R., Bonaldo, R. M., Fox, R. J., & Bellwood, D. R. (2016). Sediments and herbivory as sensitive indicators of coral reef degradation. *Ecology and Society, 21*(1). https://doi.org/10.5751/ES-08334-210129

Gordon, S. E., Goatley, C. H. R., & Bellwood, D. R. (2016). Low-quality sediments deter grazing by the parrotfish Scarus rivulatus on inner-shelf reefs. *Coral Reefs, 35*(1), 285−291. https://doi.org/10.1007/s00338-015-1374-z

Grange, N., & Allanson, B. R. (1995). The influence of freshwater inflow on the nature, amount and distribution of seston in estuaries of the Eastern Cape, South Africa. *Estuarine, Coastal and Shelf Science, 40*(4), 403−420. https://doi.org/10.1006/ecss.1995.0028

Grange, N., Whitfield, A. K., De Villiers, C. J., & Allanson, B. R. (2000). The response of two South African east coast estuaries to altered river flow regimes. *Aquatic Conservation: Marine and Freshwater Ecosystems, 10*(3), 155−177. https://doi.org/10.1002/1099-0755(200005/06)10:3<155::AID-AQC406>3.3.CO;2-Q

Greenaway, A. M., & Gordon-Smith, D. A. (2006). The effects of rainfall on the distribution of inorganic nitrogen and phosphorus in Discovery Bay, Jamaica. *Limnology & Oceanography, 51*(5), 2206–2220. https://doi.org/10.4319/lo.2006.51.5.2206

Gu, S., Qi, L., Han, S., Yuyuan, L., Guangxing, L., & Hongju, C. (2016). Impact of rainstorm on community structure of zooplankton in Jiaozhou Bay. *Hai Yang Huan Jing Ke Xue, 35*, 190–195.

Gündoğdu, S., Çevik, C., Ayat, B., Aydoğan, B., & Karaca, S. (2018). How microplastics quantities increase with flood events? An example from Mersin Bay NE Levantine coast of Turkey. *Environmental Pollution, 239*, 342–350. https://doi.org/10.1016/j.envpol.2018.04.042

Han, H., Xiao, R., Gao, G., Yin, B., Liang, S., & lv, X. (2023). Influence of a heavy rainfall event on nutrients and phytoplankton dynamics in a well-mixed semi-enclosed bay. *Journal of Hydrology, 617*. https://doi.org/10.1016/j.jhydrol.2022.128932

Han, N., Zhao, Q., Ao, H., Hu, H., & Wu, C. (2022). Horizontal transport of macro- and microplastics on soil surface by rainfall induced surface runoff as affected by vegetations. *Science of the Total Environment, 831*. https://doi.org/10.1016/j.scitotenv.2022.154989

Hayes, H. G., Kalhori, P. S., Weiss, M., Grier, S. R., Fong, P., & Fong, C. R. (2021). Storms may disrupt top-down control of algal turf on fringing reefs. *Coral Reefs, 40*(2), 269–273. https://doi.org/10.1007/s00338-020-02045-y

Haygarth, P. M., Wood, F. L., Heathwaite, A. L., & Butler, P. J. (2005). Phosphorus dynamics observed through increasing scales in a nested headwater-to-river channel study. *Science of the Total Environment, 344*(1–3), 83–106. https://doi.org/10.1016/j.scitotenv.2005.02.007

Heaney, C. D., Sams, E., Dufour, A. P., Brenner, K. P., Haugland, R. A., Chern, E., Wing, S., Marshall, S., Love, D. C., Serre, M., Noble, R., & Wade, T. J. (2012). Fecal indicators in sand, sand contact, and risk of enteric illness among beachgoers. *Epidemiology, 23*(1), 95–106. https://doi.org/10.1097/EDE.0b013e31823b504c

Herbeck, L. S., Unger, D., Krumme, U., Liu, S. M., & Jennerjahn, T. C. (2011). Typhoon-induced precipitation impact on nutrient and suspended matter dynamics of a tropical estuary affected by human activities in Hainan, China. *Estuarine, Coastal and Shelf Science, 93*(4), 375–388. https://doi.org/10.1016/j.ecss.2011.05.004

Holling, C. S. (1973). Resilience and stability of ecological systems. In *Resilience and stability of ecological systems* (pp. 245–260). Yale University Press. https://doi.org/10.12987/9780300188479-023 (1973).

Howarth, R. W., & Marino, R. (2006). Nitrogen as the limiting nutrient for eutrophication in coastal marine ecosystems: Evolving views over three decades. *Limnology & Oceanography, 51*(1part2), 364–376. https://doi.org/10.4319/lo.2006.51.1_part_2.0364

Huang, C., & Qi, Y. (1997). The abundance cycle and influence factors on red tide phenomena of Noctiluca scintillans (Dinophyceae) in Dapeng Bay, the South China Sea. *Journal of Plankton Research, 19*(3), 303–318. https://doi.org/10.1093/plankt/19.3.303

Hubertz, E. D., & Cahoon, L. B. (1999). Short-term variability of water quality parameters in two shallow estuaries of North Carolina. *Estuaries, 22*(3), 814–823. https://doi.org/10.2307/1353114

Hughes, T. P., Rodrigues, M. J., Bellwood, D. R., Ceccarelli, D., Hoegh-Guldberg, O., McCook, L., Moltschaniwskyj, N., Pratchett, M. S., Steneck, R. S., & Willis, B. (2007). Phase shifts, herbivory, and the resilience of coral reefs to climate change. *Current Biology, 17*(4), 360–365. https://doi.org/10.1016/j.cub.2006.12.049

Hyun, J. H., Mok, J. S., Cho, H. Y., Kim, S. H., Lee, K. S., & Kostka, J. E. (2009). Rapid organic matter mineralization coupled to iron cycling in intertidal mud flats of the Han River estuary, Yellow Sea. *Biogeochemistry, 92*(3), 231−245. https://doi.org/10.1007/s10533-009-9287-y

Intergovernmental Panel on Climate Change. (2014). *Climate change 2013 − the physical science basis: Working group I contribution to the fifth assessment report of the intergovernmental panel on climate change.* Cambridge: Cambridge University Press. https://doi.org/10.1017/CBO9781107415324

IPCC. (2007). Summary for policymakers. In *Climate change 2007: Impacts, adaptation and vulnerability. Contribution of working group II to the fourth assessment report of the intergovernmental panel on climate change* (pp. 7−22). Cambridge University Press.

Jedruch, A., Kwasigroch, U., Bełdowska, M., & Kuliński, K. (2017). Mercury in suspended matter of the Gulf of Gdańsk: Origin, distribution and transport at the land−sea interface. *Marine Pollution Bulletin, 118*(1−2), 354−367. https://doi.org/10.1016/j.marpolbul.2017.03.019

Jensen, M. M., Thamdrup, B., Rysgaard, S., Holmer, M., & Fossing, H. (2003). Rates and regulation of microbial iron reduction in sediments of the Baltic-North Sea transition. *Biogeochemistry, 65*(3), 295−317. https://doi.org/10.1023/A:1026261303494

Jephson, T., Carlsson, P., & Fagerberg, T. (2012). Dominant impact of water exchange and disruption of stratification on dinoflagellate vertical distribution. *Estuarine, Coastal and Shelf Science, 112*, 198−206. https://doi.org/10.1016/j.ecss.2012.07.020

Jephson, T., Fagerberg, T., & Carlsson, P. (2011). Dependency of dinoflagellate vertical migration on salinity stratification. *Aquatic Microbial Ecology, 63*(3), 255−264. https://doi.org/10.3354/ame01498

Johnson, P. W., & Sieburth, J. M. N. (1979). Chroococcoid cyanobacteria in the sea: A ubiquitous and diverse phototrophic biomass. *Limnology & Oceanography, 24*(5), 928−935. https://doi.org/10.4319/lo.1979.24.5.0928

Jørgensen, B. B., & Sørensen, J. (1985). Seasonal cycles of 02, N03- and S042- reduction in estuarine sediments: The significance of an N03- reduction maximum in spring. *Marine Ecology Progress Series, 24*, 65−74. https://doi.org/10.3354/meps024065

Joyce, E. E., Walters, W. W., Le Roy, E., Clark, S. C., Schiebel, H., & Hastings, M. G. (2020). Highly concentrated atmospheric inorganic nitrogen deposition in an urban, coastal region in the us. *Environmental Research Communications, 2*(8). https://doi.org/10.1088/2515-7620/aba637

Jr. (2005). Upper Saddle River. In *Principles of sedimentology and stratigraphy.* Prentice Hall.

Kalnejais, L. H., Martin, W. R., & Bothner, M. H. (2010). The release of dissolved nutrients and metals from coastal sediments due to resuspension. *Marine Chemistry, 121*(1), 224−235. https://doi.org/10.1016/j.marchem.2010.05.002

Kanakidou, M., Myriokefalitakis, S., Daskalakis, N., Fanourgakis, G., Nenes, A., Baker, A. R., Tsigaridis, K., & Mihalopoulos, N. (2016). Past, present, and future atmospheric nitrogen deposition. *Journal of the Atmospheric Sciences, 73*(5), 2039−2047. https://doi.org/10.1175/jas-d-15-0278.1

Kaufman, L. S. (1983). Effects of hurricane allen on reef fish assemblages near Discovery Bay, Jamaica. *Coral Reefs, 2*(1), 43−47. https://doi.org/10.1007/BF00304731

Kay, D., Kershaw, S., Lee, R., Wyer, M. D., Watkins, J., & Francis, C. (2008). Results of field investigations into the impact of intermittent sewage discharges on the microbiological quality of wild mussels (Mytilus edulis) in a tidal estuary. *Water Research, 42*(12), 033−3046. https://doi.org/10.1016/j.watres.2008.03.020

Kennison, R. L., Kamer, K., & Fong, P. (2011). Rapid nitrate uptake rates and large short-term storage capacities may explain why opportunistic green macroalgae dominate shallow eutrophic Estuaries1. *Journal of Phycology, 47*(3), 483−494. https://doi.org/10.1111/j.1529-8817.2011.00994.x

Kjerfve, B. (1994). Chapter 1 coastal lagoons. *Elsevier Oceanography Series, 60*(C), 1−8. https://doi.org/10.1016/S0422-9894(08)70006-0

Koliyavu, T., Martias, C., Singh, A., Mounier, S., Gérard, P., & Dupouy, C. (2021). In-situ variability of DOM in relation with biogeochemical and physical parameters in December 2017 in Laucala Bay (Fiji Islands) after a strong rain event. *Journal of Marine Science and Engineering, 9*(3). https://doi.org/10.3390/jmse9030241

Kossin, J. P. (2018). A global slowdown of tropical-cyclone translation speed. *Nature, 558*(7708), 104−107. https://doi.org/10.1038/s41586-018-0158-3

Kostka, J. E., Roychoudhury, A., & Van Cappellen, P. (2002). Rates and controls of anaerobic microbial respiration across spatial and temporal gradients in saltmarsh sediments. *Biogeochemistry, 60*(1), 49−76. https://doi.org/10.1023/A:1016525216426

Kragh, T., Martinsen, K. T., Kristensen, E., & Sand-Jensen, K. (2020). From drought to flood: Sudden carbon inflow causes whole-lake anoxia and massive fish kill in a large shallow lake. *Science of the Total Environment, 739*. https://doi.org/10.1016/j.scitotenv.2020.140072

Lee, C.-S., Lee, Y.-C., & Chiang, H.-M. (2016). Abrupt state change of river water quality (turbidity): Effect of extreme rainfalls and typhoons. *Science of the Total Environment, 557−558*, 91−101. https://doi.org/10.1016/j.scitotenv.2016.02.213

Lee, R., Kay, D., Wilkinson, J., Fewtrell, L., & Stapleton, C. (2003). Impact of intermittent discharges on the microbial quality of shellfish. In *Environment agency R&D technical report* (pp. 2−266).

Lewitus, A. J., Brock, L. M., Burke, M. K., DeMattio, K. A., & Wilde, S. B. (2008). Lagoonal stormwater detention ponds as promoters of harmful algal blooms and eutrophication along the South Carolina coast. *Harmful Algae, 8*(1), 60−65. https://doi.org/10.1016/j.hal.2008.08.012

Li, Y., Schichtel, B. A., Walker, J. T., Schwede, D. B., Chen, X., Lehmann, C. M. B., Puchalski, M. A., Gay, D. A., & Collett, J. L. (2016). Increasing importance of deposition of reduced nitrogen in the United States. *Proceedings of the National Academy of Sciences of the United States of America, 113*(21), 5874−5879. https://doi.org/10.1073/pnas.1525736113

Li, Y., Ye, X., Wang, A., Li, H., Chen, J., & Qiao, L. (2013). Impact of typhoon Morakot on chlorophyll a distribution on the inner shelf of the east China Sea. *Marine Ecology Progress Series, 483*, 19−29. https://doi.org/10.3354/meps10223

Lima, A. O., Lyra, G. B., Abreu, M. C., Oliveira-Júnior, J. F., Zeri, M., & Cunha-Zeri, G. (2021). Extreme rainfall events over Rio de Janeiro State, Brazil: Characterization using probability distribution functions and clustering analysis. *Atmospheric Research, 247*. https://doi.org/10.1016/j.atmosres.2020.105221

Limsakul, A., Limjirakan, S., & Sriburi, T. (2010). Observed changes in daily rainfall extreme along Thailand's coastal zones. *Applied Environmental Research, 32*(1), 49−68.

Lin, I.-I. (2012). Typhoon-induced phytoplankton blooms and primary productivity increase in the western North Pacific subtropical ocean. *Journal of Geophysical Research: Oceans, 117*(C3). https://doi.org/10.1029/2011jc007626

Lin, I., Liu, W. T., Wu, C. C., Wong, G. T. F., Hu, C., Chen, Z., Liang, W. D., Yang, Y., & Liu, K. K. (2003). New evidence for enhanced ocean primary production triggered by tropical cyclone. *Geophysical Research Letters, 30*(13), 1−51. https://doi.org/10.1029/2003GL017141

Liu, X., Zhang, Y., Han, W., Tang, A., Shen, J., Cui, Z., Vitousek, P., Erisman, J. W., Goulding, K., Christie, P., Fangmeier, A., & Zhang, F. (2013). Enhanced nitrogen deposition over China. *Nature, 494*(7438), 459−462. https://doi.org/10.1038/nature11917

Liu, Z., Wei, H., Liu, G., & Zhang, J. (2004). Simulation of water exchange in Jiaozhou Bay by average residence time approach. *Estuarine, Coastal and Shelf Science, 61*(1), 25−35. https://doi.org/10.1016/j.ecss.2004.04.009

Livingston, R. J. (2001). *Eutrophication processes in coastal systems: Origin and succession of plankton blooms and effects on secondary production in Gulf coast estuaries.* CRC Press.

Lyons, W. B., Fitzgibbon, T. O., Welch, K. A., & Carey, A. E. (2006). Mercury geochemistry of the Scioto River, Ohio: Impact of agriculture and urbanization. *Applied Geochemistry, 21*(11), 1880−1888. https://doi.org/10.1016/j.apgeochem.2006.08.005

Magalhães, C. M., Joye, S. B., Moreira, R. M., Wiebe, W. J., & Bordalo, A. A. (2005). Effect of salinity and inorganic nitrogen concentrations on nitrification and denitrification rates in intertidal sediments and rocky biofilms of the Douro River estuary, Portugal. *Water Research, 39*(9), 1783−1794. https://doi.org/10.1016/j.watres.2005.03.008

Malta, E.j., Stigter, T. Y., Pacheco, A., Dill, A. C., Tavares, D., & Santos, R. (2017). Effects of external nutrient sources and extreme weather events on the nutrient budget of a southern European coastal lagoon. *Estuaries and Coasts, 40*(2), 419−436. https://doi.org/10.1007/s12237-016-0150-9

Manhique, A. J., Reason, C. J. C., Silinto, B., Zucula, J., Raiva, I., Congolo, F., & Mavume, A. F. (2015). Extreme rainfall and floods in southern Africa in January 2013 and associated circulation patterns. *Natural Hazards, 77*(2), 679−691. https://doi.org/10.1007/s11069-015-1616-y

Margalef, R. (1978). Life-forms of phytoplankton as survival alternatives in an unstable environment. *Oceanologica Acta, 1*(4).

Martias, C., Tedetti, M., Lantoine, F., Jamet, L., & Dupouy, C. (2018). Characterization and sources of colored dissolved organic matter in a coral reef ecosystem subject to ultramafic erosion pressure (New Caledonia, Southwest Pacific). *Science of the Total Environment, 616−617*, 438−452. https://doi.org/10.1016/j.scitotenv.2017.10.261

Martin, J. H., & Fitzwater, S. E. (1988). Iron deficiency limits phytoplankton growth in the north-east pacific subarctic. *Nature, 331*(6154), 341−343. https://doi.org/10.1038/331341a0

Martino, M., Hamilton, D., Baker, A. R., Jickells, T. D., Bromley, T., Nojiri, Y., Quack, B., & Boyd, P. W. (2014). Western Pacific atmospheric nutrient deposition fluxes, their impact on surface ocean productivity. *Global Biogeochemical Cycles, 28*(7), 712−728. https://doi.org/10.1002/2013gb004794

Masson-Delmotte, V., Pörtner, H.-O., Skea, J., Zhai, P., Roberts, D., Shukla, P. R., et al. (2019). *Global warming of 1.5°C.*

McClanahan, T. R., Sala, Stickels, P. A., Cokos, B. A., Baker, A. C., Starger, C. J., & Jones. (2009). *Interaction between nutrients and herbivory in controlling algal communities and coral condition on Glover's Reef.*

Mckee, L., Eyre, B., & Hossain, S. (2000). Transport and retention of nitrogen and phosphorus in the sub-tropical Richmond River estuary, Australia - a budget approach. *School of Environmental Science and Management Papers, 50.* https://doi.org/10.1023/A:1006339910533

Meiggs, D., & Taillefert, M. (2011). The effect of riverine discharge on biogeochemical processes in estuarine sediments. *Limnology & Oceanography, 56*(5), 1797−1810. https://doi.org/10.4319/lo.2011.56.5.1797

Mendoza-Salgado, R. A., Lechuga-Devéze, C. H., & Ortega-Rubio, A. (2005). Identifying rainfall effects in an arid Gulf of California coastal lagoon. *Journal of Environmental Management, 75*(2), 183–187. https://doi.org/10.1016/j.jenvman.2004.10.008

Meng, P. J., Lee, H. J., Tew, K. S., & Chen, C. C. (2014). Effect of a rainfall pulse on phytoplankton bloom succession in a hyper-Eutrophic subtropical lagoon. *Marine and Freshwater Research, 66*(1), 60–69. https://doi.org/10.1071/MF13314

Mergler, D., Anderson, H. A., Chan, L. H. M., Mahaffey, K. R., Murray, M., Sakamoto, M., & Stern, A. H. (2007). Methylmercury exposure and health effects in humans: A worldwide concern. *AMBIO: A Journal of the Human Environment, 36*(1), 3–11. https://doi.org/10.1579/0044-7447(2007)36[3:meahei]2.0.co;2

Middelkoop, H., Daamen, K., Gellens, D., Grabs, W., Kwadijk, J. C. J., Lang, H., Parmet, B. W. A. H., Schädler, B., Schulla, J., & Wilke, K. (2001). Impact of climate change on hydrological regimes and water resources management in the Rhine Basin. *Climatic Change, 49*(1–2), 105–128. https://doi.org/10.1023/A:1010784727448

Milliman, J. D., & Meade, R. H. (1983). World-wide delivery of river sediment to the oceans. *The Journal of Geology, 91*(1), 1–21. https://doi.org/10.1086/628741

Mitchell, A. W., Bramley, R. G. V., & Johnson, A. K. L. (1997). Export of nutrients and suspended sediment during a cyclone-mediated flood event in the Herbert River catchment, Australia. *Marine and Freshwater Research, 48*(1), 79. https://doi.org/10.1071/MF96021

Mok, J. S., Kim, S. H., Kim, J., Cho, H., An, S. U., Choi, A., Kim, B., Yoon, C., Thamdrup, B., & Hyun, J. H. (2019). Impacts of typhoon-induced heavy rainfalls and resultant freshwater runoff on the partitioning of organic carbon oxidation and nutrient dynamics in the intertidal sediments of the Han River estuary, Yellow Sea. *Science of the Total Environment, 691*, 858–867. https://doi.org/10.1016/j.scitotenv.2019.07.031

Morrison, R. J., Narayan, S. P., & Gangaiya, P. (2001). Trace element studies in Laucala Bay, Suva, Fiji. *Marine Pollution Bulletin, 42*(5), 397–404. https://doi.org/10.1016/S0025-326X(00)00169-7

Morse, R. E., Shen, J., Blanco-Garcia, J. L., Hunley, W. S., Fentress, S., Wiggins, M., & Mulholland, M. R. (2011). Environmental and physical controls on the formation and transport of blooms of the dinoflagellate Cochlodinium polykrikoides Margalef in the lower Chesapeake Bay and its tributaries. *Estuaries and Coasts, 34*(5), 1006–1025. https://doi.org/10.1007/s12237-011-9398-2

Naidu, S. D., & Morrison, R. J. (1994). Contamination of Suva Harbour, Fiji. *Marine Pollution Bulletin, 29*(1–3), 126–130. https://doi.org/10.1016/0025-326X(94)90436-7

Nausch, M., Woelk, J., Kahle, P., Nausch, G., Leipe, T., & Lennartz, B. (2017). Phosphorus fractions in discharges from artificially drained lowland catchments (Warnow River, Baltic Sea). *Agricultural Water Management, 187*, 77–87. https://doi.org/10.1016/j.agwat.2017.03.006

Neves, R. A. F., Naveira, C., Miyahira, I. C., Portugal, S. G. M., Krepsky, N., & Santos, L. N. (2020). Are invasive species always negative to aquatic ecosystem services? The role of dark false mussel for water quality improvement in a multi-impacted urban coastal lagoon. *Water Research, 184*. https://doi.org/10.1016/j.watres.2020.116108

Neves, R. A. F., & Santos, L. N. (2022). Short-term effects of very heavy rainfall events on the water quality of a shallow coastal lagoon. *Hydrobiologia, 849*(17–18), 3947–3961. https://doi.org/10.1007/s10750-021-04772-x

Okin, G. S., Baker, A. R., Tegen, I., Mahowald, N. M., Dentener, F. J., Duce, R. A., Galloway, J. N., Hunter, K., Kanakidou, M., Kubilay, N., Prospero, J. M., Sarin, M., Surapipith, V., Uematsu, M., & Zhu, T. (2011). Impacts of atmospheric nutrient deposition on marine productivity: Roles of nitrogen, phosphorus, and iron. *Global Biogeochemical Cycles, 25*(2). https://doi.org/10.1029/2010GB003858

Oviatt, C., Smith, L., Krumholz, J., Coupland, C., Stoffel, H., Keller, A., McManus, M. C., & Reed, L. (2017). Managed nutrient reduction impacts on nutrient concentrations, water clarity, primary production, and hypoxia in a north temperate estuary. *Estuarine, Coastal and Shelf Science, 199*, 25−34. https://doi.org/10.1016/j.ecss.2017.09.026

Packett, R. (2017). Rainfall contributes ∼ 30% of the dissolved inorganic nitrogen exported from a southern Great Barrier Reef river basin. *Marine Pollution Bulletin, 121*(1−2), 16−31. https://doi.org/10.1016/j.marpolbul.2017.05.008

Paerl, H., & Peierls, B. L. (2008). Ecological responses of the Neuse River-Pamlico sound estuarine continuum to a period of elevated hurricane activity: Impacts of individual storms and longer term trends. *American Fisheries Society Symposium, 64*, 101−116.

Paerl, H. W. (2006). Assessing and managing nutrient-enhanced eutrophication in estuarine and coastal waters: Interactive effects of human and climatic perturbations. *Ecological Engineering, 26*(1), 40−54. https://doi.org/10.1016/j.ecoleng.2005.09.006

Paerl, H. W., Bales, J. D., Ausley, L. W., Buzzelli, C. P., Crowder, L. B., Eby, L. A., Fear, J. M., Go, M., Peierls, B. L., Richardson, T. L., & Ramus, J. S. (2001). Ecosystem impacts of three sequential hurricanes (Dennis, Floyd, and Irene) on the United States' largest lagoonal estuary, Pamlico Sound, NC. *Proceedings of the National Academy of Sciences, 98*(10), 5655−5660. https://doi.org/10.1073/pnas.101097398

Paerl, H. W., Crosswell, J. R., Van Dam, B., Hall, N. S., Rossignol, K. L., Osburn, C. L., Hounshell, A. G., Sloup, R. S., & Harding, L. W. (2018). Two decades of tropical cyclone impacts on North Carolina's estuarine carbon, nutrient and phytoplankton dynamics: Implications for biogeochemical cycling and water quality in a stormier world. *Biogeochemistry, 141*(3), 307−332. https://doi.org/10.1007/s10533-018-0438-x

Paerl, H. W., Hall, N. S., Hounshell, A. G., Luettich, R. A., Rossignol, K. L., Osburn, C. L., & Bales, J. (2019). Recent increase in catastrophic tropical cyclone flooding in coastal North Carolina, USA: Long-term observations suggest a regime shift. *Scientific Reports, 9*(1), Article 10620. https://doi.org/10.1038/s41598-019-46928-9

Paerl, H. W., Hall, N. S., Hounshell, A. G., Rossignol, K. L., Barnard, M. A., Luettich, R. A., et al. (2020). Recent increases of rainfall and flooding from tropical cyclones (TCs) in North Carolina (USA): Implications for organic matter and nutrient cycling in coastal watersheds. *Biogeochemistry, 150*(2), 197−216. https://doi.org/10.1007/s10533-020-00693-4

Paerl, R. W., Venezia, R. E., Sanchez, J. J., & Paerl, H. W. (2020). Picophytoplankton dynamics in a large temperate estuary and impacts of extreme storm events. *Scientific Reports, 10*(1), Article 22026. https://doi.org/10.1038/s41598-020-79157-6

Panton, A., Couceiro, F., Fones, G. R., & Purdie, D. A. (2020). The impact of rainfall events, catchment characteristics and estuarine processes on the export of dissolved organic matter from two lowland rivers and their shared estuary. *Science of the Total Environment, 735*. https://doi.org/10.1016/j.scitotenv.2020.139481

Pedusaar, T., Sammalkorpi, I., Hautala, A., Salujõe, J., Järvalt, A., & Pihlak, M. (2010). Shifts in water quality in a drinking water reservoir during and after the removal of cyprinids. *Hydrobiologia, 649*(1), 95−106. https://doi.org/10.1007/s10750-010-0231-x

Peierls, B. L., Christian, R. R., & Paerl, H. W. (2003). Water quality and phytoplankton as indicators of hurricane impacts on a large estuary ecosystem. *Estuaries, 26*(5), 1329−1343. https://doi.org/10.1007/BF02803635

Pereira, E., Baptista-Neto, J. A., Smith, B. J., & McAllister, J. J. (2007). The contribution of heavy metal pollution derived from highway runoff to Guanabara Bay sediments - Rio de

Janeiro/Brazil. *Anais da Academia Brasileira de Ciencias, 79*(4), 739−750. https://doi.org/
10.1590/S0001-37652007000400013

Pinckney, J. L., Millie, D. F., Vinyard, B. T., & Paerl, H. W. (1997). Environmental controls of
phytoplankton bloom dynamics in the Neuse River Estuary, North Carolina, U.S.A. *Canadian Journal of Fisheries and Aquatic Sciences, 54*(11), 2491−2501. https://doi.org/
10.1139/f97-165

Poulton, S. W., & Raiswell, R. (2002). The low-temperature geochemical cycle of iron: From
continental fluxes to marine sediment deposition. *American Journal of Science, 302*(9),
774−805. https://doi.org/10.2475/ajs.302.9.774

Pratap, A., Mani, F. S., & Prasad, S. (2020). Heavy metals contamination and risk assessment in
sediments of Laucala Bay, Suva, Fiji. *Marine Pollution Bulletin, 156*, Article 111238.
https://doi.org/10.1016/j.marpolbul.2020.111238

Pristo, M. V.de J., Dereczynski, C. P., Souza, P. R.de, & Menezes, W. F. (2018). Climatologia
de Chuvas Intensas no Município do Rio de Janeiro. *Revista Brasileira de Meteorologia,
33*(4), 615−630. https://doi.org/10.1590/0102-7786334005

Qiu, D., Zhong, Y., Chen, Y., Tan, Y., Song, X., & Huang, L. (2019). Short-term phytoplankton
dynamics during typhoon season in and near the Pearl River Estuary, South China Sea.
Journal of Geophysical Research: Biogeosciences, 124(2), 274−292. https://doi.org/
10.1029/2018jg004672

Rabalais, N. N. (2002). Nitrogen in aquatic ecosystems. *AMBIO: A Journal of the Human
Environment, 31*(2), 102−112. https://doi.org/10.1579/0044-7447-31.2.102

Rabalais, N. N., Díaz, R. J., Levin, L. A., Turner, R. E., Gilbert, D., & Zhang, J. (2010). Dynamics
and distribution of natural and human-caused hypoxia. *Biogeosciences, 7*(2),
585−619. https://doi.org/10.5194/bg-7-585-2010

Rabalais, N. N., Turner, R. E., Díaz, R. J., & Justić, D. (2009). Global change and eutrophication
of coastal waters. *ICES Journal of Marine Science, 66*(7), 1528−1537. https://doi.org/
10.1093/icesjms/fsp047

Ramos, M. C., & Martínez-Casasnovas, J. A. (2009). Impacts of annual precipitation extremes
on soil and nutrient losses in vineyards of NE Spain. *Hydrological Processes, 23*(2),
224−235. https://doi.org/10.1002/hyp.7130

Rapport, D., & Friend. (1979). *Towards a comprehensive framework for environmental statistics:
A stress-response approach.* Statistics Canada.

Räsänen, T. A., & Kummu, M. (2013). Spatiotemporal influences of ENSO on precipitation and
flood pulse in the Mekong River Basin. *Journal of Hydrology, 476*, 154−168. https://
doi.org/10.1016/j.jhydrol.2012.10.028

Rees, A. P., Hope, S. B., Widdicombe, C. E., Dixon, J. L., Woodward, E. M. S., &
Fitzsimons, M. F. (2009). Alkaline phosphatase activity in the western english channel:
Elevations induced by high summertime rainfall. *Estuarine, Coastal and Shelf Science,
81*(4), 569−574. https://doi.org/10.1016/j.ecss.2008.12.005

Ribarova, I., Ninov, P., & Cooper, D. (2008). Modeling nutrient pollution during a first flood
event using HSPF software: Iskar River case study, Bulgaria. *Ecological Modelling,
211*(1), 241−246. https://doi.org/10.1016/j.ecolmodel.2007.09.022

Ribeiro, C., & Kjerfve, B. (2002). Anthropogenic influence on the water quality in Guanabara
Bay, Rio de Janeiro, Brazil. *Regional Environmental Change, 3*(1−3), 13−19. https://
doi.org/10.1007/s10113-001-0037-5

Rissik, D., Shon, E. H., Newell, B., Baird, M. E., & Suthers, I. M. (2009). Plankton dynamics
due to rainfall, eutrophication, dilution, grazing and assimilation in an urbanized coastal
lagoon. *Estuarine, Coastal and Shelf Science, 84*(1), 99−107. https://doi.org/10.1016/
j.ecss.2009.06.009

Rözer, V., Peche, A., Berkhahn, S., Feng, Y., Fuchs, L., Graf, T., Haberlandt, U., Kreibich, H., Sämann, R., Sester, M., Shehu, B., Wahl, J., & Neuweiler, I. (2021). Impact-based forecasting for Pluvial Floods. *Earth's Future, 9*(2). https://doi.org/10.1029/2020EF001851

Rudolph, J. C., Arendt, C. A., Hounshell, A. G., Paerl, H. W., & Osburn, C. L. (2020). Use of geospatial, hydrologic, and geochemical modeling to determine the influence of wetland-derived organic matter in coastal waters in response to extreme weather events. *Frontiers in Marine Science, 7.* https://doi.org/10.3389/fmars.2020.00018

Rusuwa, B., Maruyama, A., & Yuma, M. (2006). Deterioration of cichlid habitat by increased sedimentation in the rocky littoral zone of Lake Malawi. *Ichthyological Research, 53*(4), 431–434. https://doi.org/10.1007/s10228-006-0363-1

Sahoo, S., Baliarsingh, S. K., Lotliker, A. A., Pradhan, U. K., Thomas, C. S., & Sahu, K. C. (2017). Effect of physico-chemical regimes and tropical cyclones on seasonal distribution of chlorophyll-a in the Chilika Lagoon, east coast of India. *Environmental Monitoring and Assessment, 189*(4). https://doi.org/10.1007/s10661-017-5850-5

Salamanca, A., & Rigg, J. (2017). *Adaptation to climate change in southeast Asia: Development a relational approach.*

Saniewska, D., Bełdowska, M., Bełdowski, J., Saniewski, M., Gebka, K., Szubska, M., & Wochna, A. (2018). Impact of intense rains and flooding on mercury riverine input to the coastal zone. *Marine Pollution Bulletin, 127,* 593–602. https://doi.org/10.1016/j.marpolbul.2017.12.058

Saniewska, D., Bełdowska, M., Bełdowski, J., Saniewski, M., Szubska, M., Romanowski, A., & Falkowska, L. (2014). The impact of land use and season on the riverine transport of mercury into the marine coastal zone. *Environmental Monitoring and Assessment, 186*(11), 7593–7604. https://doi.org/10.1007/s10661-014-3950-z

Santos, M., Marotta, H., & Enrich-Prast, A. (2012). Elevadas Mudanças De Curto Prazo E Heterogeneidade Intralagunar Na Emissão De Amônia De Uma Lagoa Costeira Urbana Tropical (lagoa Rodrigo De Freitas — Rio De Janeiro) À Atmosfera. *Oecologia Australis, 16,* 408–420. https://doi.org/10.4257/oeco.2012.1603.07

Sarmento, H. (2012). New paradigms in tropical limnology: the importance of the microbial food web. *Hydrobiologia, 686*(1), 1–14. https://doi.org/10.1007/s10750-012-1011-6

Seo, D. C., Yu, K., & Delaune, R. D. (2008). Influence of salinity level on sediment denitrification in a Louisiana estuary receiving diverted Mississippi River water. *Archives of Agronomy and Soil Science, 54*(3), 249–257. https://doi.org/10.1080/03650340701679075

Shi, J., Li, G., & Wang, P. (2011). Anthropogenic influences on the tidal prism and water exchanges in Jiaozhou Bay, Qingdao, China. *Journal of Coastal Research, 27,* 57–72. https://doi.org/10.2112/JCOASTRES-D-09-00011.1

Shi, W., & Wang, M. (2007). Observations of a Hurricane Katrina-induced phytoplankton bloom in the Gulf of Mexico. *Geophysical Research Letters, 34*(11). https://doi.org/10.1029/2007gl029724

Singh, A., & Aung, T. (2008). Salinity, temperature and turbidity structure in the Suva Lagoon, Fiji. *American Journal of Environmental Sciences, 4*(4), 266–275. https://doi.org/10.3844/ajessp.2008.266.275

Sinha, E., & Michalak, A. M. (2016). Precipitation dominates interannual variability of riverine nitrogen loading across the continental United States. *Environmental Science and Technology, 50*(23), 12874–12884. https://doi.org/10.1021/acs.est.6b04455

Sinha, E., Michalak, A. M., & Balaji, V. (2017). Eutrophication will increase during the 21st century as a result of precipitation changes. *Science, 357*(6349), 405–408. https://doi.org/10.1126/science.aan2409

Smetacek, V., & Cloern, J. E. (2008). On phytoplankton trends. *Science, 319*(5868), 1346−1348. https://doi.org/10.1126/science.1151330

Smith, T. B. (2008). Temperature effects on herbivory for an Indo-Pacific parrotfish in Panamá: implications for coral−algal competition. *Coral Reefs, 27*(2), 397−405. https://doi.org/10.1007/s00338-007-0343-6

Solomon, S., Qin, D., Manning, M., Chen, Z., Marquis, M., Avery, K., et al. (2007). *Climate change 2007: The physical science basis. Working group I contribution to the fourth. Assessment report of the IPCC* (Vol. 1).

Southwick, L. M., Appelboom, T. W., & Fouss, J. L. (2009). Runoff and leaching of metolachlor from Mississippi River Alluvial Soil during seasons of average and below-average rainfall. *Journal of Agricultural and Food Chemistry, 57*(4), 1413−1420. https://doi.org/10.1021/jf802468m

Souza, L. C. E., Branco, C. W. C., Domingos, P., & Bonecker, S. L. C. (2011). Zooplankton of an urban coastal lagoon: Composition and association with environmental factors and summer fish kill. *Zoologia, 28*(3), 357−364. https://doi.org/10.1590/S1984-4670 2011000300010

Statham, P. J. (2012). Nutrients in estuaries - an overview and the potential impacts of climate change. *Science of the Total Environment, 434*, 213−227. https://doi.org/10.1016/j.scitotenv.2011.09.088

Stigter, T. Y., Nunes, J. P., Pisani, B., Fakir, Y., Hugman, R., Li, Y., Tomé, S., Ribeiro, L., Samper, J., Oliveira, R., Monteiro, J. P., Silva, A., Tavares, P. C. F., Shapouri, M., Cancela da Fonseca, L., & El Himer, H. (2014). Comparative assessment of climate change and its impacts on three coastal aquifers in the Mediterranean. *Regional Environmental Change, 14*(S1), 41−56. https://doi.org/10.1007/s10113-012-0377-3

Stockner, J. G., & Antia, N. J. (1986). Algal picoplankton from marine and freshwater eco-systems: a multidisciplinary perspective. *Canadian Journal of Fisheries and Aquatic Sciences, 43*(12), 2472−2503. https://doi.org/10.1139/f86-307

Subrahmanyam, B., Rao, K. H., Srinivasa Rao, N., Murty, V. S. N., & Sharp, R. J. (2002). Influence of a tropical cyclone on chlorophyll-a concentration in the Arabian Sea. *Geophysical Research Letters, 29*(22), 22-1−22-4. https://doi.org/10.1029/2002GL015892

Swain, E. B., Jakus, P. M., Rice, G., Lupi, F., Maxson, P. A., Pacyna, J. M., Penn, A., Spiegel, S. J., & Veiga, M. M. (2007). Socioeconomic consequences of mercury use and pollution. *AMBIO: A Journal of the Human Environment, 36*(1), 45−61. https://doi.org/10.1579/0044-7447(2007)36[45:scomua]2.0.co;2

Swinton, M. W., & Boylen, C. W. (2014). Phytoplankton and macrophyte response to increased phosphorus availability enhanced by rainfall quantity. *Northeastern Naturalist, 21*(2), 234−246. https://doi.org/10.1656/045.021.0204

Tebbett, S. B., & Bellwood, D. R. (2020). Sediments ratchet-down coral reef algal turf productivity. *Science of the Total Environment, 713*. https://doi.org/10.1016/j.scitotenv.2020.136709

Tebbett, S. B., Goatley, C. H. R., & Bellwood, D. R. (2017). Fine sediments suppress detritivory on coral reefs. *Marine Pollution Bulletin, 114*(2), 934−940. https://doi.org/10.1016/j.marpolbul.2016.11.016

Tengberg, A., Almroth, E., & Hall, P. (2003). Resuspension and its effects on organic carbon recycling and nutrient exchange in coastal sediments: in situ measurements using new experimental technology. *Journal of Experimental Marine Biology and Ecology, 285−286*, 119−142. https://doi.org/10.1016/S0022-0981(02)00523-3

Tester, P. A., Varnam, S. M., Culver, M. E., Eslinger, D. L., Stumpf, R. P., Swift, R. N., Yungel, J. K., Black, M. D., & Litaker, R. W. (2003). Airborne detection of ecosystem

The ecological implications of land reclamation are profound and can result in a cascade of environmental challenges. The alteration of hydrodynamic patterns and water exchange capacities can disrupt the habitats of various marine organisms, impacting their feeding and breeding patterns. This, in combination with other stressors, can lead to a decline in biodiversity and the loss of important coastal habitats such as wetlands and mangroves. The resulting ecological vulnerability exacerbates issues related to environmental quality, putting additional pressure on ecosystems (Liang et al., 2015).

(3) Multiple anthropogenic pressures encompass overfishing practices alongside excessive agricultural activities (Liang et al., 2015).

Overfishing, driven by extreme fishing practices, has emerged as a critical threat to marine ecosystems. This practice, characterized by the relentless pursuit of marine resources beyond sustainable levels, has triggered alarming declines in marine living resources. The consequences are profound and far-reaching, affecting not only the targeted species but also the intricate balance of entire marine ecosystems. As certain species are fished to the brink of collapse, the intricate web of predator—prey relationships is disrupted, with cascading impacts throughout the food chain (Smith et al., 2010; Worm et al., 2006, pp. 787—791).

With a significant increase in population density projected for coastal areas over the next few decades, achieving sustainable eco-environmental development and optimizing coastal zone utilization under anthropogenic pressure is a critical and meaningful topic (Geng et al., 2022). Consequently, this study provides a comprehensive review on strategies for achieving sustainable management of coastal eco-environment amid multiple anthropogenic pressures. The study encompasses three key perspectives: the management concepts of coastal zone eco-environment management (Section 2), evaluation and management models for coastal zone eco-environment (Section 3), and prediction models for assessing the impact on coastal zone eco-environment. Additionally, it introduces the origins, development, and application of concepts and models related to managing coastal zone eco-environment under human-induced pressure. This research serves as a foundation for expanding research ideas and methodologies pertaining to managing coastal zone ecology under anthropogenic pressures.

2. Coastal zone eco-environment management concepts

2.1 Integrated coastal zone management

Integrated Coastal Zone Management (ICZM) refers to integrated land and sea management practices (Armstrong & Ryner, 1981). The San Francisco Bay, Conservation and Development Commission, established by the US government in 1965, marked a significant milestone toward the ICZM implementation. In 1972, The *Coastal Zone Management Act (CZAM)* became one of America's earliest pieces of legislation encouraging states to develop comprehensive ICZM plans. Since then, various countries have introduced their laws and regulations on coastal zone management, such as "North Sea Oil and Gas: Guidelines for Coastal Zone Planning" promulgated by the UK during the 1970s; Implementation Order of Public Water Surface Management Act formulated by South Korea; Marine and Underwater Land Act promulgated by Australia in 1973. Since 1980, the concept of integrated coastal zone management has been widely embraced by coastal countries worldwide. Jiangsu Province in China was a pioneer in enacting the Interim Regulations on Coastal Zone Management in 1985.

In 2006, UNESCO launched Marine Spatial Planning (MSP) to implement ecosystem-based marine management. With increasing global attention to integrated coastal zone management, an ecosystem-based approach is gradually being adopted. Recently, more emphasis has been placed on carefully considering natural conditions, such as resources, ecology, and social conditions like political economy in coastal zones, to efficiently utilize coastal resources and energy while improving and protecting regional environment (Sorensen, 2002). In 2015, the United Nations General Assembly adopted the 2030 Agenda for Sustainable Development during its 70th session. It recognized sustainable development in coastal areas as a pivotal objective encompassing environmental, economic, and social dimensions.

ICZM could be used in coastal eco-environment event solutions, which involve balance ecological and socioeconomic concerns, address climate change impacts, collaboration across disciplines, and engage local communities. ICZM could be used to address the multiple demands and potential conflicts in coastal zones. ICZM could be used as a framework for marine biodiversity conversation. Bin et al. (2009) proposed a series of strategies for implementing ICZM to conserve marine biodiversity in Quanzhou Bay, Fujian, China. These strategies encompass defining management boundaries for marine biodiversity conservation, developing assessment models to evaluate the impacts of human activities on coastal marine biodiversity, formulating plans and management approaches for marine biodiversity conservation, and establishing monitoring systems. Mahony et al. (2020) investigated the potential of ICZM for implementing climate adaptation measures, focusing on local ICZM initiatives and emphasizing processes, principles, and stakeholder engagement. The optimization of ICZM to support climate adaptation and enhance resilience is crucial. Evidence suggests that the ICZM model, based on local partnerships, can serve as a facilitation mechanism by promoting capacity building, facilitating knowledge exchange and learning, thereby supporting the implementation of national climate policies at the local level.

Collaborative governance has emerged as a prominent paradigm in environmental management, particularly in the context of coastal management. The extensive literature on ICZM emphasizes the coordination of policy sectors, involvement of nonstate actors, and integration of scientific knowledge from diverse disciplines. In this study, Walsh (2019) investigates the relationship between knowledge integration and collaborative practice by examining the management of coastal change at the German Wadden Sea coast. Taking Algeria as a case study, Khelil et al. (2019) evaluated the institutional capacities required for effective implementation of ICZM principles. The findings highlight that bolstering stakeholder coordination and engagement, particularly with local communities and NGOs, emerges as a pivotal factor in achieving comprehensive ICZM adoption.

2.2   Land-sea coordination

The complex interaction between land and sea makes land-sea integration vital for solving eco-environmental problems in the coastal zone and achieving sustainable development goals. Given the interaction characteristics and key spatial elements between land and sea, it is urgent to refine research content related to integrated eco-environmental management between land and sea while enhancing capacity for such efforts (Yue et al., 2023). As a practical approach to developing and protecting coastlines, China's research capacity for the overall planning of lands and seas' is continuously improving with changing times. The scope of land-sea planning has expanded from initial economic synergies (LSC) to multidimensional coordination of resources, environment, and social economy (Yang & Sun, 2014) through coastal zone spatial planning (Li & Ye, 2020; Morf et al., 2022).

In 2011, the outline of "12th Five-Year Plan" emphasized the need to "adhere to land-sea coordination, develop marine development strategies, enhance our capacity for marine exploitation, control, and comprehensive management" (Xinhua News Agency, 2011). In 2012, the 18th National Congress of the Communist Party of China (CPC) efforts to "enhance our ability to exploit marine resources, foster the growth of the maritime economy, protect eco-environment" (Xinhua News Agency, 2012). In 2017, the report presented at the 19th National Congress of CPC explicitly highlighted that China should adhere to LSC. This provided guidance for promoting LSC (Xinhua News Agency, 2017). In 2021, "14th Five-Year Plan" embodies the LSC as "adhere to land-sea coordination, develop marine economy, and build a maritime power" (Xinhua News Agency, 2021). The Ministry of Ecology and Environment and other six departments jointly issued the "*14th Five-Year Plan for Marine Ecological Environment Protection*" (hereinafter referred to as the Plan) in 2022, which integrated planning, design, and promotion of basin and marine eco-environmental protection under the "14th Five-Year Plan" (Ministry of Ecology and Environment of the People's Republic of China, 2022). The Plan implements pertinent strategies across five dimensions: Firstly, it aims to enhance precise pollution control measures and continuously advance the environmental quality of offshore waters. Secondly, it emphasizes the simultaneous protection and restoration of marine ecosystems to elevate their quality and stability. Thirdly, it focuses on effectively responding to marine environmental emergencies and ecological disasters while strengthening emergency response capacity building. Fourthly, comprehensive measures are taken to address prominent eco-environmental issues in the marine environment. Lastly, collaborative efforts are made to promote climate change response alongside marine eco-environmental protection, thereby enhancing ocean resilience in adapting to climate change.

The Plan proposes key tasks and support measures in four aspects: firstly, enhancing marine eco-environmental protection laws and regulations, as well as the responsibility system; secondly, promoting scientific and technological innovation; thirdly, implementing the vision of a maritime community with a shared future; and fourthly, clarifying organizational safeguard measures such as strengthening leadership, increasing investment guarantee, conducting strict supervision and assessment, and enhancing publicity and guidance. Therefore, the LSC mechanism is expected to improve further.

The research concept of LSC has gradually evolved from the initial focus on economic synergy and interactive development between land and sea to encompass multidimensional coordination of resources, ecology, economy, disaster prevention, rights, and interests, as well as spatial planning. The research scope of LSC primarily encompasses the establishment and evaluation of an index system, the development of a management framework, and the formulation of a management and control system.

Yang and Sun (2014) developed a comprehensive LSC index system encompassing resources, environment, economy, and society within the marine and terrestrial systems. They conducted a detailed investigation on the current status of LSC in cities surrounding the Bohai Sea, China. Ma et al. (2020) established an index system for assessing the carrying capacity of resources and environment in Tangshan City, China, from three perspectives: resources, environment, and socio-economic factors, which was conducted against the backdrop of LSC. Yao et al. (2021) presented a framework for managing marine eco-environment within the context of LSC. They proposed key tasks such as establishing marine eco-environment management zones, implementing a collaborative emission control system between land and sea areas, and enhancing supervision mechanisms for marine ecosystems.

Li and Ye (2020) categorized the control and support system for coastal zone spatial planning into two aspects: zoning control and use classification. Zoning control encompasses the attributes of coastal zone protection, restoration, development, and utilization. Use classification determines land use permitted by the control system based on principles of full coverage, noncrossing, and nonoverlapping for different zones.

Lin et al. (2018) proposed a comprehensive approach to spatial control in urban coastal zones that integrates flexibility and rigidity with LSC, focusing on key elements of land space, utilization modes of sea space, pollution control from land sources in marine environment, and conservation of ecological shorelines and marine resources.

Current research on integrated land-sea management focuses on integrating factors to understand socio-economic driving forces' impact on system state evolution under stress, ecological effects resulting from multiple pressures and management decisions, as well as feedback effects from eco-environmental implications on socio-economic systems. A comprehensive regulatory scheme that combines top-down responsibility sharing with stakeholder coordination and bottom-up participatory management should be established (General Office of the CPC Central Committee and General Office of the China State Council, 2017). However, challenges remain regarding how best to rationally coordinate coastal zone spatial elements while prioritizing which elements require attention while effectively measuring integrated elements (Li, Tian, et al., 2022, Li, Wu, et al., 2022, Li, Xu, et al., 2022, Li, Yu, et al., 2022).

Currently, land use and environmental management regulation in coastal zones lack an overall plan for basins, estuaries, or nearshore areas; furthermore, existing frameworks lack analysis functions for complex systems with multiple factors or uncertainties within coastal zones, making scientific research ineffective at supporting effective decision-making (Chen et al., 2022).

A comprehensive, precise, and quantitative assessment of the dynamic changes in the coupled social-eco-environmental system is crucial for identifying early warning points/thresholds and improving the research framework for integrated land-sea eco-environmental management (Yu, 2010). Due to the presence of objective factors such as limited conditions and insufficient supporting elements for LSC, it is imperative to extract the advantages and experiences from various relevant professions in order to advance and incorporate the thoughts and control requirements of each profession into LSC. Proactively addressing market, temporal, personnel, and technological deviations resulting from reforms is essential to meet the heightened demands of land and sea coordination in this new era.

2.3 Collaborative governance

Collaborative governance refers to a dynamic process wherein, when confronted with complex issues that transcend the capacity of a single public department, a central force leads and unifies all stakeholders to engage in division of labor and collaboration based on relatively formal rules. Through information and resource interaction and integration, problems are resolved (Huang & Wang, 2022). The study of collaborative governance can be broadly categorized into four aspects: collaborative field, collaborative subject, collaborative concept, and collaborative method. By adapting and coordinating among subjects, collaborative governance can transition from being scattered, partial, and fragmented to becoming centralized, holistic, and integrated through interconnections between the system's elements as well as interactions among these elements within the system's environment. This results in an orderly cooperative landscape characterized by thorough coordination (Zhao & Wang, 2018). Therefore, the concept of collaborative governance exhibits obvious systematism along with integrity and complementarity (Huang & Wang, 2022).

In the context of LSC, collaborative governance encompasses the integration of soil-water-biological linkages. In their study on the impact of intergovernmental collaborative governance on transdomain environmental pollution, Rao and Zhao (2022) discovered that effective intergovernmental collaborative governance at different scales exhibits distinct action paths and characteristics. Effective interprovincial collaborative governance involves two functional pathways: interest-driven and authority-driven, whereas effective interprovincial collaborative governance relies solely on an assessment-driven pathway, with the establishment of an assessment mechanism being a crucial element.

Huang, Li, et al. (2022) conducted a comprehensive analysis of the pollution status and interaction mechanism between marine and soil eco-environment through collaborative governance. They identified issues related to the ecological restoration process and policy convergence and proposed the establishment of a collaborative governance mechanism focusing on restoration subjects, responsibilities, resources, and funds. This approach aims to effectively control the land-sea eco-environment in a scientific, efficient, and systematic manner. Li, Yu, et al. (2022) reviewed the current research status, technical challenges, and development trends in monitoring and modeling of land-sea-air environmental media as well as interface processes. They suggested studying multiobjective treatment optimization schemes for coastal environmental pollution while establishing a monitoring-model-assessment-decision system for land-sea air nitrogen and phosphorus pollution to accelerate collaborative efforts in addressing these pollutants.

In recent years, scholars and management decision-makers have advocated for a collaborative approach to addressing offshore pollution through systematic thinking and methods; however, effective measures are still lacking. Future studies should prioritize the scientific frontier of the evolution of the eco- environment at the sea—land interface and promote interdisciplinary research in science engineering management to foster sustainable development of the social economy and environment in coastal zones based on the crucial needs of land—sea overall planning and integrated watershed and offshore water management (Li, Tian, et al., 2022).

2.4 System theory and complex systems

Integrating science and management requires scientific tools to analyze the complex coastal zone system that interacts with humans. The application of system theory provides a perspective on interconnections and interactions between elements within a structure, revealing common nature and internal regularity (Li, Xu, et al., 2022).

Biologist Bertalanffy initially developed system theory during World War II (Bertalanffy, 1976). The systemic perspective unveils the inherent nature and internal regularity of the interconnectedness and interaction among entities and objects (Hofkirchner & Schafranek, 2011). With the advancement of system theory, scholars hold diverse viewpoints on system research, leading to varying definitions of system theory (Fan & Chen, 2022).

The overarching definition of system theory used to depict the shared characteristics of all systems is as follows: a system is an organic entirety comprising distinct elements arranged in a specific structure with designated functions and purposes. This definition encompasses four fundamental concepts: system, element, structure, and function, signifying interactions between elements themselves, elements within a system, as well as between the entire system and its environment (Yu, 2010).

Scholars hold varying perspectives on system research, yet they concur that a system is an organic entity with specific functions comprised of multiple elements arranged in a particular structure. The general characteristics of all systems include integrity, interrelatedness among interacting and interdependent elements, and a hierarchy consisting of subsystems or elements. The system itself can be regarded as a subsystem of a more extensive system. The core idea or system principle of system theory is the whole concept of the system (Yu, 2010).

The complex systems can be comprehended from three perspectives: (1) determining the multilayered framework structure within the system; (2) identifying relatively separable subsystems that interact with each other; and (3) recognizing that different combinations of variables within the system lead to distinct management strategies, and complex systems are greater than the sum of their parts.

Under multiple anthropogenic pressures, socio-economic and eco-environmental factors in coastal zones constitute a complex system that possesses natural characteristics and interacts with human beings, making eco-environmental management decisions more intricate. In managing eco-environmental complex systems, there may be challenges such as multiobjective or multifactor management needs, high uncertainty, and difficulty maintaining social-ecological scales consistency (Xu et al., 2013).

Integrating existing knowledge from natural and social sciences interdisciplinarily can help connect socio-economic and eco-environmental systems' management in specific regions/cultures to emerging global sustainability needs while translating interdisciplinary knowledge into eco-environmental management strategies. Transition management concepts and transformative governance are necessary for effectively linking concepts, frameworks, theories, and models (Bruckmeier, 2014).

2.5 Nature-based Solutions

According to the definition of the International Union for Conservation of Nature (IUCN) (Resolution WCC-2016-Res-069), "Nature-based Solutions leverage nature and the power of healthy ecosystems to protect people, optimize infrastructure and safeguard a stable and biodiverse" (IUCN, 2016). The concept of Nature-based Solutions (NbS) was formally introduced by the World Bank in 2008 (World Bank, 2008). Subsequently, it was incorporated and defined by the International Union for Conservation of Nature (IUCN) within the United Nations Framework Convention on Climate Change in 2009 (UNFCCC, 2009). In 2019, a global alliance dedicated to NbS was established at the United Nations Climate Action Summit held in New York. China and New Zealand jointly released the *Declaration on Nature-Based Solutions for Climate Change Mitigation and Adaptation*, proposing the utilization of natural environment as means to mitigate and adapt to climate change.

Due to the comprehensive nature of NbS, different organizations such as the World Bank, IUCN, and the European Commission, along with various experts and scholars, often interpret NbS from diverse perspectives (Eggermont et al., 2015; European Commission, 2021). However, in scaling up NbS practices, there is a risk of generalization, misuse, and abuse that can lead to unforeseen ecological damage (Nesshöver et al., 2017). To standardize the concept of NbS and summarize its practical achievements for unleashing its potential on a large scale while making it mainstream in addressing global challenges, in July 2020, IUCN released "IUCN Global Standard for Nature-based Solutions" (IUCN Commission on Ecosystem Management (CEM), 2020) along with "Guidance for using the IUCN Global Standard for Nature-based Solutions" (IUCN, 2020).

The "Guidance for using the IUCN Global Standard for Nature-based Solutions" proposed eight standards: societal challenges, design at scale, biodiversity net gain, economic feasibility, inclusive governance, balance trade-offs, adaptive management, mainstreaming and sustainability. Therefore, adaptive management is also one of the global standards of NbS.

Restoration of ocean and coastal critical zones represents a promising avenue for future research and implementation of natural solutions. Currently, China's strategies for NbS in the marine environment primarily encompass: designating ecological red-lines along the entire coast; establishing a comprehensive network of marine protected areas; strictly regulating land reclamation activities; undertaking initiatives such as "Blue Bay Remediation" and "Ecological Islands and Reefs" to actively restore coastal wetlands and nearshore ecosystems; advancing blue carbon scientific research; developing a robust blue carbon standard system; and fostering international collaboration on blue carbon development. While these NbS approaches have effectively enhanced climate change mitigation and adaptation capabilities in the oceanic environment, certain limitations remain. The NbS working group proposes to restore 50% of degraded coastal ecosystems by 2030 while safeguarding 30% of oceans. This can be achieved through natural systems that regulate pollutants and protect ecosystem services, thereby reducing exposure and vulnerability in high-risk coastal areas.

In the coming decade, China's recommendations for implementing ocean NbS encompass: (1) exerting strict control over urban spatial development and land reclamation, optimizing the layout to enhance regional microclimate and bolster capacity for addressing and mitigating climate change; (2) advancing NbS models for climate change mitigation and adaptation in marine environment, establishing pertinent funding mechanisms to ensure food security, alleviate climate change impacts, and achieve sustainable socio-economic development; (3) strengthening protection and restoration of marine ecosystems, promoting rational utilization of marine resources, reversing trends of ecosystem degradation, safeguarding marine biodiversity, and enhancing carbon sequestration capacity; (4) considering oceanic NbS approaches across multiple scales such as the United Nations Framework Convention on Climate Change (United Nations, 1992a), Convention on Biological Diversity (United Nations, 1992b), Sustainable Development Goals (United Nations, 2015); and (5) enhance international collaboration in ocean-related fields, ratify accords on mitigating climate change's impact on the ocean, and incorporate NbS for oceans as a significant element of the Belt and Road Initiative.

In several European and American countries, NbS has emerged as a well-established management tool. Based in the United States, Naturally Resilient Communities has developed an interactive network tool grounded on natural solutions (Communities Naturally-Resilient, n.d.). This innovative tool facilitates local decision-makers, planners, and engineers to identify NbS opportunities and provide tailored solutions for diverse scales and challenges (Communities Naturally-Resilient, n.d.). In Europe, robust online platforms are also dedicated to sharing knowledge on NbS, such as ThinkNature (Think Nature, n.d.) and oppla (Repository EU, n.d.).

2.6   Research on adaptive management

2.6.1   Adaptive management background

In eco-environment management, changes are driven by advancements in scientific understanding and the broader evolution of society and politics. Establishing correct goals and selecting appropriate management methods are effective strategies for achieving desired outcomes. Compared to general management, eco-environmental management is more complex due to the intricate nature of socio-ecological systems.

In ICZM, researchers usually need to spend a lot of time collecting interdisciplinary data based on different research needs, and a lack of data and incomplete information can hinder the smooth progress of research. At the same time, in the process of coastal zone management and practice, scientists, managers, and decision-makers usually have barriers to information exchange due to their different roles (Portman, 2014). To this end, it is necessary to build a coastal zone management platform based on the coastal zone database as a way for researchers to access research data and information, as a communication medium between scientists, managers, and decision-makers, and to enhance public participation in integrated coastal zone management.

The establishment of a database platform for coastal areas requires data information. On April 26, 2001, a symposium on how to manage data integration in the water was held in Pensacola Beach, Florida, USA, where discussions were held on various ways partnerships could improve information management systems. Hale et al. (2003) proposed and answered the question "Why do we need coastal data partners" by asking and answering the following questions to the panel members: (1) question: Why do we need to share and integrate data? answer: Environmental issues require data from multiple sources; Data sharing has led to more and better sources of information; (2) question: What are the barriers to data sharing and integration? answer: technical barriers; uneven level; social, cultural, and organizational barriers; (3) question: How can data partnerships help overcome these barriers? answer: overcoming technical obstacles; Efforts were made to close the level gap; overcoming social, cultural, and organizational barriers; (4) question: How can we make the data partnership work? answer: to create data-sharing requirements; use collaborative leadership to create conditions for open communication; production of human resources and material support.

There are various types and structures of relevant data for ICZM. Currently, several challenges hinder the effective integration and utilization of data in ICZM, including scattered data distribution, slow development of data products, and lack of supporting information platforms to integrate and summarize relevant data for ICZM. Consequently, the maximum potential value of data remains unrealized. In 2003, Yan (2003), aiming at the environmental problems of Jiaozhou Bay (JZB), established the hydrodynamic model and pollutant transport model, developed the coastal zone information management system by using geographic information system (GIS) technology from the perspective of physical oceanography, and proposed the conceptual model of the ICZM system of JZB.

In 2019, Yuan et al. (2019) developed a regional ocean database in response to the lack of databases in regional marine and coastal zone systems. The database contains three data sources, including field monitoring data, satellite data, and modeling data. It uses winter temperature and salt distribution in JZB, chlorophyll concentration distribution in the JZB area, and the impact of the JZB Bridge on tidal dynamics to illustrate the use of data.

Vaitis et al. (2022) proposed to provide high-quality spatial data to stakeholders of MSP through the construction of spatial data infrastructure. Noble et al. (2021) took the eastern coastal area of Australia as an example. They proposed integrating social-ecological data with GIS fuzzy set modeling to support the overall resilience of marine protected area spatial planning. Conti et al., (2018) proposed the establishment of a local spatial data infrastructure (SDI) for collecting, managing, and delivering coastal information.

3. Coastal zone eco-environment evaluation and management models

In managing coastal zone eco-environment, models are often necessary for evaluating its status under anthropogenic pressure. The available models for evaluation and management introduced in this chapter were shown below in Table 13.1.

The DPSIR model is a theoretical framework for measuring the evaluation index system of environment and sustainable development. Its components include revealing the dynamic causal relationship between social economy and government response to eco-environmental changes, emphasizing human factors. The European Environment Agency (EEA) proposed the DPSIR framework to comprehensively analyze and describe environmental issues and their relationship with social development (Europen Environment Agency, 2020). DPSIR is adopted in conducting research and evaluation on coastal sustainable development, such as cross-social and ecosystem assessment of coastal development and sustainability (Bidone & Lacerda, 2004), management of sustainable tourism in coastal areas, mangrove, crab fisheries in coastal areas (Baldwin et al., 2016), and postcoastal zone change evaluation (Karageorgis et al., 2006), among others.

The GRA is widely utilized in the Gray Model (GM) theory (Zhou & Zeng, 2020). It measures the strength, magnitude, and order of relationships based on the similarity or dissimilarity of development trends among various factors within a system (Cheng et al., 2019). Correlation quantifies the degree of association between factors in different systems or objects that may vary over time. In system development, if two factors exhibit similar change trends with higher synchronicity, their correlation degree is higher; otherwise, it is lower.

GRA typically involves the following steps: (1) determining the parent sequence (reference value) and feature sequence while formatting data accordingly; (2) conducting nondimensional processing of data; (3) solving for gray relational values between the parent sequence and feature sequence; (4) calculating correlation degree values; and (5) sorting correlation degree values to conclude. In coastal zone

Table 13.1 The evaluation and management models and their references.

Evaluation and management models	References
DPSIR (driving-pressure-state-impact-response)	Europen Environment Agency (2020)
Gray relational analysis (GRA)	Zhou and Zeng (2020)
Coupling coordination degree (CCD)	Cheng et al. (2019)
Geographic information system (GIS)	Chang (2018)
SWOT (strengths, weaknesses, opportunities, and threats) analysis	Stewart (1963)
Principal component analysis (PCA)	Pearson (1901)
Analytic hierarchy process (AHP)	Saaty (1977)

management, the GRA finds extensive application in areas related to coastal zone management, such as studying coordinated industrial development within coastal zones (Li, Wu, et al., 2022), evaluating eco-environmental carrying capacity (Shen et al., 2017), and assessing correlations between eco-environmental quality and urban construction land use (Li et al., 2021).

The CCD model is utilized to analyze the level of coordination development. The term "coupling degree" refers to the interaction between two or more systems, reflecting the extent of interdependence and constraints among these systems. On the other hand, the concept of "coordination degree" pertains to the level of favorable coupling within this interactive relationship, which effectively reflects the quality of coordination. The interactive relationship of coordinated development among subsystems in a complex system can be defined as the degree of coupling coordination. The construction and calculation of its evaluation index system can reflect the extent to which subsystems mutually influence and interact and the level of coupling and coordination among them (Cheng et al., 2019; Shu et al., 2015; Xun & Yu, 2022).

A comprehensive, precise, and quantitative evaluation of the dynamic changes in the coupled social—ecological system is crucial for identifying early warning points or thresholds and improving the research framework for integrated land-sea eco-environmental management, thereby achieving science-management integration (Li, Tian, et al., 2022). The CCD could be used for coastal sustainable development by combing social economy and eco-environment. Cheng et al. (2023) analyzed marine economic-ecological-social system (MEESS) of 11 coastal regions of China to get findings of the tendency of MEESS comprehensive level. Liu et al. (2023) explored the CCD between the marine economy and urban resilience to improving urban strength and the quality of the maritime economy. Gao et al. (2022) explored the CCD of coastal four-dimensional index, which included economic development, resource utilization, eco-environment, and social livelihood.

GIS is a computer system utilized for acquiring, storing, querying, analyzing, and displaying spatial data (Chang, 2018). In the early 1960s, Tomlinson (2003) developed the first Canada Geographic Information System (CGIS) for storing, manipulating, and analyzing data for the Federal Department of Forestry and Rural Development in Ottawa, Ontario, Canada. In the 1980s, with the advent of personal computers, the application scope of GIS expanded significantly. This era witnessed the emergence of both commercial and free GIS software packages such as GRASS, MaoInfo, Trans-CAD, and Smallworld. Notably, Environmental Systems Research Institute (Esri) introduced ARC/INFO, which seamlessly integrated point, line, and polygon spatial features into a robust database management system to associate feature attributes efficiently. In the 1990s, graphical user interfaces (GUIs), along with the widespread availability of software and hardware, further broadened the application scope of GIS due to the introduction of public digital data sources (Chang, 2018). In recent years, continuous advancements in GIS research have led to two prominent development trends: firstly, as a core spatial technology integrating various other spatial data sources like satellite imagery and GPS data; secondly, by connecting GIS with network services, mobile technology social media platforms, and cloud computing (Chang, 2018).

The GIS possesses robust spatial management capabilities, facilitating the dynamic management of spatial features. In the realm of coastal zone eco-environment research, ArcGIS is frequently employed as an effective tool for management purposes. GIS technology can be introduced in pollution source analysis, pollution load assessment, water quality evaluation, water quality simulation, environmental capacity analysis, data management, and planning schemes to achieve dynamic pollution source management while establishing a connection between these sources' spatial location and attributes.

The ArcGIS software could analyze the land use/cover change data in the coastal zone while investigating its driving mechanisms using the GRA (Wu et al., 2018). Besides, ArcGIS software also could be used to evaluate the river basin, which is characterized by a fragile eco-environment. To assess the coastal eco-environment, an evaluation model incorporating vegetation coverage, topographic relief, slope, and soil salinization was developed using ArcGIS' spatial analysis function (Wang & Xi, 2016). For coastal eco-environment under anthropogenic pressures, ArcGIS could be used to assess and map coastal human activities (Wu et al., 2016), spatial conflicts analysis of aquaculture with coastal fisheries (Bergh et al., 2023), assess the environmental vulnerability resulting in climate change and anthropogenic pressures for the conservation planning and adaptation measures (Thirumurthy et al., 2022).

The SWOT analysis, initially developed by Stewart (1963), has emerged as one of the most extensively utilized strategic tools on a global scale (Puyt et al., 2023). The SWOT analysis comprehensively assesses the internal and external competitive environment and conditions closely associated with the research subject. It involves identifying and evaluating major internal strengths and weaknesses, external opportunities, and threats. By investigating to list the advantages, disadvantages, opportunities, and threats faced by the development of the coastal zone under multiple anthropogenic pressures in a matrix format, a series of corresponding conclusions through systematic analysis could be derived (Huang, Wei, et al., 2022). For coastal eco-environment analysis, the SWOT analysis could be used for the evaluation of coastal zone management in coastal policy-making (Panigrahi & Mohanty, 2012) and strategic planning and sustainable development of islands (Gkoltsiou & Mougiakou, 2021).

PCA is a statistical technique that employs an orthogonal transformation to convert a set of potentially correlated variables into a set of linearly uncorrelated variables known as principal components (Pearson, 1901). When conducting statistical analysis on topics involving multiple variables, numerous variables can introduce complexity. Researchers often seek to reduce the number of variables while maximizing information gain. In many cases, variable correlation indicates overlapping information about the topic. PCA eliminates redundant or closely related variables from the original set and generates as few new components as possible, ensuring pairwise uncorrelatedness among these components while preserving substantial original information. The PCA was applied to coastal issues, such as coastal flood vulnerability for climate adaption policy (Wu, 2021) and climate vulnerability mapping of the coastal region (Uddin et al., 2019).

AHP is a widely recognized and influential decision-making method that was introduced by T. L. Saaty, an esteemed American operations research scientist, in the 1970s

(Saaty, 1977). This innovative approach has gained significant popularity across various fields due to its ability to effectively handle complex decision problems. AHP is a systematic and hierarchical analysis method that integrates qualitative and quantitative analyses. It begins by establishing a hierarchical structure model, dividing the decision event into various levels such as target level, criterion level, and factor level. The relevant factors to be considered in the studied event are placed at their appropriate levels, clearly expressing their relationships through a hierarchy diagram. Based on this foundation, analysis and calculations are conducted layer by layer according to the hierarchical relationship and influence magnitude of each factor, ultimately yielding quantitative results for overall target analysis (Zhang et al., 2022). This approach has been extensively employed in multifactor decision analysis of complex events. Feng et al. (2014) employed AHP to assess the appropriateness of reclamation for facilitating future sustainable development in the coastal region of Lianyungang, China. The study conducted by Mandal et al. (2023) integrated AHP with machine learning techniques to evaluate and classify multiple-hazards risk index in the coastal regions of Sundarban, India.

4. Coastal zone eco-environment prediction models

The marine eco-environment is highly intricate and susceptible to numerous influencing factors. Utilizing a marine eco-environment prediction model can enhance the accuracy of predicting and evaluating the marine ecosystem. Many models could be used for coastal eco-environmental prediction. According to different classification methods, there are time series analysis and machine learning models introduced in this chapter (Table 13.2).

Table 13.2 The prediction models and references.

Models classification	Prediction models	References
Time series analysis	Autoregression (AR)	Slutsky (1927), Yule (1921), (1926), (1927)
	Moving average (MA)	Slutsky (1927), Yule (1921), (1926), (1927)
	Autoregressive moving average (ARMA)	Nerlove (2018)
	Vector autoregression (VAR)	Sims (1980)
	Gray model (GM)	Deng (1982)
Machine learning models	Markov analysis method	Basharin et al. (2004)
	Artificial neural networks (ANN)	Kang (2019)

4.1   Time series analysis

The time series analysis model refers to a sequential arrangement of random data, which can be organized based on spatial, temporal, or other relevant physical parameters. It enables the prediction of future trends and developmental patterns by utilizing comprehensive information. This analytical approach focuses on monitoring variable changes over time and forecasting specific element variations within defined timeframes. Consequently, the alterations in the marine eco-environment can be perceived as a collection of time series (Liu, 2014).

Russian statistician and economist Slutsky (1927) and British statistician Yule (1921, 1926, 1927) almost simultaneously proposed methods to characterize time series through AR or MA processes or combined ARMA processes (Nerlove, 2018, pp. 608−615).

AR involves establishing a regression equation for prediction by leveraging the interdependence between values in the historical time series of the target variable across different periods, known as self-correlation. Specifically, it utilizes the time series of a variable as the dependent variable and employs another or multiple independent variables derived from the same time series to analyze their correlations and establish a predictive regression equation.

The MA method can be employed as a data smoothing technique, enabling the calculation of an average for a specific number of items in a time series to capture its long-term trend. When periodic and irregular fluctuations significantly impact the time series values, making it challenging to discern the underlying development trend, the moving average method can effectively mitigate these influences and facilitate analysis and prediction of the series' long-term trajectory. The MA method encompasses three variations: simple moving average, weighted moving average, and trend-moving average.

ARMA is an important method to study time series. It is a hybrid of AR and MA model. It has the characteristics of wide application range and small prediction error.

VAR is a nonstructural equations model proposed by Sims in 1980 (Sims, 1980). This model, although not grounded in economic theory, adopts a multiple equation framework. Each equation within the model regresses endogenous variables on lagged terms of all endogenous independent variables and estimates the dynamic relationships among all endogenous variables. It is commonly employed for predicting interrelated time series systems and analyzing the dynamic impact of random perturbations on the variable system (Stock & Watson, 2001).

The GM, proposed by Chinese scholar Deng (1982), has found widespread application across various fields. This model is particularly effective for predicting systems characterized by small sample sizes, discrete forms of data, information scarcity, and uncertainty. Its key advantage lies in its ability to achieve high levels of accuracy with relatively limited amounts of input data while also being capable of solving nonlinear problems (Wu, 2021; Xie & Chen, 2019). Given these attributes, the GM is well-suited for eco-environmental forecasting. Specifically, the classical GM (GM (1,1)) represents a one-variable order GM (Deng, 1985). Chen (2017) used this model to forecast and evaluate coordinated development between regional social economy and eco-environment.

4.2 Machine learning models

Machine learning models can be categorized as either supervised or unsupervised. There are two subcategories within supervised learning: regression models and classification models. In a regression model, the output is continuous, encompassing techniques such as linear regression, decision trees (DTs), random forests (RFs), and neural networks. On the other hand, in a classification model, the outputs are discrete and include methods like logistic regression (LR), support vector machine (SVM), and naive Bayes. Unlike supervised learning approaches that rely on labeled results for training data, unsupervised learning aims to infer and discover patterns from unlabeled input data. The main methods of unsupervised learning consist of clustering and dimensionality reduction techniques. Clustering involves grouping logarithmic data points without predefined labels using K-means, hierarchical, mean-shift, or density-based clustering techniques. Dimensionality reduction focuses on reducing the number of random variables by obtaining a set of principal variables through principal component analysis (PCA) methods. In the field of coastal zone management, machine learning can be employed in various capacities, such as utilizing predictive models based on machine learning algorithms to forecast marine water quality (Deng et al., 2021), estimate coastal chlorophyll-a (Niu et al., 2023), and map the shoreline position (Fogarin et al., 2023) and urban expansion in coastal cities (Singh et al., 2023). Below are detailed examples from machine learning models.

The Markov analysis method (Basharin et al., 2004) is applicable for analyzing changes in complex systems, including biological, physical, chemical, and other factors affecting coastal areas and the marine eco-environment. The Markov analysis can accurately predict future alterations by examining known changes in these factors (Kang, 2019). The Markov analysis could usually be improved and combined with other methods. The combination of the Extreme Gradient Boosting (XGBoost) algorithm and multicriteria evaluation-cellular automata-Markov (MCE-CA-Markov) model could be used to predict the landscape pattern of the coastal area (Hao et al., 2022). The CA-Markov model could be used to predict coastal land use prediction (Rahman & Ferdous, 2021).

The ANN model is a sophisticated system that simulates the human brain's neural network, enabling it to effectively handle nonlinear relationships between entities. It can approximate arbitrary nonlinear functions within a finite subset by employing neurogenic action functions in a simple combination. Furthermore, its topology structure allows for processing continuous or intermittent input information and generating corresponding output (Kang, 2019). Predicting the marine eco-environment is a highly complex problem due to the intricate, nonlinear, and uncertain relationships between various factors. To establish an ANN model for the marine eco-environment, it is necessary to screen these factors and conduct numerous experiments to obtain optimal model parameters that can accurately predict future changes (Wu et al., 2000).

In practical eco-environmental forecasting, single or combined models can be utilized. Liu, Wang, et al. (2018) integrated the GM with the Markov model to examine the accuracy of the Gray Markov prediction model, which outperformed the weighted plus growth rate moving average method. Gao (2008) coupled the GM with artificial neural network (ANN) theory to establish a shallow lake group prediction model for Baiyangdian Lake's water quantity, quality, and ecology. Zhou and Wang (2018) employed a pressure-state-response (PSR) framework to construct a comprehensive eco-environmental quality evaluation index system. Fannassi et al. (2023) integrated machine learning, including ANN, DT, LR, RF, and SVM, to select a better performance prediction model as an appropriate tool predicting and mapping the coastal vulnerability index (CVI). They combined it with the coefficient of variation method and GM thoroughly evaluate and forecast the Yangtze River Economic Belt's eco-environmental quality.

5. Conclusions

As a critical interface between land and ocean, the coastal zone sustains 70% of the world's population as the most dynamic arena for human activities and natural processes. The sustainable utilization of coastal resources is essential to human survival, economic development, and environmental well-being. However, anthropogenic factors have increasingly impacted the coastal environment in recent decades due to intensified engineering construction and pollution discharge. These pressures have led to issues such as eutrophication in coastal waters. With projected increases in population density along coastlines over the coming decades, achieving sustainable development of ecology and the environment under multiple anthropogenic pressures remains a pressing challenge.

Starting from the multiple anthropogenic pressures the coastal zone faces, this chapter provides a comprehensive overview of the primary anthropogenic impacts on the ecology and environment of this area. The chapter examines representative management concepts for eco-environmental management in the face of multiple anthropogenic pressures within the coastal zone, which includes ICZM, LSC, collaborative governance, system theory and complex systems, NbS, adaptive management, and coastal zone eco-environment database.

The chapter also organizes representative management evaluation methods for eco-environmental management in the face of multiple anthropogenic pressures within the coastal zone. The evaluation and management model reviewed in this chapter mainly includes DPSIR, GRA, CCD, GIS, SWOT, and PCA. The prediction model introduced in this chapter mainly includes GM, Markov analysis, and ANN model (Kang, 2019).

As the interaction between anthropogenic pressure and the coastal eco-environment becomes increasingly intricate, effective management of the coastal eco-environment under such pressure necessitates the utilization of relevant concepts and the selection of appropriate assessment and prediction methods to achieve adaptive management objectives. Moreover, sustainable development in coastal areas requires collaborative efforts among governments, communities, and stakeholders to formulate viable plans that strike a harmonious balance between coastal development and environmental preservation while meeting humanity's needs without compromising delicate coastal ecosystems. As an interface between human activities and the natural environment, the coastal zone requires a delicate equilibrium to effectively accomplish sustainable development goals through integrated management while simultaneously addressing many benefits and challenges.

References

Armstrong, J. M., & Ryner, P. C. (1981). *Ocean management: A new perspective*. Michigan: Ann Arbor Science.

Baldwin, C., Lewison, R. L., Lieske, S. N., Beger, M., Hines, E., Dearden, P., Rudd, M. A., et al. (2016). Using the DPSIR framework for transdisciplinary training and knowledge elicitation in the Gulf of Thailand. *Ocean and Coastal Management, 134*, 163−172.

Barrows, C. W., Swartz, M. B., Hodges, W. L., Allen, M. F., Rotenberry, J. T., Li, B.-L., Scott, T. A., et al. (2005). A framework for monitoring multiple-species conservation plans. *Journal of Wildlife Management, 69*, 1333−1345.

Basharin, G. P., Langville, A. N., & Naumov, V. A. (2004). The life and work of A.A. Markov. *Linear Algebra and its Applications, 386*(1−3 Suppl. L), 3−26.

Bergh, Ø., Beck, A. C., Tassetti, A. N., Olsen, E., Thangstad, T. H., Gonzalez-Mirelis, G., Grati, F., et al. (2023). Analysis of spatial conflicts of large scale salmonid aquaculture with coastal fisheries and other interests in a Norwegian fjord environment, using the novel GIS-tool SEAGRID and stakeholder surveys. *Aquaculture, 574*(May). https://doi.org/10.1016/j.aquaculture.2023.739643

Bertalanffy, L. Von (1976). *General system theory*. George Braziller.

Beverton, R. J. H., & Holt, S. J. (2012). *On the dynamics of exploited fish populations*. Springer Dordrecht. https://doi.org/10.1007/978-94-011-2106-4. available at:.

Bidone, E. D., & Lacerda, L. D. (2004). The use of DPSIR framework to evaluate sustainability in coastal areas. Case study: Guanabara Bay basin, Rio de Janeiro, Brazil. *Regional Environmental Change, 4*(1), 5−16.

Bin, C., Hao, H., Weiwei, Y., Senlin, Z., Jinkeng, W., & Jinlong, J. (2009). Ocean & coastal management marine biodiversity conservation based on integrated coastal zone management (ICZM) d a case study in Quanzhou Bay , Fujian , China. *Ocean and Coastal Management, 52*(12), 612−619.

Bruckmeier, K. (2014). Problems of cross-scale coastal management in Scandinavia. *Regional Environmental Change, 14*(6), 2151−2160.

Buenau, K., & Anderson, M. (2016). Adaptive management. In M. J. Kennish (Ed.), *Encyclopedia of earth sciences series* (pp. 1−2). Netherlands, Dordrecht: Springer.

Chang, K. (2018). *Introduction to geographic information systems*. McGraw-Hill Education.

Chen, D., & Sun, Y. (2021). The construction of the adaptive management model of marine eco-economical system and its realization path. *Journal of Ocena University of China, 1*.

Chen, K., Gao, Y., Wu, K., & Huang, H. (2022). Integrated coastal zone management in China: System, practies and problems. *Journal of Applied Oceanography, 41*(3). https://doi.org/10.3969/J.ISSN.2095-4972.2022.03.017

Chen, Y. (2017). Assessment for the coordination between socio-economy and eco-environment based on GM (1,1) - a case study of Zhongshan City, China (in Chinese). *Environmental and Development, 29*(03). https://doi.org/10.16647/j.cnki.cn15-1369/X.2017.03.143

Cheng, H., Xu, Q., & Guo, Y. (2019). Temporal and spatial evolution of the coupling coordinated development between tourism resources development and ecological environment in China(in Chinese). *Economic Geography, 39*(7).

Cheng, J., Zhang, X., & Gao, Q. (2023). Analysis of the spatio-temporal changes and driving factors of the marine economic—ecological—social coupling coordination: A case study of 11 coastal regions in China. *Ecological Indicators, 153*(February), Article 110392.

Communities Naturally-Resilient. *Nature-based solutions strategy.* Available at https://nrcsolutions.org/strategies/. (Accessed 3 August 2023).

Conti, L. A., Fonseca Filho, H., Turra, A., & Amaral, A. C. Z. (2018). Building a local spatial data infrastructure (SDI) to collect, manage and deliver coastal information. *Ocean and Coastal Management, 164*(January), 136—146.

Deng, J. (1982). Control problems of grey system (in Chinese). *Syetems and Control Letters, 1*(5).

Deng, J. (1985). *Grey control system.* Wuhan: Huazhong University of Science and Technology Press.

Deng, T., Chau, K. W., & Duan, H. F. (2021). Machine learning based marine water quality prediction for coastal hydro-environment management. *Journal of Environmental Management, 284*(October 2020), Article 112051.

Eggermont, H., Balian, E., Azevedo, J. M. N., Beumer, V., Brodin, T., Claudet, J., Fady, B., et al. (2015). Nature-based solutions: New influence for environmental management and research in Europe. *GAIA - Ecological Perspectives for Science and Society, 24*(4), 243—248.

European Commission. (2021). *Communication from the commission to the European parliament, the Council, the European economic and social committee and the committee of the regions*, 2021. COM. *550 Final*, available at: https://eur-lex.europa.eu/legal-content/EN/TXT/?uri=CELEX:52021DC0550.

Europen Environment Agency. (2020). *The DPSIR framework.*

Fan, D., & Chen, L. (2022). Systematology is the bridge from system science to marxist philosophy-reviews of systematology:philosophy of systems science(in Chinese). *Journal of System Science, 30*(1).

Fannassi, Y., Ennouali, Z., Hakkou, M., Benmohammadi, A., Al-Mutiry, M., Elbisy, M. S., & Ali Masria. (2023). Prediction of coastal vulnerability with machine learning techniques, Mediterranean coast of Tangier-Tetouan, Morocco. *Estuarine, Coastal and Shelf Science, 291*, 108422.

Feng, L., Zhu, X., & Sun, X. (2014). Assessing coastal reclamation suitability based on a fuzzy-AHP comprehensive evaluation framework: A case study of Lianyungang , China. *Marine Pollution Bulletin, 89*(1—2), 102—111.

Fogarin, S., Zanetti, M., Dal Barco, M. K., Zennaro, F., Furlan, E., Torresan, S., Pham, H. V., et al. (2023). Combining remote sensing analysis with machine learning to evaluate short-term coastal evolution trend in the shoreline of Venice. *Science of the Total Environment, 859*(October 2022), Article 160293.

Gao, F. (2008). *Variation and prediction of the ecological environment of baiyangdian wetland.* Agricultural University of Hebei.

Gao, J., An, T., Shen, J., Zhang, K., Yin, Y., Zhao, R., He, G., et al. (2022). Development of a land-sea coordination degree index for coastal regions of China. *Ocean and Coastal Management, 230*(September), Article 106370.

Ge, H., Chen, K., & Wang, D. (2020). Study on the theoretical framework of ecological adaptive managment downstream of the dam. *Northwest Hydropower, 6*.

General Office of the CPC Central Committee and General Office of the China State Council. (2017). *General Office of the CPC central committee and general office of the state council issued several opinions on delineating and strictly observing the red line for ecological protection(in Chinese)*. Xinhua News Agency. available at: https://www.gov.cn/zhengce/2017-02/07/content_5166291.htm.

Geng, J., Wang, J., Huang, J., Zhou, D., Bai, J., Wang, J., Zhang, H., et al. (2022). Quantification of the carbon emission of urban residential buildings: The case of the Greater Bay area cities in China. *Environmental Impact Assessment Review, 95*(February), Article 106775.

Gkoltsiou, A., & Mougiakou, E. (2021). The use of Islandscape character assessment and participatory spatial SWOT analysis to the strategic planning and sustainable development of small islands. The case of Gavdos. *Land Use Policy, 103*(March 2020), Article 105277.

Gregory, R., Ohlson, D., & Arvai, J. (2006). Deconstructing adaptive management: Criteria for applications to environmental management. *Ecological Applications, 16*(6), 2411−2425.

Hale, S. S., Miglarese, A. H., Bradley, M. P. T., Belton, T. J., Cooper, L. D., Michael, T., Frame, C. A. F., Harwell, L. M., et al. (2003). Managing troubled data: Coastal data partnerships smooth data integration. *Environmental Monitoring and Assessment, 81*, 133−148.

Hao, L., He, S., Zhou, J., Zhao, Q., & Lu, X. (2022). Prediction of the landscape pattern of the Yancheng Coastal Wetland, China, based on XGBoost and the MCE-CA-Markov model. *Ecological Indicators, 145*(99), Article 109735.

Hendrickx, G. G., Antolínez, J. A. A., & Herman, P. M. J. (2023). Predicting the response of complex systems for coastal management. *Coastal Engineering, 182*, Article 104289.

Hofkirchner, W., & Schafranek, M. (2011). General system theory. In C. Hooker (Ed.), *Philosophy of complex systems* (Vol 10, pp. 177−194). North-Holland, Amsterdam.

Holling, C. S. (1978). *Adaptive environmental assessment and management.* Chinchester, USA: Wiley. https://doi.org/10.1093/forestscience/26.3.435. available at:.

Holling, C. S., & Meffe, G. K. (1996). Command and control and the pathology of natural resource management. *Conservation Biology, 10*(2), 328−337.

Holling, C. S. (1973). Resilience and stability of ecological systems. *Annual Review of Ecology, Evolution, and Systematics, 4*, 1−23.

Huan, L. (2015). *The physiological and biochemical basis of ulva prolifera response to stresses (in Chinese)*. University of Chinese Academy of Sciences.

Huang, B., & Wang, Z. (2022). Current situation, hot spots and future prospects of collaborative governance research at home and abroad in the past two decades: Based on CiteSpace visual comparative analysis (in Chinese). *Leadership Science, 07*, 129−134.

Huang, Y., Li, Z., & Chen, M. (2022). Study on the collaborative governance mechanism of marine ecological restoration and soil remediation under the background of land-se coordination (in Chinese). *Ocean Development and Management, 39*(05). https://doi.org/10.20016/j.cnki.hykfygl.20220426.001

Huang, Y., Wei, S., Le, Y., Chen, X., Sun, B., & Li, M. (2022). A SWOT analysis of Heishui River national wetland park in Daxin of Guangxi(in Chinese). *Wetland Science and Manageent, 18*(2).

Ichikawa, S., Onaka, S., Izumi, M., & Uda, T. (2020). Adaptive coastal management for gravel beach nourishment in Tuvalu based on monitoring results. In *APAC 2019 - proceedings of the 10th international conference on Asian and Pacific Coasts* (pp. 435−440). Springer.

IUCN Commission on Ecosystem Management (CEM). (2020). *IUCN global standard for nature-based solutions* (1st ed.). https://doi.org/10.2305/IUCN.CH.2020.08.en Gland, Switzerland, available at:.

IUCN. (2020). *Guidance for using the IUCN global standard for nature-based solutions: First editions, guidance for using the IUCN global standard for nature-based solutions: First editions* https://doi.org/10.2305/iucn.ch.2020.09.en. available at:.

IUCN. (2016). *Nature-based solutions*. available at: https://www.iucn.org/our-work/nature-based-solutions. (Accessed 3 August 2023).

Kang, X. (2019). *Study on marine ecological environment prediction system*. Zhejiang Ocean University.

Karageorgis, A. P., Kapsimalis, V., Kontogianni, A., Skourtos, M., Turner, K. R., & Salomons, W. (2006). Impact of 100-year human interventions on the deltaic coastal zone of the inner thermaikos gulf (Greece): A DPSIR framework analysis. *Environmental Management, 38*(2), 304−315.

Khelil, N., Larid, M., Grimes, S., Le Berre, I., & Peuziat, I. (2019). Challenges and opportunities in promoting integrated coastal zone management in Algeria: Demonstration from the Algiers coast. *Ocean and Coastal Management, 168*, 185−196.

Li, J., Tian, P., Li, C., & Gong, H. (2022). Land-sea economic relations and land spaceutilization based onland-sea coordination: Research status, problems and future priorities (in Chinese). *Journal of Natural Resources, 37*(4), 924−941.

Li, R., Xu, C., Li, Y., & Hu, H. (2022). Progressn of international research on coastal resilience and implications for China(in Chinese). *Resources Science, 44*(2).

Li, S., Yu, D., Kong, J., Chen, J., Wu, S., Wang, Y., & Chen, N. (2022). Coastal land-ocean-atmosphere cooperative management of nitrogen and phosphrus pollution: Monitoring, modeling and decision-making(in Chinese). *Environmental Engineering, 40*(6).

Li, W., Xu, J., & Sun, C. (2021). Spatial and temporal evolution of Shenzhen's eco-environment and urban sprawl based on grey correlation degree(in Chinese). *Ecology and Environmental Science, 30*(4).

Li, X., & Ye, G. (2020). Study on zoning and use classification of coastal zone planning fromo the perspective of land space planning-a case of Jiaozhou Bay in Qingdao (in Chinese). *Urbanism and Architecture, 17.*

Li, Y., Wu, J., & Liang, Y. (2022). Empirical study on grey correlation degree of three industries in Guangdong-Hong Kong-Macao Greater Bay area. *Journal of South China University of Technology, 24*(2).

Liang, S. kang, Pearson, S., Wu, W., Ma, Y., jie, Q., lu, L., Wang, X. H., Li, J. mei, et al. (2015). Research and integrated coastal zone management in rapidly developing estuarine harbours: A review to inform sustainment of functions in Jiaozhou bay, China. *Ocean and Coastal Management, 116*, 470−477.

Lin, X., Wang, L., & Wen, C. (2018). Coastal spatial control under land-sea coordination_The case of coastal zone planning of Xiamen (in Chinese). *Urban Planning Forum, 4.* https://doi.org/10.16361/j.upf.201804009

Liu, T. (2014). *Assessment and forecast of shandong province based on the time series model and the change point model(in Chinese)*. Shandong University.

Liu, P., Wang, M., Ma, L., & Shi, S. (2018). Comparison of prediction accuracy between gray markov forecasting model and weighted plus growth rate moving average method(in Chinese). *Statistics and Decision, 34*(22).

Liu, Y., Han, L., Pei, Z., & Jiang, Y. (2023). Evolution of the coupling coordination between the marine economy and urban resilience of major coastal cities in China. *Marine Policy, 148*(June 2022), Article 105456.

Liu, Y., Zhou, S., Hu, J., & Ma, Z. (2018). Progress in environmental stress research of aquatic animals (in Chinese). *Journal of Tianjing Agricultural University, 25*(4). https://doi.org/10.19640/j.cnki.jtau.2018.04.016

Luijendijk, A., Hagenaars, G., Ranasinghe, R., Baart, F., Donchyts, G., & Aarninkhof, S. (2018). The state of the world's beaches. *Scientific Reports, 8*(1), 1−11.

Ma, Y., Yan, J., Sha, J., Wang, F., Cui, W., Ding, L., & Ma, L. (2020). Comprehensive evaluation on carrying of resources and environment under the background of land-sea coordination in Tangshan city (in Chinese). *China Mining Magazine, 29*(1).

Mahony, C. O., Gray, S., Gault, J., & Cummins, V. (2020). ICZM as a framework for climate change adaptation action — experience from Cork Harbour , Ireland. *Marine Policy, 111*, Article 102223.

Mandal, P., Maiti, A., Paul, S., Bhattacharya, S., & Paul, S. (2023). ScienceDirect mapping the multi-hazards risk index for coastal block of Sundarban , India using AHP and machine learning algorithms. *Tropical Cyclone Research and Review, 11*(4), 225—243.

Ministry of Ecology and Environment of the People's Republic of China. (2022). *The ministry of ecology and environment and other six departments jointly issued the '14th five-year plan for marine ecological environment protection' (in Chinese)*. Ministry of Ecology and Environment of the People's Republic of China.

Morf, A., Moodie, J., Cedergren, E., Eliasen, S. Q., Gee, K., Kull, M., Mahadeo, S., et al. (2022). Challenges and enablers to integrate land-sea-interactions in cross-border marine and coastal planning: Experiences from the Pan baltic scope collaboration. *Planning Practice and Research, 37*(3), 333—354.

Nerlove, M. (2018). *Autoregressive and moving-average time-series processes BT - the new palgrave dictionary of economics*. London: Palgrave Macmillan UK.

Nesshöver, C., Assmuth, T., Irvine, K. N., Rusch, G. M., Waylen, K. A., Delbaere, B., Haase, D., et al. (2017). The science, policy and practice of nature-based solutions: An interdisciplinary perspective. *The Science of the Total Environment, 579*, 1215—1227.

Niu, J., Feng, Z., He, M., Xie, M., Lv, Y., Zhang, J., Sun, L., et al. (2023). Incorporating marine particulate carbon into machine learning for accurate estimation of coastal chlorophyll-a. *Marine Pollution Bulletin, 192*(May), Article 115089.

Noble, M. M., Harasti, D., Pittock, J., & Doran, B. (2021). Using GIS fuzzy-set modelling to integrate social-ecological data to support overall resilience in marine protected area spatial planning: A case study. *Ocean and Coastal Management, 212*(July), Article 105745.

O'Donnell, T. K., & Galat, D. L. (2008). Evaluating success criteria and project monitoring in river enhancement within an adaptive management framework. *Environmental Management, 41*(1), 90—105.

Ostrom, E. (2007). A diagnostic approach for going beyond panaceas. *Proceedings of the National Academy of Sciences of the United States of America, 104*(39), 15181—15187.

Panigrahi, J. K., & Mohanty, P. K. (2012). Effectiveness of the Indian coastal regulation zones provisions for coastal zone management and its evaluation using SWOT analysis. *Ocean and Coastal Management, 65*, 34—50.

Pearson, K. (1901). On lines and planes of closest fit to systems of points in space. *Philosophical Magazine, Series 6, 2*(11), 559—572.

Pfaff, M. C., Hart-Davis, M., Smith, M. E., & Veitch, J. (2022). A new model-based coastal retention index (CORE) identifies bays as hotspots of retention, biological production and cumulative anthropogenic pressures. *Estuarine, Coastal and Shelf Science, 273*(May). https://doi.org/10.1016/j.ecss.2022.107909

Portman, M. E. (2014). Visualization for planning and management of oceans and coasts. *Ocean and Coastal Management, 98*, 176—185.

Puyt, R. W., Lie, F. B., & Wilderom, C. P. M. (2023). The origins of SWOT analysis. *Long Range Planning, 56*, Article 102304.

Rahman, M. T. U., & Ferdous, J. (2021). Spatio-temporal variation and prediction of land use based on CA-Markov of southwestern coastal district of Bangladesh. *Remote Sensing Applications: Society and Environment, 24*(August), Article 100609.

Ramesh, R., Chen, Z., Cummins, V., Day, J., D'Elia, C., Dennison, B., Forbes, D. L., et al. (2015). Land-Ocean interactions in the coastal zone: Past, present & future. *Anthropocene, 12*(2015), 85−98.

Rao, C., & Zhao, S. (2022). The influencing factors and action approaches for the effect of intergovernmental collaboration in governance of cross-domain evironmental pollution-A qualitive comprarative analysis based on twelve cases (in Chese). *Journal of Central China Normal University (Humanities and Social Sciences), 61*(4). https://doi.org/10.19992/j.cnki.1000-2456.2022.04.006

Repository EU. *Oppla.* Available at https://oppla.eu/about. (Accessed 3 August 2023).

Saaty, T. L. (1977). A scaling method for priorities in hierarchical structures. *Journal of Mathematical Psychology, 15*(3), 234−281.

Shen, S., Niu, E., & Meng, B. (2017). Evaluation of ecological environment carrying capacity in coastal waters of Liaoning based on Grey relation. *Journal of Dalian Maritime University, 43*(3).

Shu, X., Gao, Y., Zhang, Y., & Yang, C. (2015). Study on the coupling relationship and coordinative development between tourism industry and eco-civilization City(in Chinese). *China Population,Resources and Environment, 25*(3).

Sims, C. A. (1980). Macroeconomics and reality. *Econometrica, 48*(1). https://doi.org/10.2307/1912017

Singh, S., Rao, M. J., Baranval, N. K., Kumar, K. V., & Kumar, Y. V. (2023). Geoenvironment factors guided coastal urban growth prospect (UGP) delineation using heuristic and machine learning models. *Ocean and Coastal Management, 236*(October 2022), Article 106496.

Slutsky, E. (1927). The summation of random causes as the source of cyclic processes. *Trans. Econometrica, 5*, 105−146.

Smith, M. D., Roheim, C. A., Crowder, L. B., Halpern, B. S., Turnipseed, M., Anderson, J. L., Asche, F., et al. (2010). Sustainability and global seafood. *327*(February), 784−787.

Sorensen, J. (2002). *Baseline 2000 background report: The status of integrated coastal management as an international practice (second iteration).* Urban Harbors Institute Publications.

Stewart, R. F. (1963). *A framework for business palnning (No.162),Long range planning service.*

Stock, J. H., & Watson, M. W. (2001). Vector autoregressions. *The Journal of Economic Perspectives, 15*(4), 101−115.

Sun, L. (2008). *Coastal ecosystem health assessment and prediction research of Jiaozhou bay(in Chinese).* Ocean University of China.

Sun, M. (2012). *Ecosystem health assessment and stress factor analysis on Zhuhai coastal marine areas.* Ocean University of China *(in Chinese).*

Think Nature. *Plantform for nature-based solutions.* Available at https://www.think-nature.eu/. (Accessed 25 July 2022).

Thirumurthy, S., Jayanthi, M., Samynathan, M., Duraisamy, M., Kabiraj, S., & Anbazhahan, N. (2022). Multi-criteria coastal environmental vulnerability assessment using analytic hierarchy process based uncertainty analysis integrated into GIS. *Journal of Environmental Management, 313*(March), Article 114941.

Tomlinson, R. F. (2003). *Thinking about GIS: geographic information system planning for managers.* Esri Press.

Uddin, M. N., Saiful Islam, A. K. M., Bala, S. K., Islam, G. M. T., Adhikary, S., Saha, D., Haque, S., et al. (2019). Mapping of climate vulnerability of the coastal region of Bangladesh using principal component analysis. *Applied Geography, 102*(May 2016), 47−57.

UNFCCC. (2009). *COP15: Copenhagen accord - draft decision, fccc/cp/2009/L.7.* available at: papers3://publication/uuid/01CE7DB2-A61E-4B51-8E75-B5621C2DA364.

United Nations. (1992a). *United Nations framework convention* (Vol 62220).

United Nations. (1992b). Convention on biological diversity. *Encyclopedia of biodiversity* (2nd ed.) https://doi.org/10.1016/B978-0-12-384719-5.00418-4 available at:.

United Nations. (2015). *The 17 goals.* Department of Economic and Social Affairs. available at: https://sdgs.un.org/goals.

Vaitis, M., Kopsachilis, V., Tataris, G., Michalakis, V. I., & Pavlogeorgatos, G. (2022). The development of a spatial data infrastructure to support marine spatial planning in Greece. *Ocean and Coastal Management, 218,* Article 106025.

Walsh, C. (2019). Integration of expertise or collaborative practice?: Coastal management and climate adaptation at the Wadden Sea. *Ocean and Coastal Management, 167*(January 2018), 78−86.

Walters, C. J. (1986). *Adaptive management of renewable resources.* New York: MacMillan Publishing Company.

Wang, C. (2009). *Study the pattern for integrated coastal zone management based on ecosystem approach:A case study of Jiaozhou bay(in Chinese).* Ocean University of China.

Wang, G., & Cai, W. (2020). Review of research on ecosystem adaptive management based on bibliometric analysis(in Chinese). *Journal of Liaoning Normal University (Natural Science Edition), 43*(3).

Wang, X., & Xi, R. (2016). Assessment of eco-environment vulnerability in Weigan River basin based on GIS (in Chinese). *Ecological Science, 35*(4). https://doi.org/10.14108/j.cnki. 1008-8873.2016.04.023

Wei, J., Lin, L., Pan, X., Liu, M., & Li, M. (2020). Reserach progress about the effects of different environmental stress factors on algae in molecular biology (in Chinese). *Journal of Yangtze River Scientific Research Institue, 37*(4).

Weinstein, M. P., Balletto, J. H., Teal, J. M., & Ludwig, D. F. (1996). Success criteria and adaptive management for a large-scale wetland restoration project. *Wetlands Ecology and Management, 4*(2), 111−127.

Williams, B. K. (2011). Adaptive management of natural resources-framework and issues. *Journal of Environmental Management, 92*(5), 1346−1353.

Williams, B. K., Szaro, R. C., & Shapiro, C. D. (2009). *Adaptive management: The U.S. Department of the interior technical guide.* available at: https://api.semanticscholar.org/ CorpusID:197455508.

Winther, J. G., Dai, M., Rist, T., Hoel, A. H., Li, Y., Trice, A., Morrissey, K., et al. (2020). Integrated ocean management for a sustainable ocean economy. *Nature Ecology and Evolution, 4*(11), 1451−1458.

World Bank. (2008). *Biodiversity, climate change, and adaptation : Nature-based solutions from the World Bank Portfolio, Washington, DC.* Available at: http://hdl.handle.net/10986/ 6216.

Worm, B., Barbier, E. B., Beaumont, N., Duffy, J. E., Folke, C., Halpern, B. S., Jackson, J. B. C., et al. (2006). *Impacts of biodiversity loss on ocean ecosystem services.* November.

Wu, C., Wang, Q., Dong, Z., & Chen, W. (2018). Land use/cover change and its driving forces in coastal zone of Fujian Province (in Chinese). *Bulletin of Soil and Water Conservation, 38*(3). https://doi.org/10.13961/j.cnki.stbctb.2018.03.051

Wu, H., Lin, Z., & Gao, S. (2000). The application of artifical neural networks in the resources and environment. *Resources and Environment in the Yangtze Basin, 9*(2).

Wu, T. (2021). Quantifying coastal flood vulnerability for climate adaptation policy using principal component analysis. *Ecological Indicators, 129,* Article 108006.

Wu, Z., Yu, Z., Song, X., Li, Y., Cao, X., & Yuan, Y. (2016). A methodology for assessing and mapping pressure of human activities on coastal region based on stepwise logic decision process and GIS technology. *Ocean and Coastal Management, 120*, 80–87.

Xie, M., & Chen, X. (2019). Advances in the application of bibliometrics-based grey system theory in fisheries science (in Chinese). *Transactions of Oceanology and Limnology, 05*. https://doi.org/10.13984/j.cnki.cn37-1141.2019.05.015

Xinhua News Agency. (2011). *Outline of the 12th five-year plan for national economic and social development (in Chinese)*, available at: https://www.gov.cn/2011lh/content_1825 838.htm.

Xinhua News Agency. (2012). *Report of the 18th national congress of the communist party of China (in Chinese)*, available at: http://www.beijingreview.com.cn/18da/txt/2012-11/19/content_502297.htm.

Xinhua News Agency. (2017). *Report of the 19th national congress of the communist party of China (in Chinese)*, available at: http://news.cnr.cn/native/gd/20171027/t20171027_524003098.shtml.

Xinhua News Agency. (2021). *The 14th five-year plan for national economic and social development of the People's Republic of China and the outline of the long-range goals for 2035 (in Chinese)*, available at: https://www.gov.cn/xinwen/2021-03/13/content_5592681.htm.

Xu, G., Kang, M., & Shi, Y. (2013). A review of adaptive management reserach on natrual resources. *Journal of Natural Resources, 28*(10).

Xun, W., & Yu, Z. (2022). On coupling relationship and coordination path between land development intensity and carrying capacity of resource and environment in Shenyang from the perspective of green development(in Chinese). *Scientific and Technological Management of Land and Resources, 39*(1).

Yan, J. (2003). *Research on comprehensive coastal zone management in Jiaozhou bay(in Chinese)*. Ocean University of China.

Yang, Y., & Sun, C. (2014). Assessment of land-sea coordination in the Bohai sea ring area and spatial-temporal differences (in Chinese). *Resources Science, 36*(4).

Yao, R., Zhang, X., Yan, D., Xu, M., Ma, L., & Zhao, Y. (2021). Marine ecological environment management system based on land and sea coordination (in Chinese). *Chinese Journal of Environmental Management, 13*(05). https://doi.org/10.16868/j.cnki.1674-6252.2021.05.079. available at:.

Yu, Y. (2010). *Integrated coastal area and river basin management in Jiaozhou bay*. Ocean University of China.

Yuan, Y., Jalón-Rojas, I., Wang, X. H., & Song, D. (2019). Design, construction, and application of a regional ocean database: A case study in Jiaozhou Bay, China. *Limnology and Oceanography: Methods, 17*(3), 210–222.

Yuan, Y., Song, D., Wu, W., Liang, S., Wang, Y., & Ren, Z. (2016). The impact of anthropogenic activities on marine environment in Jiaozhou Bay, Qingdao, China: A review and a case study. *Regional Studies in Marine Science, 8*, 287–296.

Yue, W., Hou, B., Ye, G., & Wang, Z. (2023). China's land-sea coordination practice in territorial spatial planning. *Ocean and Coastal Management, 237*(October 2022), Article 106545.

Yule, G. U. (1927). On a method for investigating periodicities in disturbed series with special reference to Wolfer's sunspot numbers. *Philosophical Transactions of the Royal Society of London, 226*, 267–298.

Yule, G. U. (1921). On the time-correlation problem, with especial reference to the variate-difference correlation method. *Journal of the Royal Statistical Society, 84*(4), 497–537.

Yule, G. U. (1926). Why do we sometimes get nonsense-correlations between time-series?–A study in sampling and the nature of time-series. *Journal of the Royal Statistical Society, 89*(1), 1–63.

Zhang, M., Zhang, J., Liu, Z., & Ding, D. (2022). Evaluation of coastal rectification and repairing engineering effect based on AHP and fuzzy mathematics in the Jinshi Beach in Dalian(in Chinese). *The Journal of Ocean Technology, 41*(5).

Zhao, X., & Wang, P. (2018). The reform and remolding of the governance concept of xiong'an newarea under the background of coordinated development of the BeijingTianjin-Hebei region (in Chinese). *Administrative Tribune, 25*(02), 31–39.

Zhou, W., & Zeng, B. (2020). A Research review of grey relational degree model. *Statistics and Decision, 36*(15), 29–34.

Zhou, Z., & Wang, J. (2018). Comprehensive evaluation and forecasting of ecological environmental quality in the Yangtze economic zone. *Journal of Shandong Normal University(Natural Science), 33*(4). https://doi.org/10.3969/j.issn.1001-4748.2018.04.014

Conclusions

14

Lulu Qiao[1] and Xiao Hua Wang[2]
[1]College of Marine Geosciences, Ocean University of China, Key Lab of Submarine Geosciences and Prospecting Techniques, Ministry of Education, Qingdao, China; [2]The Sino-Australian Research Consortium for Coastal Management, School of Science, University of New South Wales, Canberra, ACT, Australia

Coastal zones are the significant areas being affected by land—sea interaction and global climate change. Thus, they have been for a long-time target areas and hotbeds for the multidisciplinary study. In this book, we presented the recent trend in observations and modeling of estuarine and coastal dynamics, with a focus on hydrodynamics, sediment transport dynamics, and the coastal feedback to human impacts. A total of 12 different perspectives discussed topics including sea surface temperature (SST), coastal bathymetry inversion, coastal eddies, estuarine circulation, hydrologic and sediment investigation methods and numerical modeling, hydro and sediment dynamics related to the coastline changes and engineering constructions, boundary upwelling, nutrient and chlorophyll dynamics related to the intense precipitation, and environmental management under multiple anthropogenic pressures and climate change.

SST has been recognized as an essential climate variable. In Chapter 2, Haifeng Zhang summarized the features and measurements' qualities of several in situ SST platform types, taking the ones around the Australian and adjacent waters as examples. They pointed out that nearly all ship SST sensors were placed at a depth of 7−10 m, those on mooring float buoys were at ∼ 1 m depth, and satellites can only measure the optimal source skin SST. The observation depth may result in SST diurnal variation (DV) signals, especially during the daytime under warm and calm conditions.

Satellite remote sensing is not the sole method for investigating SST, it can also be employed to acquire coastal bathymetry. In Chapter 3, Nan Xu and Yue Ma outlined the procedures for generating a bathymetric map in shallow waters using novel ICESat-2 lidar data and Sentinel-2 multispectral imagery. They recommended an enhanced DBSCAN method, incorporating corrections for surface fluctuation effects, as well as local linear band models and/or band ratio models for depth mapping. ICESat-2 data were also utilized to extract wave parameters of sea surfaces, enabling successful bathymetry estimation irrespective of water clarity.

For the interaction between coastal water and open seas, eddies and estuarine processes play significant roles. Coastal and near-coastal eddies arise in proximity to coastlines and driven by the complex interactions of tides, currents, and coastal features. In Chapter 4, Zhibing Li presented that these eddies bridged coastal and open ocean realms, facilitating the exchange of water masses, nutrients, and marine life. However, the detection of these eddies with fleeting existence and complicated

Current Trends in Estuarine and Coastal Dynamics. https://doi.org/10.1016/B978-0-443-21728-9.00014-4

dynamic systems has become a challenge contrasting with the eddies in the vast open ocean. Li also noted that dynamics processes of eddies in the estuaries also fostered frequent exchanges of materials and energy.

In Chapter 5, Ziyu Xiao and collaborators established a three-dimensional coastal ocean model to reproduce the intratidal and spring-neap variations of current velocity, salinity, and residual currents in the Sydney Harbor estuary. They showed that the presence of a horizontal salinity gradient served as the primary driving force for the estuarine circulation. Residual flows can be generated due to the spring-neap and flood-ebb asymmetric stratification occurring within these tidal cycles. This is different with the mechanism of high precipitation-induced stratification reported by most previous studies.

To describe the hydrodynamic characteristics and sediment movement in the estuary and coastal seas, geochemical analysis, numerical models and field investigations are effective methodologies, which were introduced in Chapters 6 and 7 by Zhixin Cheng, Yongzhi Wang, and their colleagues, respectively. By analyzing the mineralogical and elemental composition of sediments, geochemical analyses are adept at identifying unique signatures from different sediment sources. Rooted in mathematical and physical tenets, numerical modeling can empower researchers to simulate diverse sediment transport scenarios and even forecast potential sediment movement pathways. Based on a variety of instruments and equipment, marine surveys can display the spatial and temporal distributions in these vital hydrodynamic and environmental elements. Besides research vessels, satellites, underwater devices, and buoys can also conduct large-area monitor, cross-sectional surveys, and continuous observations.

Several case studies were conducted in Hangzhou Bay, Yalu River Estuary and Yangshan Harbor in China and Darwin Harbor and Batemans Bay in Australia. In Chapter 8, Li Li and collaborators concluded the increased/advanced tidal amplitude/phase and intensified deposition/erosion of sediment near the southern/northern bank are caused by the coastline changes of the inner Hangzhou Bay from the 16th Century to 2010. Based on the comparative studies of Yalu River Estuary, Darwin Harbor and Batemans Bay in Chapter 9, Gang Yang and colleagues pointed out that the two-layer estuarine circulation can influence the sediment transport and contributed to the formation of turbidity maximum zones. Furthermore, the strong gradient of wave radiation stress can generate wave-driven flow and then control the sediment dynamics and morphological evolution in coastal embayment with shallow waters and small tidal range such as Batemans Bay. To investigate the hydrodynamic response to the construction of Yangshan Deepwater Harbor, Wenyun Guo and collaborators demonstrated that the tidal amplitudes change varied in space and the tidal asymmetries were dominated by the development of shallow-water tides and the strengthening of tidal choking after the harbor construction.

Marine ecosystem and its influencing factors have always been the focus of coastal research, since its close relationship with coastal biodiversity and marine economy. Upwelling and precipitation are two of the important factors that significantly affect the ecosystem variation. In Chapter 11, Fanglou Liao reviewed the recent studies on upwelling and concluded that the combined indexes of wind stress and SST data could provide a more comprehensive understanding of upwelling dynamics. Furthermore, the robust upwelling systems in the subsurface waters of major western boundary currents cannot be ignored. Considering the vertical transport of nutrients and bottom cold water, upwelling regions have been recognized as the potential refuges for marine organisms in light of the ongoing warming trends.

In response to future warming, the frequency and intensity of global precipitation are expected to rise. Coastal environments exhibit heightened susceptibility to the impacts of intense precipitation events. In Chapter 12, Guandong Gao and Rushui Xiao anticipated that forthcoming instances of intense precipitation would further contribute to the deterioration of water quality in various areas, leading to an increase in the frequency or intensity of algal blooms and the occurrence of hypoxic conditions. They suggested further investigation into the long-term impacts on water quality and phytoplankton communities to increase the resilience of marine ecosystems. In addition to ocean and climate factors, intensified human activities such as engineering construction and pollutant discharge have compromised the health of coastal zones in recent years.

To identify key coastal management issues and offer strategies to address the environmental problems facing coastal oceans, Wen Wu and Liu Wan provided a comprehensive overview of the primary anthropogenic pressures affecting the ecological environment of coastal zones in Chapter 13. The chapter introduced the evaluation and management methods for the coastal zone ecological environment including Driving-Pressure-State-Impact-Response (DPSIR) framework analysis, the Gray Relational Analysis (GRA), and the Coupling Coordination Degree (CCD). They also presented a prediction method for assessing changes in the coastal zone eco-environment based on time series analysis and machine learning models. Finally, this chapter is concluded by summarizing future research directions for managing the coastal zone ecological environment under multiple anthropogenic pressures and climate change.

Index

'*Note:* Page numbers followed by "f" indicate figures and "t" indicate table.'

Printed in the United States
by Baker & Taylor Publisher Services